Library of America, a nonprofit organization,
champions our nation's cultural heritage
by publishing America's greatest writing in
authoritative new editions and providing resources
for readers to explore this rich, living legacy.

LOREN EISELEY
COLLECTED ESSAYS

VOLUME ONE

LOREN EISELEY

COLLECTED ESSAYS ON EVOLUTION, NATURE, AND THE COSMOS

VOLUME ONE

The Immense Journey
The Firmament of Time
The Unexpected Universe
Uncollected Prose

William Cronon, *editor*

THE LIBRARY OF AMERICA

This paper meets the requirements of
ANSI/NISO Z39.48–1992 (Permanence of Paper).

Distributed to the trade in the United States
by Penguin Random House Inc.
and in Canada by Penguin Random House Canada Ltd.

Library of Congress Control Number: 2016935164
ISBN 978–1–59853–506–8

First Printing
The Library of America—285

Manufactured in the United States of America

Loren Eiseley:
Collected Essays on Evolution, Nature, and the Cosmos
is published with support from

THE GOULD FAMILY FOUNDATION

and will be kept in print by its gift to
the Guardians of American Letters Fund,
established by the Library of America
to ensure that every volume in the series
will be permanently available.

Contents

THE IMMENSE JOURNEY

The Slit 5
The Flow of the River 13
The Great Deeps 21
The Snout 33
How Flowers Changed the World 41
The Real Secret of Piltdown 51
The Maze 61
The Dream Animal 69
Man of the Future 81
Little Men and Flying Saucers 91
The Judgment of the Birds 103
The Bird and the Machine 113
The Secret of Life 123

THE FIRMAMENT OF TIME

How the World Became Natural 135
How Death Became Natural 151
How Life Became Natural 167
How Man Became Natural 183
How Human Is Man? 199
How Natural Is "Natural"? 219
Bibliography 236

THE UNEXPECTED UNIVERSE

The Ghost Continent 243
The Unexpected Universe 259
The Hidden Teacher 275
The Star Thrower 289
The Angry Winter 307
The Golden Alphabet 325
The Invisible Island 343
The Inner Galaxy 361
The Innocent Fox 375

The Last Neanderthal 387
Bibliography 401

UNCOLLECTED PROSE
Autumn—A Memory......................... 407
Riding the Peddlers 409
The Mop to K. C. 415
Neanderthal Man and the Dawn of Human
    Paleontology 422
FROM *The Lost Notebooks* 435

Chronology................................. 459
Note on the Texts 470
Notes..................................... 476
Index..................................... 491

THE IMMENSE JOURNEY

DEDICATED TO THE MEMORY OF

CLYDE EDWIN EISELEY,

who lies in the grass of the prairie frontier
but is not forgotten by his son

The author wishes to thank the editors of the *American Scholar*, *Harper's Magazine*, and the *Scientific American* for permission to reprint material which appeared separately in those publications. He would like, also, to express his gratitude to the Wenner-Gren Foundation for Anthropological Research for providing the leisure from professional duties during which a major number of the essays in this book were written.

"Man can not afford to be a naturalist, to look at Nature directly, but only with the side of his eye. He must look through and beyond her."

HENRY DAVID THOREAU

"Unless all existence is a medium of revelation, no particular revelation is possible. . . ."

WILLIAM TEMPLE

The Slit

S OME LANDS ARE flat and grass-covered, and smile so evenly
up at the sun that they seem forever youthful, untouched
by man or time. Some are torn, ravaged and convulsed like the
features of profane old age. Rocks are wrenched up and exposed to view; black pits receive the sun but give back no light.

It was to such a land I rode, but I rode to it across a sunlit,
timeless prairie over which nothing passed but antelope or a
wandering bird. On the verge where that prairie halted before
a great wall of naked sandstone and clay, I came upon the Slit.
A narrow crack worn by some descending torrent had begun
secretly, far back in the prairie grass, and worked itself deeper
and deeper into the fine sandstone that led by devious channels
into the broken waste beyond. I rode back along the crack to a
spot where I could descend into it, dismounted, and left my
horse to graze.

The crack was only about body-width and, as I worked my
way downward, the light turned dark and green from the
overhanging grass. Above me the sky became a narrow slit of
distant blue, and the sandstone was cool to my hands on either
side. The Slit was a little sinister—like an open grave, assuming
the dead were enabled to take one last look—for over me the
sky seemed already as far off as some future century I would
never see.

I ignored the sky, then, and began to concentrate on the
sandstone walls that had led me into this place. It was tight
and tricky work, but that cut was a perfect cross section
through perhaps ten million years of time. I hoped to find at
least a bone, but I was not quite prepared for the sight I finally
came upon. Staring straight out at me, as I slid farther and
deeper into the green twilight, was a skull embedded in the
solid sandstone. I had come at just the proper moment when it
was fully to be seen, the white bone gleaming there in a kind
of ashen splendor, water worn, and about to be ground away
in the next long torrent.

It was not, of course, human. I was deep, deep below the
time of man in a remote age near the beginning of the reign of

mammals. I squatted on my heels in the narrow ravine, and we stared a little blankly at each other, the skull and I. There were marks of generalized primitiveness in that low, pinched brain case and grinning jaw that marked it as lying far back along those converging roads where, as I shall have occasion to establish elsewhere, cat and man and weasel must leap into a single shape.

It was the face of a creature who had spent his days following his nose, who was led by instinct rather than memory, and whose power of choice was very small. Though he was not a man, nor a direct human ancestor, there was yet about him, even in the bone, some trace of that low, snuffling world out of which our forebears had so recently emerged. The skull lay tilted in such a manner that it stared, sightless, up at me as though I, too, were already caught a few feet above him in the strata and, in my turn, were staring upward at that strip of sky which the ages were carrying farther away from me beneath the tumbling debris of falling mountains. The creature had never lived to see a man, and I, what was it I was never going to see?

I restrained a panicky impulse to hurry upward after that receding sky that was outlined above the Slit. Probably, I thought, as I patiently began the task of chiseling into the stone around the skull, I would never again excavate a fossil under conditions which led to so vivid an impression that I was already one myself. The truth is that we are all potential fossils still carrying within our bodies the crudities of former existences, the marks of a world in which living creatures flow with little more consistency than clouds from age to age.

As I tapped and chiseled there in the foundations of the world, I had ample time to consider the cunning manipulability of the human fingers. Experimentally I crooked one of the long slender bones. It might have been silica, I thought, or aluminum, or iron—the cells would have made it possible. But no, it is calcium, carbonate of lime. Why? Only because of its history. Elements more numerous than calcium in the earth's crust could have been used to build the skeleton. Our history is the reason—we came from the water. It was there the cells took the lime habit, and they kept it after we came ashore.

It is not a bad symbol of that long wandering, I thought

again—the human hand that has been fin and scaly reptile foot and furry paw. If a stone should fall (I cocked an eye at the leaning shelf above my head and waited, fatalistically) let the bones lie here with their message, for those who might decipher it, if they come down late among us from the stars.

Above me the great crack seemed to lengthen.

Perhaps there is no meaning in it at all, the thought went on inside me, save that of journey itself, so far as men can see. It has altered with the chances of life, and the chances brought us here; but it was a good journey—long, perhaps—but a good journey under a pleasant sun. Do not look for the purpose. Think of the way we came and be a little proud. Think of this hand—the utter pain of its first venture on the pebbly shore.

Or consider its later wanderings.

I ceased my tappings around the sand-filled sockets of the skull and wedged myself into a crevice for a smoke. As I tamped a load of tobacco into my pipe, I thought of a town across the valley that I used sometimes to visit, a town whose little inhabitants never welcomed me. No sign points to it and I rarely go there any more. Few people know about it and fewer still know that in a sense we, or rather some of the creatures to whom we are related, were driven out of it once, long ago. I used to park my car on a hill and sit silently observant, listening to the talk ringing out from neighbor to neighbor, seeing the inhabitants drowsing in their doorways, taking it all in with nostalgia—the sage smell on the wind, the sunlight without time, the village without destiny. We can look, but we can never go back. It is prairie-dog town.

"Whirl is king," said Aristophanes, and never since life began was Whirl more truly king than eighty million years ago in the dawn of the Age of Mammals. It would come as a shock to those who believe firmly that the scroll of the future is fixed and the roads determined in advance, to observe the teetering balance of earth's history through the age of the Paleocene. The passing of the reptiles had left a hundred uninhabited life zones and a scrambling variety of newly radiating forms. Unheard-of species of giant ground birds threatened for a moment to dominate the earthly scene. Two separate orders of life contended at slightly different intervals for the pleasant grasslands—for the seeds and the sleepy burrows in the sun.

Sometimes, sitting there in the mountain sunshine above prairie-dog town, I could imagine the attraction of that open world after the fern forest damp or the croaking gloom of carboniferous swamps. There by a tree root I could almost make him out, that shabby little Paleocene rat, eternal tramp and world wanderer, father of all mankind. He ruffed his coat in the sun and hopped forward for a seed. It was to be a long time before he would be seen on the grass again, but he was trying to make up his mind. For good or ill there was to be one more chance, but that chance was fifty million years away.

Here in the Paleocene occurred the first great radiation of the placental mammals, and among them were the earliest primates—the zoological order to which man himself belongs. Today, with a few unimportant exceptions, the primates are all arboreal in habit except man. For this reason we have tended to visualize all of our remote relatives as tree dwellers. Recent discoveries, however, have begun to alter this one-sided picture. Before the rise of the true rodents, the highly successful order to which present-day prairie dogs and chipmunks belong, the environment which they occupy had remained peculiarly open to exploitation. Into this zone crowded a varied assemblage of our early relatives.

"In habitat," comments one scholar, "many of these early primates may be thought of as the rats of the Paleocene. With the later appearance of true rodents, the primate habitat was markedly restricted." The bone hunters, in other words, have succeeded in demonstrating that numerous primates reveal a remarkable development of rodent-like characteristics in the teeth and skull during this early period of mammalian evolution. The movement is progressive and distributed in several different groups. One form, although that of a true primate, shows similarities to the modern kangaroo rat, which is, of course, a rodent. There is little doubt that it was a burrower.

It is this evidence of a lost chapter in the history of our kind that I used to remember on the sunny slope above prairie-dog town, and that enables me to say in a somewhat figurative fashion that we were driven out of it once ages ago. We are not, except very remotely as mammals, related to prairie dogs. Nevertheless, through several million years of Paleocene time, the primate order, instead of being confined to trees, was

experimenting to some extent with the same grassland bur-
rowing life that the rodents later perfected. The success of
these burrowers crowded the primates out of this environment
and forced them back into the domain of the branches. As a
result, many primates, by that time highly specialized for a
ground life, became extinct.

In the restricted world of the trees, a "refuge area," as the
zoologist would say, the others lingered on in diminished
numbers. Our ancient relatives, it appeared, were beaten in
their attempt to expand upon the ground; they were dying out
in the temperate zone, and their significance as a widespread
and diversified group was fading. The shabby pseudo-rat I had
seen ruffling his coat to dry after the night damps of the reptile
age, had ascended again into the green twilight of the rain
forest. The chatterers with the ever-growing teeth were his
masters. The sunlight and the grass belonged to them.

It is conceivable that except for the invasion of the rodents,
the primate line might even have abandoned the trees. We
might be there on the grass, you and I, barking in the high-
plains sunlight. It is true we came back in fifty million years
with the cunning hands and the eyes that the tree world gave
us, but was it victory? Once more in memory I saw the high
blue evening fall sleepily upon that village, and once more
swung the car to leave, lifting, as I always did, a figurative lan-
tern to some ambiguous crossroads sign within my brain. The
pointing arms were nameless and nameless were the distances
to which they pointed. One took one's choice.

I ceased my daydreaming then, squeezed myself out of the
crevice, shook out my pipe, and started chipping once more,
the taps sounding along the inward-leaning walls of the Slit
like the echo of many footsteps ascending and descending. I
had come a long way down since morning; I had projected
myself across a dimension I was not fitted to traverse in the
flesh. In the end I collected my tools and climbed painfully up
through the colossal debris of ages. When I put my hands on
the surface of the crack I looked all about carefully in a sudden
anxiety that it might not be a grazing horse that I would see.

He had not visibly changed, however, and I mounted in
some slight trepidation and rode off, having a memory for a
camp—if I had gotten a foot in the right era—which should lie

somewhere over to the west. I did not, however, escape totally from that brief imprisonment.

Perhaps the Slit, with its exposed bones and its far-off vanishing sky, has come to stand symbolically in my mind for a dimension denied to man, the dimension of time. Like the wistaria on the garden wall he is rooted in his particular century. Out of it—forward or backward—he cannot run. As he stands on his circumscribed pinpoint of time, his sight for the past is growing longer, and even the shadowy outlines of the galactic future are growing clearer, though his own fate he cannot yet see. Along the dimension of time, man, like the rooted vine in space, may never pass in person. Considering the innumerable devices by which the mindless root has evaded the limitations of its own stability, however, it may well be that man himself is slowly achieving powers over a new dimension—a dimension capable of presenting him with a wisdom he has barely begun to discern.

Through how many dimensions and how many media will life have to pass? Down how many roads among the stars must man propel himself in search of the final secret? The journey is difficult, immense, at times impossible, yet that will not deter some of us from attempting it. We cannot know all that has happened in the past, or the reason for all of these events, any more than we can with surety discern what lies ahead. We have joined the caravan, you might say, at a certain point; we will travel as far as we can, but we cannot in one lifetime see all that we would like to see or learn all that we hunger to know.

The reader who would pursue such a journey with me is warned that the essays in this book have not been brought together as a guide but are offered rather as a somewhat unconventional record of the prowlings of one mind which has sought to explore, to understand, and to enjoy the miracles of this world, both in and out of science. It is, without doubt, an inconsistent record in many ways, compounded of fear and hope, for it has grown out of the seasonal jottings of a man preoccupied with time. It involves, I see now as I come to put it together, the four ancient elements of the Greeks: mud and the fire within it we call life, vast waters, and something—space, air, the intangible substance of hope which at the last proves unanalyzable by science, yet out of which the human dream is made.

Forward and backward I have gone, and for me it has been an immense journey. Those who accompany me need not look for science in the usual sense, though I have done all in my power to avoid errors in fact. I have given the record of what one man thought as he pursued research and pressed his hands against the confining walls of scientific method in his time. It is not, I must confess at the outset, an account of discovery so much as a confession of ignorance and of the final illumination that sometimes comes to a man when he is no longer careful of his pride. In the last three chapters of the book I have tried to put down such miracles as can be evoked from common earth. But men see differently. I can at best report only from my own wilderness. The important thing is that each man possess such a wilderness and that he consider what marvels are to be observed there.

Finally, I do not pretend to have set down, in Baconian terms, a true, or even a consistent model of the universe. I can only say that here is a bit of my personal universe, the universe traversed in a long and uncompleted journey. If my record, like those of the sixteenth-century voyagers, is confused by strange beasts or monstrous thoughts or sights of abortive men, these are no more than my eyes saw or my mind conceived. On the world island we are all castaways, so that what is seen by one may often be dark or obscure to another.

The Flow of the River

IF THERE IS magic on this planet, it is contained in water. Its least stir even, as now in a rain pond on a flat roof opposite my office, is enough to bring me searching to the window. A wind ripple may be translating itself into life. I have a constant feeling that some time I may witness that momentous miracle on a city roof, see life veritably and suddenly boiling out of a heap of rusted pipes and old television aerials. I marvel at how suddenly a water beetle has come and is submarining there in a spatter of green algae. Thin vapors, rust, wet tar and sun are an alembic remarkably like the mind; they throw off odorous shadows that threaten to take real shape when no one is looking.

Once in a lifetime, perhaps, one escapes the actual confines of the flesh. Once in a lifetime, if one is lucky, one so merges with sunlight and air and running water that whole eons, the eons that mountains and deserts know, might pass in a single afternoon without discomfort. The mind has sunk away into its beginnings among old roots and the obscure tricklings and movings that stir inanimate things. Like the charmed fairy circle into which a man once stepped, and upon emergence learned that a whole century had passed in a single night, one can never quite define this secret; but it has something to do, I am sure, with common water. Its substance reaches everywhere; it touches the past and prepares the future; it moves under the poles and wanders thinly in the heights of air. It can assume forms of exquisite perfection in a snowflake, or strip the living to a single shining bone cast up by the sea.

Many years ago, in the course of some scientific investigations in a remote western county, I experienced, by chance, precisely the sort of curious absorption by water—the extension of shape by osmosis—at which I have been hinting. You have probably never experienced in yourself the meandering roots of a whole watershed or felt your outstretched fingers touching, by some kind of clairvoyant extension, the brooks of snow-line glaciers at the same time that you were flowing toward the Gulf over the eroded debris of worn-down mountains. A poet, MacKnight Black, has spoken of being "limbed . . . with waters gripping pole and pole." He had the idea, all right, and it is obvious that these sensations are not unique, but they are hard to come by; and the sort of extension of the senses that people will accept when they put their ear against a sea shell, they will smile at in the confessions of a bookish professor. What makes it worse is the fact that because of a traumatic experience in childhood, I am not a swimmer, and am inclined to be timid before any large body of water. Perhaps it was just this, in a way, that contributed to my experience.

As it leaves the Rockies and moves downward over the high plains towards the Missouri, the Platte River is a curious stream. In the spring floods, on occasion, it can be a mile-wide roaring torrent of destruction, gulping farms and bridges. Normally, however, it is a rambling, dispersed series of streamlets flowing erratically over great sand and gravel fans that are, in part, the remnants of a mightier Ice Age stream bed. Quicksands and shifting islands haunt its waters. Over it the prairie suns beat mercilessly throughout the summer. The Platte, "a mile wide and an inch deep," is a refuge for any heat-weary pilgrim along its shores. This is particularly true on the high plains before its long march by the cities begins.

The reason that I came upon it when I did, breaking through a willow thicket and stumbling out through ankle-deep water to a dune in the shade, is of no concern to this narrative. On various purposes of science I have ranged over a good bit of that country on foot, and I know the kinds of bones that come gurgling up through the gravel pumps, and the arrowheads of shining chalcedony that occasionally spill out of water-loosened sand. On that day, however, the sight of sky and willows and the weaving net of water murmuring a little in the shallows on its way to the Gulf stirred me, parched as I was with miles of walking, with a new idea: I was going to float. I was going to undergo a tremendous adventure.

The notion came to me, I suppose, by degrees. I had shed my clothes and was floundering pleasantly in a hole among some reeds when a great desire to stretch out and go with this gently insistent water began to pluck at me. Now to this bronzed, bold, modern generation, the struggle I waged with timidity while standing there in knee-deep water can only seem farcical; yet actually for me it was not so. A near-drowning accident in childhood had scarred my reactions; in addition to the fact that I was a nonswimmer, this "inch-deep river" was treacherous with holes and quicksands. Death was not precisely infrequent along its wandering and illusory channels. Like all broad wastes of this kind, where neither water nor land quite prevails, its thickets were lonely and untraversed. A man in trouble would cry out in vain.

I thought of all this, standing quietly in the water, feeling the sand shifting away under my toes. Then I lay back in the floating position that left my face to the sky, and shoved off. The sky wheeled over me. For an instant, as I bobbed into the main channel, I had the sensation of sliding down the vast tilted face of the continent. It was then that I felt the cold needles of the alpine springs at my fingertips, and the warmth of the Gulf pulling me southward. Moving with me, leaving its taste upon my mouth and spouting under me in dancing springs of sand, was the immense body of the continent itself, flowing like the river was flowing, grain by grain, mountain by mountain, down to the sea. I was streaming over ancient sea beds thrust aloft where giant reptiles had once sported; I was wearing down the face of time and trundling cloud-wreathed ranges into oblivion. I touched my margins with the delicacy of a crayfish's antennae, and felt great fishes glide about their work.

I drifted by stranded timber cut by beaver in mountain fastnesses; I slid over shallows that had buried the broken axles of prairie schooners and the mired bones of mammoth. I was streaming alive through the hot and working ferment of the sun, or oozing secretively through shady thickets. I *was* water and the unspeakable alchemies that gestate and take shape in water, the slimy jellies that under the enormous magnification of the sun writhe and whip upward as great barbeled fish mouths, or sink indistinctly back into the murk out of which they arose. Turtle and fish and the pinpoint chirpings of individual frogs are all watery projections, concentrations—as man himself is a concentration—of that indescribable and liquid brew which is compounded in varying proportions of salt and sun and time. It has appearances, but at its heart lies water, and as I was finally edged gently against a sand bar and dropped like any log, I tottered as I rose. I knew once more the body's revolt against emergence into the harsh and unsupporting air, its reluctance to break contact with that mother element which still, at this late point in time, shelters and brings into being nine tenths of everything alive.

As for men, those myriad little detached ponds with their own swarming corpuscular life, what were they but a way that water has of going about beyond the reach of rivers? I, too, was a microcosm of pouring rivulets and floating driftwood gnawed by the mysterious animalcules of my own creation. I was three fourths water, rising and subsiding according to the hollow knocking in my veins: a minute pulse like the eternal pulse that lifts Himalayas and which, in the following systole, will carry them away.

Thoreau, peering at the emerald pickerel in Walden Pond, called them "animalized water" in one of his moments of strange insight. If he had been possessed of the geological knowledge so laboriously accumulated since his time, he might have gone further and amusedly detected in the planetary rumblings and eructations which so delighted him in the gross habits of certain frogs, signs of that dark interior stress which has reared sea bottoms up to mountainous heights. He might have developed an acute inner ear for the sound of the surf on Cretaceous beaches where now the wheat of Kansas rolls. In any case, he would have seen, as the long trail of life was unfolded by the fossil hunters, that his animalized water had changed its shapes eon by eon to the beating of the earth's dark millennial heart. In the swamps of the low continents, the amphibians had flourished and had their day; and as the long skyward swing—the isostatic response of the crust—had come about, the era of the cooling grasslands and mammalian life had come into being.

A few winters ago, clothed heavily against the weather, I wandered several miles along one of the tributaries of that same Platte I had floated down years before. The land was stark and ice-locked. The rivulets were frozen, and over the marshlands the willow thickets made such an array of vertical lines against the snow that tramping through them produced strange optical illusions and dizziness. On the edge of a frozen backwater, I stopped and rubbed my eyes. At my feet a raw prairie wind had swept the ice clean of snow. A peculiar green object caught my eye; there was no mistaking it.

Staring up at me with all his barbels spread pathetically, frozen solidly in the wind-ruffled ice, was a huge familiar face. It was one of those catfish of the twisting channels, those dwellers in the yellow murk, who had been about me and beneath me on the day of my great voyage. Whatever sunny dream had kept him paddling there while the mercury plummeted downward and that Cheshire smile froze slowly, it would be hard to say. Or perhaps he was trapped in a blocked channel and had simply kept swimming until the ice contracted around him. At any rate, there he would lie till the spring thaw.

At that moment I started to turn away, but something in the bleak, whiskered face reproached me, or perhaps it was the river calling to her children. I termed it science, however—a convenient rational phrase I reserve for such occasions—and decided that I would cut the fish out of the ice and take him home. I had no intention of eating him. I was merely struck by a sudden impulse to test the survival qualities of high-plains fishes, particularly fishes of this type who get themselves immured in oxygenless ponds or in cut-off oxbows buried in winter drifts. I blocked him out as gently as possible and dropped him, ice and all, into a collecting can in the car. Then we set out for home.

Unfortunately, the first stages of what was to prove a remarkable resurrection escaped me. Cold and tired after a long drive, I deposited the can with its melting water and ice in the basement. The accompanying corpse I anticipated I would either dispose of or dissect on the following day. A hurried glance had revealed no signs of life.

To my astonishment, however, upon descending into the basement several hours later, I heard stirrings in the receptacle and peered in. The ice had melted. A vast pouting mouth ringed with sensitive feelers confronted me, and the creature's gills labored slowly. A thin stream of silver bubbles rose to the surface and popped. A fishy eye gazed up at me protestingly.

"A tank," it said. This was no Walden pickerel. This was a yellow-green, mud-grubbing, evil-tempered inhabitant of floods and droughts and cyclones. It was the selective product of the high continent and the waters that pour across it. It had outlasted prairie blizzards that left cattle standing frozen up-right in the drifts.

"I'll get the tank," I said respectfully.

He lived with me all that winter, and his departure was to-tally in keeping with his sturdy, independent character. In the spring a migratory impulse or perhaps sheer boredom struck him. Maybe, in some little lost corner of his brain, he felt, far off, the pouring of the mountain waters through the sandy coverts of the Platte. Anyhow, something called to him, and he went. One night when no one was about, he simply jumped out of his tank. I found him dead on the floor next morning. He had made his gamble like a man—or, I should say, a fish. In the proper place it would not have been a fool's gamble. Fishes in the drying shallows of intermittent prairie streams who feel their confinement and have the impulse to leap while there is yet time may regain the main channel and survive. A million ancestral years had gone into that jump, I thought as I looked at him, a million years of climbing through prairie sunflowers and twining in and out through the pillared legs of drinking mammoth.

"Some of your close relatives have been experimenting with air breathing," I remarked, apropos of nothing, as I gathered him up. "Suppose we meet again up there in the cottonwoods in a million years or so."

I missed him a little as I said it. He had for me the kind of lost archaic glory that comes from the water brotherhood. We were both projections out of that timeless ferment and locked as well in some greater unity that lay incalculably beyond us. In many a fin and reptile foot I have seen myself passing by—some part of myself, that is, some part that lies unrealized in the momentary shape I inhabit. People have occasionally written me harsh letters and castigated me for a lack of faith in man when I have ventured to speak of this matter in print. They distrust, it would seem, all shapes and thoughts but their own. They would bring God into the compass of a shopkeep-er's understanding and confine Him to those limits, lest He proceed to some unimaginable and shocking act—create per-haps, as a casual afterthought, a being more beautiful than man. As for me, I believe nature capable of this, and having been part of the flow of the river, I feel no envy—any more than the frog envies the reptile or an ancestral ape should envy man.

Every spring in the wet meadows and ditches I hear a little shrilling chorus which sounds for all the world like an endlessly reiterated "We're here, we're here, we're here." And so they are, as frogs, of course. Confident little fellows. I suspect that to some greater ear than ours, man's optimistic pronouncements about his role and destiny may make a similar little ringing sound that travels a small way out into the night. It is only its nearness that is offensive. From the heights of a mountain, or a marsh at evening, it blends, not too badly, with all the other sleepy voices that, in croaks or chirrups, are saying the same thing.

After a while the skilled listener can distinguish man's noise from the katydid's rhythmic assertion, allow for the offbeat of a rabbit's thumping, pick up the autumnal monotone of crickets, and find in all of them a grave pleasure without admitting any to a place of preëminence in his thoughts. It is when all these voices cease and the waters are still, when along the frozen river nothing cries, screams or howls, that the enormous mindlessness of space settles down upon the soul. Somewhere out in that waste of crushed ice and reflected stars, the black waters may be running, but they appear to be running without life toward a destiny in which the whole of space may be locked in some silvery winter of dispersed radiation.

It is then, when the wind comes straitly across the barren marshes and the snow rises and beats in endless waves against the traveler, that I remember best, by some trick of the imagination, my summer voyage on the river. I remember my green extensions, my catfish nuzzlings and minnow wrigglings, my gelatinous materializations out of the mother ooze. And as I walk on through the white smother, it is the magic of water that leaves me a final sign.

Men talk much of matter and energy, of the struggle for existence that molds the shape of life. These things exist, it is true; but more delicate, elusive, quicker than the fins in water, is that mysterious principle known as "organization," which leaves all other mysteries concerned with life stale and insignificant by comparison. For that without organization life does not persist is obvious. Yet this organization itself is not strictly the product of life, nor of selection. Like some dark and passing shadow within matter, it cups out the eyes' small windows or spaces the notes of a meadow lark's song in the interior of a mottled egg. That principle—I am beginning to suspect—was there before the living in the deeps of water.

The temperature has risen. The little stinging needles have given way to huge flakes floating in like white leaves blown from some great tree in open space. In the car, switching on the lights, I examine one intricate crystal on my sleeve before it melts. No utilitarian philosophy explains a snow crystal, no doctrine of use or disuse. Water has merely leapt out of vapor and thin nothingness in the night sky to array itself in form. There is no logical reason for the existence of a snowflake any more than there is for evolution. It is an apparition from that mysterious shadow world beyond nature, that final world which contains—if anything contains—the explanation of men and catfish and green leaves.

The Great Deeps

THERE IS A night world that few men have entered and from whose greatest depths none have returned alive—the abyssal depths of the sea. Darwin's associates dreamed of its hovering and intangible shapes as possibly those of the lost Paleozoic world. The great naturalist himself pleaded with outbound voyagers: "Urge the use of the dredge in the tropics; how little or nothing we know of the limit of life downward in the hot seas."

Anything that has been supposedly dead for a hundred million years—anything that no living eye has beheld except in the chalks of vanished geological epochs—is monstrous when you find it alive and pulsing in your hand. But this was the experience of Sir Charles Thomson, one of the first explorers of the North Atlantic sea bed. Few men in the years since have laid hands or eyes upon living denizens of the fossil kingdom, and this was an adventure no man could forget. The discovery influenced indirectly the formation of the world's largest oceanographic expedition and made Sir Charles its leader. Speaking of his find, long afterwards, he said: "It was like a little round red cake. And like a little round red cake, it began to pant there in my hand. Curious undulations were passing through it and I had to summon up all my resolution before handling the weird little monster."

Now to the ordinary man that little round red cake would have been a sea urchin, and whether it panted would have meant nothing at all except that it was alive. Nevertheless, the man in the street would have been wrong. The fact *was* monstrous and the little red sea urchin more startling still. Even the "panting" had significance. No living sea urchin had ever been observed in such a performance. The known forms were all too rigid. The undulations of this little beast were a sure sign of its relationship to a more leathery and flexible ancestral group.

As a living fossil it had been dredged out of the North Atlantic sea bed almost a solid mile below the surface. A mile today is not a great depth compared with the six-mile depression of the

Tuscarora Deep, but in the sixties of the last century—Sir Charles Thomson's time—it was below the level at which life was generally supposed to exist. Anything below three hundred fathoms was Azoic, lifeless—so wrote Edward Forbes, the first great oceanographer of the eighteen forties. Like many pioneers he was destined to be proved wrong, yet looking back, it is possible to sympathize. The cold, the dark, the pressure of those unknown depths was frightening to contemplate. The human mind shied subconsciously away from the notion that even here sentient beings had groped their way down into the primeval slime of the sea floor. It was the world of the abyss, supposedly as lifeless as the earth's first midnight.

Today we know that the abyss is haunted. Through it drift luminous jack-o'-lantern faces with wolf-trap mouths and meager bodies, as though a head floating in that enormous darkness were more important than a body, which could almost be dispensed with in the lean economy of the night. It is a world of delicately groping yard-long antennae, or of great staring eyes that can pick up remote pinpoints of light and follow them through the restless luminescence of a firefly darkness. To Sir Charles Thomson, however, the abyss was more than haunted. It was the world of the past.

The fascination of lost worlds has long preoccupied humanity. It is inevitable that transitory man, student of the galaxies and computer of light-years, should entertain nostalgic yearnings for some island outside of time, some Avalon untouched by human loss. Even the scholar has not been averse to searching for the living past on islands or precipice-guarded plateaus. Jefferson repeated the story of a trapper who had heard the mammoth roaring in the Virginia woods; in 1823 a South American traveler imaginatively viewed through his spyglass mastodons grazing in remote Andean valleys.

Nevertheless when the explorers had penetrated the last woodland, gazed on the last new animal—something they had pretty well accomplished by the middle of the nineteenth century—the past had been nowhere found. Only the great waters remained, the planetary expanse that, since the days of Thorfinn the Skull Cleaver, has received and, on occasion, swallowed the restless sails of men. Its surface was known, but its depths

remained unplumbed. The treasure of countless piracies, the
dead of innumerable battles had gone down into the green
gloom of the mermaids' kingdom. In return, men had had
only the legendary glimpse of a white arm at evening or the
voice of a siren singing from some isle that would be gone at
daybreak. Later, as men's youthful imaginations faded, only
the rumor of sea monsters—serpents or archaic water beasts—
survived from the abyss.

From a belief that the great deeps were lifeless, scholars examining the growths on submarine cables and the scrapings
brought up by newly devised dredges began to visualize something like Conan Doyle's *Lost World* in reverse. By 1870 this
conception had two aspects: first, a theory that the ocean
depths were populated by the living marine fossils of past
geological ages which had here escaped the disasters that had
destroyed their kind in the shallow seas of the earlier world;
second—and reflecting the materialistic philosophy which was
beginning to arise under the stimulus of the Darwinian theory
—a belief that widespread on the floor of the abyssal plain lay
the "*Urschleim*," a protoplasmic half-living matter representing
that transition between the living and the nonliving out of
which more complex life had, in the course of time, developed.
The abyss, in other words, was thought to contain not only
the living record of the past, but the ultimate secret of life itself; Creation might still be in process. Sir Charles Thomson in
one enthusiastic statement in his *Depths of the Sea* even ventured to maintain: "The [depth] range of the various groups in
modern seas corresponds remarkably with their vertical range
in ancient strata." Down at the bottom, of course, lay that living undifferentiated primordial ooze as deep in the sea as it lay
deep in time.

As the number of deep-sea soundings increased, as men
slowly grasped the antiquity of that dark, cold world that is
called the abyssal plain, a new idea arose: the notion, as I have
hinted, of a lost world in reverse, a midnight city of refuge in
which the present mingled and lived on with the past. It was,
of course, the world of the uttermost depths, the place without
light since the beginning, and whose extent no continent
above the waters could ever fill.

Of all the worlds of life the abyss alone remains unaltered. It is the one place on the planet where conditions remain as they have been since the beginning, where the five-mile pressures have not altered, where no suns have ever shone, where the cold is the same at the poles as at the equator, where the seasons are unchanging, where there is no wind and no wave to stir the ooze above which the glass sponges rise on graceful stems, or the abyssal sea squirts float like little balloons on strings above the mud. This is the sole world on the planet which we can enter only by a great act of the imagination. There has been, perhaps, only one greater imaginative effort—the attempt of nineteenth-century biology, intoxicated by its own successes, to observe on the sea floor life in the process of becoming, to glimpse in the abyssal oozes the crossing between life and death.

The story begins with the laying of the first Atlantic cable in the sixties of the last century. It involves one of the most peculiar and fantastic errors ever committed in the name of science. It is useless to blame this error upon one man because many leading figures of the day participated in what was, and remains, one of the most curious cases of self-delusion ever indulged in by scholars. It was the product of an overconfident materialism, a vainglorious assumption that the secrets of life were about to be revealed.

Haeckel in Germany and Huxley in England were proceeding to show that as one passed below the stage of nucleated single-celled organisms one arrived at a simple stirring of the abyssal slime wherein something that was neither life nor non-life oozed and fed without cellular individuality.

This soft, gelatinous matter had been taken from the ocean bed during dredging operations. Examined and pronounced upon by Professor Huxley, it was given the name of *Bathybius haeckelii* in honor of his great German colleague. Speaking before the Royal Geographical Society in 1870, Huxley confidently maintained that *Bathybius* formed a living scum or film on the sea bed extending over thousands of square miles. Moreover, he expanded, it probably formed a continuous sheet of living matter girdling the whole surface of the earth.

Sir Charles Thomson shared this view, commenting that the "organism" showed "no trace of differentiation of organs" and consisted apparently "of an amorphous sheet of a protein compound, irritable to a low degree and capable of assimilating food . . . a diffused formless protoplasm." Haeckel conceived of these formless "monera" as arising from non-living matter, their vital phenomena being traceable to "physicochemical causes." Here was the "*Urschleim*" with a vengeance, the seething, unindividualized ooze whose potentialities included the butterfly and the rose. Man was mud and mud was man. Mechanism was the order of the day.

Unfortunately for this beautiful theory wistfully remembered by one writer as "explaining so much," *Bathybius* proved to be what the microscopists call an artifact; that is, it did not exist. A certain unfeeling Mr. Buchanan of the *Challenger* Expedition discovered, as he tried to investigate the nature of *Bathybius*, that he could produce all the characters of that indescribable animal by the simple process of adding strong alcohol to sea water. It was not necessary to drink the potion. One simply examined a specimen under the lens and observed that sulphate of lime was precipitated in the form of a gelatinous ooze which clung around particles as though ingesting them, thus lending a superficial protoplasmic appearance to the solution.

Mr. Huxley's original specimen had apparently been treated in this manner when it was sent to him. Huxley took the episode in good grace, but it was a severe blow to the materialists. The structureless protoplasmic "*Urschleim*" was a projective dream of scientists striving to build an evolutionary family tree upon existing organisms. Being nineteenth-century zoologists they unfortunately forgot the world of microscopic plant life, its basic position in the nourishment of living things, and the fact that it must have sunlight in order to perform its mysterious green miracles.

The abyss, it was now to be learned, whatever might roam its waters or slither wetly through its midnights, was not the original abode of life. If there was a past on the black plain far beneath us, if indeed the strange life of remote eras lingered there, it was not stacked with the layered neatness of geological strata as some oceanographers had imagined. The floating heads with their starveling bodies, the squid which emitted clouds of luminescent ink and vanished in their own bright explosions, were all a part of one of life's strangest qualities—its eternal dissatisfaction with what is, its persistent habit of reaching out into new environments and, by degrees, adapting itself to the most fantastic circumstances.

Once long ago as a child I can remember removing the cover from an old well. I was alone at the time and I can still anticipate, with a slight crawling of my scalp, the sight I inadvertently saw as I peered over the brink and followed a shaft of sunlight many feet down into the darkness. It touched, just touched in passing, a rusty pipe which projected across the well space some twenty feet above the water. And there, secretive as that very underground whose mystery had lured me into this adventure, I saw, passing surely and unhurriedly into the darkness, a spidery thing of hair and many legs. I set the rotting cover of boards back into place with a shiver, but that unidentifiable creature of the well has stayed with me to this day.

For the first time I must have realized, I think, the frightening diversity of the living; something that did not love the sun was down there, something that could walk through total darkness upon slender footholds over evil waters, something that had come down there by preference from above. It was in this way that the oceanic abyss was entered: by preference from above. Life did not arise on the bottom; the muds of the deep waters did not compound it. Instead, with its own pale lanterns or with the delicate, strawlike feelers of blindness, it has groped its way down into the dark.

The four-year voyage of the *Challenger* under the auspices of the British Admiralty, beginning in 1872, was the most ambitious project to investigate the ocean depths that men had ever attempted. The vessel was equipped with floating laboratories and a staff of naturalists. She traveled sixty-nine thousand nautical miles, took hundreds of soundings, and the observations of her staff of investigators occupy fifty huge volumes.

When the *Challenger* left port oceanography was still essentially a speculative science. Her biological director, Sir Charles Wyville Thomson, the same zoologist who had dredged the little red sea urchin out of the North Atlantic, believed along with many of his colleagues that the deep recesses of the ocean, unchanging through the ages, would reveal "living fossils," actual missing links in the history of life. Thomas Huxley, then at the height of his powers, proclaimed with characteristic vigor:

> *It may be confidently assumed that . . . the things brought up will . . . be zoological antiquities which in the tranquil and little changed depths of the ocean have escaped the causes of destruction at work in the shallows and represent the predominant population of a past age.*

This view was enthusiastically shared by the great Swiss naturalist, Louis Agassiz, who contended that in deep waters "we should expect to find representatives of earlier geological periods." Agassiz went even further and observed that it was the deep waters which today most closely approximated the conditions under which life had originally emerged. It was, he said, the depths of the ocean alone which could place animals under a pressure such as he believed corresponded to the heavy atmosphere of a young world.

These were the excited dreams of science in 1872 as the *Challenger* steamed out of port. Sixty-nine thousand miles and four years later her weary scientists came home. They had rocked sickeningly in all seas, had dragged with cumbersome and ill-devised apparatus the very bowels of Creation. They had handled rare forms of life, looked on things denied to ordinary men, and, above all, they had laid the foundations of a true science of the sea. Nevertheless, their eyes were empty.

The great globe-girdling carpet of the living ooze was gone—that evolutionary base in which the German scholars had seen "an infinite capacity for improvement in every conceivable direction." "Our ardor," wearily confessed Moseley, the coral specialist, "abated somewhat . . . as the same tedious animals kept appearing from the depths in all parts of the world."

In the beginning even the cabin boys had crowded to see what four miles of rope would bring up from the bottom. Gradually, however, as the novelty wore off, the spectators became fewer. Even members of the scientific staff were not always present, particularly when the dredge arrived during the dinner hour.

The great hopes of the beginning were fading in disappointment, but Moseley gives an unforgettable picture of Sir Charles
Thomson's sturdy persistence and enthusiasm in the face of the
collapse of his theories. "To the last," he writes, "every cuttle-
fish which came up in our deep sea net was squeezed to see if
it had a belemnite's bone in its back, and trilobites were eagerly
looked out for." Either of these events would have found the
world of the Paleozoic floundering alive on the deck of the
Challenger. To the despair of Sir Charles they never appeared.
It is true that here and there a few animals were recovered that
were believed to be extinct and to exist only as fossils, but
these were only such discoveries as might be expected when
any vast unexplored region is first investigated, whether it be
land or sea.

The secret and remote abysses were yielding not the protected
remnants of the very earliest world, but a scattering of later
antique types along with a more modern abyssal fauna obviously related to, and descended from, the swarming creatures
of the shallow seas and upper waters. Such ancient forms as
survive in the abyss represent adaptations and migrations that
took place in antiquity from the continental shoals far above.
In that sense the midnight timeless city does indeed exist, for
in those depths the ages overlap and some few elements of the
older world, losing out in competition with more highly
evolved and modern types, have chosen to slip by degrees into
the freezing cold of the abyss. Here in the unchanging mud
and comforting darkness they have survived. After them in
time have come others, groping into that enormous cellar with
lanterns or light-magnifying eyes—clever adaptations possible
to squids and higher vertebrates.

Even among the mammals, the great sperm whale has come
sounding down into the fearful pressures of the kraken's world,
the last of all to enter, and capable of enduring only moments
on what is actually the upper edge of the abyss. If it is a place
of refuge it is also, we know now, a famine world. There is no
vegetable life below there. All that lives preys on others or on
the dead raining down from above. This is the reason for the
curiously abbreviated bodies of many of the fishes and their

enormous jaws; this is the reason why we know that life came relatively late to the abyss.

According to the biochemists the conditions under which cellular life is possible are very restricted, nor have they changed in any marked degree since life began. At first glance this statement seems absurd. Life has crept upward from the waters, it crawls in the fields, it penetrates the air, it is not unknown even in the frozen wastes of the Antarctic. Surely this enormous diversity is the very reverse of restriction.

The answer, of course, lies in that modest little phrase "the conditions of cellular life." All of the tremendous differences between living forms have been achieved only by the elaboration of devices for the maintenance of that inner nourishing liquidity in which cells can live and grow within a certain narrow range of tolerance. Not for nothing has the composition of mammalian blood led to our description as "walking sacks of sea water." Not for nothing did the great French physiologist Bernard comment that "the stability of the interior environment is the condition of free life."

The drifting cell masses of the early ocean lived in a nutrient solution. Salt and sun and moisture were accessible without great mechanical elaboration. It was the reaching out that changed this pattern, the reaching out that forced the cells to bring the sea ashore with them, to elaborate in their own bodies the very miniature of that all-embracing sea from which they came. It was the reaching out, that magnificent and age-long groping that only life—blindly and persistently among stones and the indifference of the entire inanimate universe—can continue to endure and prolong.

Men have worked in many places. They have seen this sea-born protoplasm creeping upward in the shape of lichens, among the howling winds of snow-clad mountains. They have seen it in the delicate "snowshoe" feet of desert lizards devised for running over sand. From some unknown spot, most probably along the shoals above the continental shelf, it has reached out into lakes and grasslands, edged stealthily into deserts, learned even to endure the heat of boiling springs or to hatch eggs, like the emperor penguin, in the blizzards by the southern pole. It has similarly found its way into the downward coursing streams of the abyss. It has solved the pressures of the ocean bottom as it has survived the rarefied air of the highest mountains. In these difficult surroundings life thins a little; the inventions that support it grow more difficult to produce and the intrusions are apt to be late, because life has experimented last in these bleak planetary wastelands.

Nevertheless the reaching out that began a billion years ago is still in process. The cells, so carefully transferring their limited range of endurance through astounding extremes of heat and frost and pressure, show no inclination toward content. Content is a word unknown to life; it is also a word unknown to man.

In 1949, on the White Sands proving grounds, a Wac Corporal rocket reached an altitude of 250 miles and, on the verge of outer space, paused and fell back. Somehow I like to think of those rockets, pounding year after year at that ocean of air, roaring away into an immensity from which, before long, one will not come back. Sometimes, walking in the star-sprinkled evenings, I think of that almost forgotten theory of Arrhenius that the spores of life came originally from outer space.

Perhaps that explains it, I think wistfully—life reaching out, groping for a billion years, life desperate to go home.

The nineteenth-century mechanists, at least, did not find our origins in the abyss, and every bubble of the chemist's broth has left the secret of life as inscrutably remote as ever. The ingredients are known; they are to be had on any drug-store shelf. You can take them yourself and pour them and wait hopefully for the resulting slime to crawl. It will not. The beautiful pulse of streaming protoplasm, that unknown organization of an unstable chemistry which makes up the life process, will not begin. Carbon, nitrogen, hydrogen, and oxygen you have mixed, and the same dead chemicals they remain.

Shape of sea water and carbon rings, yet simultaneously a perplexed professor on a village street, I look up across the moon and Venus—outward, outward into that blue-white glitter beyond the galaxy. And as I look and shiver I feel the voice in every fiber of my being: Have we come from elsewhere? By these our instruments shall we go home? Whatever the beginning, and by whatever mechanical extensions, life is about to cross into the open domain of space. Has not the great 200-inch reflector upon Mount Palomar already spied out the prospect?

A billion years have gone into the making of that eye; the water and the salt and the vapors of the sun have built it; things that squirmed in the tide silts have devised it. Light-year beyond light-year, deep beyond deep, the mind may rove by means of it, hanging above the bottomless and surveying impartially the state of matter in the white-dwarf suns.

Yet whenever I see a frog's eye low in the water warily ogling the shoreward landscape, I always think inconsequentially of those twiddling mechanical eyes that mankind manipulates nightly from a thousand observatories. Someday, with a telescopic lens an acre in extent, we are going to see something not to our liking, some looming shape outside there across the great pond of space.

Whenever I catch a frog's eye I am aware of this, but I do not find it depressing. I stand quite still and try hard not to move or lift a hand since it would only frighten him. And standing thus it finally comes to me that this is the most enormous extension of vision of which life is capable: the projection of itself into other lives. This is the lonely, magnificent power of humanity. It is, far more than any spatial adventure, the supreme epitome of the reaching out.

The Snout

I HAVE LONG BEEN an admirer of the octopus. The cephalo-
pods are very old, and they have slipped, protean, through
many shapes. They are the wisest of the mollusks, and I have
always felt it to be just as well for us that they never came
ashore, but—there are other things that have.

There is no need to be frightened. It is true some of the
creatures are odd, but I find the situation rather heartening than
otherwise. It gives one a feeling of confidence to see nature still
busy with experiments, still dynamic, and not through nor satisfied because a Devonian fish managed to end as a two-legged
character with a straw hat. There are other things brewing and
growing in the oceanic vat. It pays to know this. It pays to know
there is just as much future as there is past. The only thing that
doesn't pay is to be sure of man's own part in it.

There are things down there still coming ashore. Never make
the mistake of thinking life is now adjusted for eternity. It gets
into your head—the certainty, I mean—the human certainty, and
then you miss it all: the things on the tide flats and what they
mean, and why, as my wife says, "they ought to be watched."

The trouble is we don't know what to watch for. I have a
friend, one of these Explorers Club people, who drops in now
and then between trips to tell me about the size of crocodile
jaws in Uganda, or what happened on some back beach in
Arnhem Land.

"They fell out of the trees," he said. "Like rain. And into the
boat."

"Uh?" I said, noncommittally.

"They did *so*," he protested, "and they were hard to catch."

"Really—" I said.

"We were pushing a dugout up one of the tidal creeks in
northern Australia and going fast when *smacko* we jam this
mangrove bush and the things come tumbling down.

"What were they doing sitting up there in bunches? I ask
you. It's no place for a fish. Besides that they had a way of sidling off with those popeyes trained on you. I never liked it.
Somebody ought to keep an eye on them."

33

"Why?" I asked.

"I don't know why," he said impatiently, running a rough, square hand through his hair and wrinkling his forehead. "I just mean they make you feel that way, is all. A fish belongs in the water. It ought to stay there—just as we live on land in houses. Things ought to know their place and stay in it, but those fish have got a way of sidling off. As though they had mental reservations and weren't keeping any contracts. See what I mean?"

"I see what you mean," I said gravely. "They ought to be watched. My wife thinks so too. About a lot of things."

"She does?" He brightened. "Then that's two of us. I don't know why, but they give you that feeling."

He didn't know why, but I thought that I did.

It began as such things always begin—in the ooze of unnoticed swamps, in the darkness of eclipsed moons. It began with a strangled gasping for air.

The pond was a place of reek and corruption, of fetid smells and of oxygen-starved fish breathing through laboring gills. At times the slowly contracting circle of the water left little windrows of minnows who skittered desperately to escape the sun, but who died, nevertheless, in the fat, warm mud. It was a place of low life. In it the human brain began.

There were strange snouts in those waters, strange barbels nuzzling the bottom ooze, and there was time—three hundred million years of it—but mostly, I think, it was the ooze. By day the temperature in the world outside the pond rose to a frightful intensity; at night the sun went down in smoking red. Dust storms marched in incessant progression across a wilderness whose plants were the plants of long ago. Leafless and weird and stiff they lingered by the water, while over vast areas of grassless uplands the winds blew until red stones took on the polish of reflecting mirrors. There was nothing to hold the land in place. Winds howled, dust clouds rolled, and brief erratic torrents choked with silt ran down to the sea. It was a time of dizzying contrasts, a time of change.

On the oily surface of the pond, from time to time a snout thrust upward, took in air with a queer grunting inspiration, and swirled back to the bottom. The pond was doomed, the water was foul, and the oxygen almost gone, but the creature would not die. It could breathe air direct through a little accessory lung, and it could walk. In all that weird and lifeless landscape, it was the only thing that could. It walked rarely and under protest, but that was not surprising. The creature was a fish.

In the passage of days the pond became a puddle, but the Snout survived. There was dew one dark night and a coolness in the empty stream bed. When the sun rose next morning the pond was an empty place of cracked mud, but the Snout did not lie there. He had gone. Down stream there were other ponds. He breathed air for a few hours and hobbled slowly along on the stumps of heavy fins.

It was an uncanny business if there had been anyone there to see. It was a journey best not observed in daylight, it was something that needed swamps and shadows and the touch of the night dew. It was a monstrous penetration of a forbidden element, and the Snout kept his face from the light. It was just as well, though the face should not be mocked. In three hundred million years it would be our own.

There was something fermenting in the brain of the Snout. He was no longer entirely a fish. The ooze had marked him. It takes a swamp-and-tide-flat zoologist to tell you about life; it is in this domain that the living suffer great extremes, it is here that the water-failures, driven to desperation, make starts in a new element. It is here that strange compromises are made and new senses are born. The Snout was no exception. Though he breathed and walked primarily in order to stay in the water, he was coming ashore.

He was not really a successful fish except that he was managing to stay alive in a noisome, uncomfortable, oxygen-starved environment. In fact the time was coming when the last of his kind, harried by more ferocious and speedier fishes, would slip off the edge of the continental shelf, to seek safety in the sunless abysses of the deep sea. But the Snout was a fresh-water Crossopterygian, to give him his true name, and cumbersome and plodding though he was, something had happened back of his eyes. The ooze had gotten in its work.

It is interesting to consider what sort of creatures we, the remote descendants of the Snout, might be, except for that green quagmire out of which he came. Mammalian insects perhaps we should have been—solid-brained, our neurones wired for mechanical responses, our lives running out with the perfection of beautiful, intricate, and mindless clocks. More likely we should never have existed at all. It was the Snout and the ooze that did it. Perhaps there also, among rotting fish heads and blue, night-burning bog lights, moved the eternal mystery, the careful finger of God. The increase was not much. It was two bubbles, two thin-walled little balloons at the end of the Snout's small brain. The cerebral hemispheres had appeared.

Among all the experiments in that dripping, ooze-filled world, one was vital: the brain had to be fed. The nerve tissues are insatiable devourers of oxygen. If they do not get it, life is gone. In stagnant swamp waters, only the development of a highly efficient blood supply to the brain can prevent disaster. And among those gasping, dying creatures, whose small brains winked out forever in the long Silurian drought, the Snout and his brethren survived.

Over the exterior surface of the Snout's tiny brain ran the myriad blood vessels that served it; through the greatly enlarged choroid plexuses, other vessels pumped oxygen into the spinal fluid. The brain was a thin-walled tube fed from both surfaces. It could only exist as a thing of thin walls permeated with oxygen. To thicken, to lay down solid masses of nervous tissue such as exist among the fishes in oxygenated waters was to invite disaster. The Snout lived on a bubble, two bubbles in his brain.

It was not that his thinking was deep; it was only that it had to be thin. The little bubbles of the hemispheres helped to spread the area upon which higher correlation centers could be built, and yet preserve those areas from the disastrous thickenings which meant oxygen death to the swamp dweller. There is a mystery about those thickenings which culminate in the so-called solid brain. It is the brain of insects, of the modern fishes, of some reptiles and all birds. Always it marks the appearance of elaborate patterns of instinct and the end of thought. A road has been taken which, anatomically, is well-nigh irretraceable; it does not lead in the direction of a high order of consciousness.

Wherever, instead, the thin sheets of gray matter expand upward into the enormous hemispheres of the human brain, laughter, or it may be sorrow, enters in. Out of the choked Devonian waters emerged sight and sound and the music that rolls invisible through the composer's brain. They are there still in the ooze along the tideline, though no one notices. The world is fixed, we say: fish in the sea, birds in the air. But in the mangrove swamps by the Niger, fish climb trees and ogle uneasy naturalists who try unsuccessfully to chase them back to the water. There are things still coming ashore.

The door to the past is a strange door. It swings open and things pass through it, but they pass in one direction only. No man can return across that threshold, though he can look down still and see the green light waver in the water weeds.

There are two ways to seek the doorway: in the swamps of the inland waterways and along the tide flats of the estuaries where rivers come to the sea. By those two pathways life came ashore. It was not the magnificent march through the breakers and up the cliffs that we fondly imagine. It was a stealthy advance made in suffocation and terror, amidst the leaching bite of chemical discomfort. It was made by the failures of the sea.

Some creatures have slipped through the invisible chemical barrier between salt and fresh water into the tidal rivers, and later come ashore; some have crept upward from the salt. In all cases, however, the first adventure into the dreaded atmosphere seems to have been largely determined by the inexorable crowding of enemies and by the retreat further and further into marginal situations where the oxygen supply was depleted. Finally, in the ruthless selection of the swamp margins, or in the scramble for food on the tide flats, the land becomes home.

Not the least interesting feature of some of the tide-flat emergents is their definite antipathy for the full tide. It obstructs their food-collecting on the mud banks and brings their enemies. Only extremes of fright will drive them into the water for any period.

I think it was the great nineteenth-century paleontologist Cope who first clearly enunciated what he called the "law of the unspecialized," the contention that it was not from the most highly organized and dominant forms of a given geological era that the master type of a succeeding period evolved, but that instead the dominant forms tended to arise from more lowly and generalized animals which were capable of making new adaptations, and which were not narrowly restricted to a given environment.

There is considerable truth to this observation, but, for all that, the idea is not simple. Who is to say without foreknowledge of the future which animal is specialized and which is not? We have only to consider our remote ancestor, the Snout, to see the intricacies into which the law of the unspecialized may lead us.

If we had been making zoological observations in the Paleozoic Age, with no knowledge of the strange realms life was to penetrate in the future, we would probably have regarded the Snout as specialized. We would have seen his air-bladder lung, his stubby, sluggish fins, and his odd ability to wriggle overland as specialized adaptations to a peculiarly restricted environmental niche in stagnant continental waters. We would have thought in water terms and we would have dismissed the Snout as an interesting failure off the main line of progressive evolution, escaping from his enemies and surviving successfully only in the dreary and marginal surroundings scorned by the swift-finned teleost fishes who were destined to dominate the seas and all quick waters.

Yet it was this poor specialization—this bog-trapped failure —whose descendants, in three great movements, were to dominate the earth. It is only now, looking backward, that we dare to regard him as "generalized." The Snout was the first vertebrate to pop completely through the water membrane into a new dimension. His very specializations and failures, in a water sense, had preadapted him for a world he scarcely knew existed.

The day of the Snout was over three hundred million years ago. Not long since I read a book in which a prominent scientist spoke cheerfully of some ten billion years of future time remaining to us. He pointed out happily the things that man might do throughout that period. Fish in the sea, I thought again, birds in the air. The climb all far behind us, the species fixed and sure. No wonder my explorer friend had had a momentary qualm when he met the mudskippers with their mental reservations and lack of promises. There is something wrong with our world view. It is still Ptolemaic, though the sun is no longer believed to revolve around the earth.

We teach the past, we see farther backward into time than any race before us, but we stop at the present, or, at best, we project far into the future idealized versions of ourselves. All that long way behind us we see, perhaps inevitably, through human eyes alone. We see ourselves as the culmination and the end, and if we do indeed consider our passing, we think that sunlight will go with us and the earth be dark. We are the end. For us continents rose and fell, for us the waters and the air were mastered, for us the great living web has pulsated and grown more intricate.

To deny this, a man once told me, is to deny God. This puzzled me. I went back along the pathway to the marsh. I went, not in the past, not by the bones of dead things, not down the lost roadway of the Snout. I went instead in daylight, in the Now, to see if the door was still there, and to see what things passed through.

I found that the same experiments were brewing, that up out of that ancient well, fins were still scrambling toward the sunlight. They were small things, and which of them presaged the future I could not say. I saw only that they were many and that they had solved the oxygen death in many marvelous ways, not always ours.

I found that there were modern fishes who breathed air, not through a lung but through their stomachs or through strange chambers where their gills should be, or breathing as the Snout once breathed. I found that some crawled in the fields at nightfall pursuing insects, or slept on the grass by pond sides and who drowned, if kept under water, as men themselves might drown.

Of all these fishes the mudskipper *Periophthalmus* is perhaps the strangest. He climbs trees with his fins and pursues insects; he snaps worms like a robin on the tide flats; he sees as land things see, and above all he dodges and evades with a curious popeyed insolence more suggestive of the land than of the sea. Of a different tribe and a different time he is, nevertheless, oddly reminiscent of the Snout.

But not the same. There lies the hope of life. The old ways are exploited and remain, but new things come, new senses try the unfamiliar air. There are small scuttlings and splashings in the dark, and out of it come the first croaking, illiterate voices of the things to be, just as man once croaked and dreamed darkly in that tiny vesicular forebrain.

Perpetually, now, we search and bicker and disagree. The eternal form eludes us—the shape we conceive as ours. Perhaps the old road through the marsh should tell us. We are one of many appearances of the thing called Life; we are not its perfect image, for it has no image except Life, and life is multitudinous and emergent in the stream of time.

How Flowers Changed the World

IF IT HAD BEEN possible to observe the Earth from the far side of the solar system over the long course of geological epochs, the watchers might have been able to discern a subtle change in the light emanating from our planet. That world of long ago would, like the red deserts of Mars, have reflected light from vast drifts of stone and gravel, the sands of wandering wastes, the blackness of naked basalt, the yellow dust of endlessly moving storms. Only the ceaseless marching of the clouds and the intermittent flashes from the restless surface of the sea would have told a different story, but still essentially a barren one. Then, as the millennia rolled away and age followed age, a new and greener light would, by degrees, have come to twinkle across those endless miles.

This is the only difference those far watchers, by the use of subtle instruments, might have perceived in the whole history of the planet Earth. Yet that slowly growing green twinkle would have contained the epic march of life from the tidal oozes upward across the raw and unclothed continents. Out of the vast chemical bath of the sea—not from the deeps, but from the element-rich, light-exposed platforms of the continental shelves—wandering fingers of green had crept upward along the meanderings of river systems and fringed the gravels of forgotten lakes.

In those first ages plants clung of necessity to swamps and watercourses. Their reproductive processes demanded direct access to water. Beyond the primitive ferns and mosses that enclosed the borders of swamps and streams the rocks still lay vast and bare, the winds still swirled the dust of a naked planet. The grass cover that holds our world secure in place was still millions of years in the future. The green marchers had gained a soggy foothold upon the land, but that was all. They did not reproduce by seeds but by microscopic swimming sperm that had to wriggle their way through water to fertilize the female cell. Such plants in their higher forms had clever adaptations for the use of rain water in their sexual phases, and survived with increasing success in a wet land environment. They now seem part of man's normal environment. The truth is, however, that there is nothing very "normal" about nature. Once upon a time there were no flowers at all.

A little while ago—about one hundred million years, as the geologist estimates time in the history of our four-billion-year-old planet—flowers were not to be found anywhere on the five continents. Wherever one might have looked, from the poles to the equator, one would have seen only the cold dark monotonous green of a world whose plant life possessed no other color.

Somewhere, just a short time before the close of the Age of Reptiles, there occurred a soundless, violent explosion. It lasted millions of years, but it was an explosion, nevertheless. It marked the emergence of the angiosperms—the flowering plants. Even the great evolutionist, Charles Darwin, called them "an abominable mystery," because they appeared so suddenly and spread so fast.

Flowers changed the face of the planet. Without them, the world we know—even man himself—would never have existed. Francis Thompson, the English poet, once wrote that one could not pluck a flower without troubling a star. Intuitively he had sensed like a naturalist the enormous interlinked complexity of life. Today we know that the appearance of the flowers contained also the equally mystifying emergence of man.

If we were to go back into the Age of Reptiles, its drowned swamps and birdless forests would reveal to us a warmer but, on the whole, a sleepier world than that of today. Here and there, it is true, the serpent heads of bottom-feeding dinosaurs might be upreared in suspicion of their huge flesh-eating compatriots. Tyrannosaurs, enormous bipedal caricatures of men, would stalk mindlessly across the sites of future cities and go their slow way down into the dark of geologic time.

In all that world of living things nothing saw save with the intense concentration of the hunt, nothing moved except with the grave sleepwalking intentness of the instinct-driven brain. Judged by modern standards, it was a world in slow motion, a cold-blooded world whose occupants were most active at noonday but torpid on chill nights, their brains damped by a slower metabolism than any known to even the most primitive of warm-blooded animals today.

A high metabolic rate and the maintenance of a constant body temperature are supreme achievements in the evolution of life. They enable an animal to escape, within broad limits, from the overheating or the chilling of its immediate surroundings, and at the same time to maintain a peak mental efficiency. Creatures without a high metabolic rate are slaves to weather. Insects in the first frosts of autumn all run down like little clocks. Yet if you pick one up and breathe warmly upon it, it will begin to move about once more.

In a sheltered spot such creatures may sleep away the winter, but they are hopelessly immobilized. Though a few warm-blooded mammals, such as the woodchuck of our day, have evolved a way of reducing their metabolic rate in order to undergo winter hibernation, it is a survival mechanism with drawbacks, for it leaves the animal helplessly exposed if enemies discover him during his period of suspended animation. Thus bear or woodchuck, big animal or small, must seek, in this time of descending sleep, a safe refuge in some hidden den or burrow. Hibernation is, therefore, primarily a winter refuge of small, easily concealed animals rather than of large ones.

A high metabolic rate, however, means a heavy intake of energy in order to sustain body warmth and efficiency. It is for this reason that even some of these later warm-blooded mammals existing in our day have learned to descend into a slower, unconscious rate of living during the winter months when food may be difficult to obtain. On a slightly higher plane they are following the procedure of the cold-blooded frog sleeping in the mud at the bottom of a frozen pond.

The agile brain of the warm-blooded birds and mammals demands a high oxygen consumption and food in concentrated forms, or the creatures cannot long sustain themselves. It was the rise of the flowering plants that provided that energy and changed the nature of the living world. Their appearance parallels in a quite surprising manner the rise of the birds and mammals.

Slowly, toward the dawn of the Age of Reptiles, something over two hundred and fifty million years ago, the little naked sperm cells wriggling their way through dew and raindrops had given way to a kind of pollen carried by the wind. Our present-day pine forests represent plants of a pollen-disseminating variety. Once fertilization was no longer dependent on exterior water, the march over drier regions could be extended. Instead of spores simple primitive seeds carrying some nourishment for the young plant had developed, but true flowers were still scores of millions of years away. After a long period of hesitant evolutionary groping, they exploded upon the world with truly revolutionary violence.

The event occurred in Cretaceous times in the close of the Age of Reptiles. Before the coming of the flowering plants our own ancestral stock, the warm-blooded mammals, consisted of a few mousy little creatures hidden in trees and underbrush. A few lizard-like birds with carnivorous teeth flapped awkwardly on ill-aimed flights among archaic shrubbery. None of these insignificant creatures gave evidence of any remarkable talents. The mammals in particular had been around for some millions of years, but had remained well lost in the shadow of the mighty reptiles. Truth to tell, man was still, like the genie in the bottle, encased in the body of a creature about the size of a rat.

As for the birds, their reptilian cousins the Pterodactyls, flew farther and better. There was just one thing about the birds that paralleled the physiology of the mammals. They, too, had evolved warm blood and its accompanying temperature control. Nevertheless, if one had been seen stripped of his feathers, he would still have seemed a slightly uncanny and unsightly lizard.

Neither the birds nor the mammals, however, were quite what they seemed. They were waiting for the Age of Flowers. They were waiting for what flowers, and with them the true encased seed, would bring. Fish-eating, gigantic leather-winged reptiles, twenty-eight feet from wing tip to wing tip, hovered over the coasts that one day would be swarming with gulls.

Inland the monotonous green of the pine and spruce forests with their primitive wooden cone flowers stretched every-where. No grass hindered the fall of the naked seeds to earth. Great sequoias towered to the skies. The world of that time has a certain appeal but it is a giant's world, a world moving slowly like the reptiles who stalked magnificently among the boles of its trees.

The trees themselves are ancient, slow-growing and im-mense, like the redwood groves that have survived to our day on the California coast. All is stiff, formal, upright and green, monotonously green. There is no grass as yet; there are no wide plains rolling in the sun, no tiny daisies dotting the meadows underfoot. There is little versatility about this scene; it is, in truth, a giant's world.

A few nights ago it was brought home vividly to me that the world has changed since that far epoch. I was awakened out of sleep by an unknown sound in my living room. Not a small sound—not a creaking timber or a mouse's scurry—but a sharp, rending explosion as though an unwary foot had been put down upon a wine glass. I had come instantly out of sleep and lay tense, unbreathing. I listened for another step. There was none.

Unable to stand the suspense any longer, I turned on the light and passed from room to room glancing uneasily behind chairs and into closets. Nothing seemed disturbed, and I stood puzzled in the center of the living room floor. Then a small button-shaped object upon the rug caught my eye. It was hard and polished and glistening. Scattered over the length of the room were several more shining up at me like wary little eyes. A pine cone that had been lying in a dish had been blown the length of the coffee table. The dish itself could hardly have been the source of the explosion. Beside it I found two ribbon-like strips of a velvety-green. I tried to place the two strips together to make a pod. They twisted resolutely away from each other and would no longer fit.

I relaxed in a chair, then, for I had reached a solution of the midnight disturbance. The twisted strips were wistaria pods that I had brought in a day or two previously and placed in the dish. They had chosen midnight to explode and distribute their multiplying fund of life down the length of the room. A plant, a fixed, rooted thing, immobilized in a single spot, had devised a way of propelling its offspring across open space. Immediately there passed before my eyes the million airy troopers of the milkweed pod and the clutching hooks of the sandburs. Seeds on the coyote's tail, seeds on the hunter's coat, thistledown mounting on the winds—all were somehow triumphing over life's limitations. Yet the ability to do this had not been with them at the beginning. It was the product of endless effort and experiment.

The seeds on my carpet were not going to lie stiffly where they had dropped like their antiquated cousins, the naked seeds on the pine-cone scales. They were travelers. Struck by the thought, I went out next day and collected several other varieties. I line them up now in a row on my desk—so many little capsules of life, winged, hooked or spiked. Every one is an angiosperm, a product of the true flowering plants. Contained in these little boxes is the secret of that far-off Cretaceous explosion of a hundred million years ago that changed the face of the planet. And somewhere in here, I think, as I poke seriously at one particularly resistant seedcase of a wild grass, was once man himself.

When the first simple flower bloomed on some raw upland late in the Dinosaur Age, it was wind pollinated, just like its early pine-cone relatives. It was a very inconspicuous flower because it had not yet evolved the idea of using the surer attraction of birds and insects to achieve the transportation of pollen. It sowed its own pollen and received the pollen of other flowers by the simple vagaries of the wind. Many plants in regions where insect life is scant still follow this principle today. Nevertheless, the true flower—and the seed that it produced—was a profound innovation in the world of life.

In a way, this event parallels, in the plant world, what happened among animals. Consider the relative chance for survival of the exteriorly deposited egg of a fish in contrast with the fertilized egg of a mammal, carefully retained for months in the mother's body until the young animal (or human being) is developed to a point where it may survive. The biological wastage is less—and so it is with the flowering plants. The primitive spore, a single cell fertilized in the beginning by a swimming sperm, did not promote rapid distribution, and the young plant, moreover, had to struggle up from nothing. No one had left it any food except what it could get by its own unaided efforts.

By contrast, the true flowering plants (angiosperm itself means "encased seed") grew a seed in the heart of a flower, a seed whose development was initiated by a fertilizing pollen grain independent of outside moisture. But the seed, unlike the developing spore, is already a fully equipped *embryonic plant* packed in a little enclosed box stuffed full of nutritious food. Moreover, by featherdown attachments, as in dandelion or milkweed seed, it can be wafted upward on gusts and ride the wind for miles; or with hooks it can cling to a bear's or a rabbit's hide; or like some of the berries, it can be covered with a juicy, attractive fruit to lure birds, pass undigested through their intestinal tracts and be voided miles away.

The ramifications of this biological invention were endless. Plants traveled as they had never traveled before. They got into strange environments heretofore never entered by the old spore plants or stiff pine-cone-seed plants. The well-fed, carefully cherished little embryos raised their heads everywhere. Many of the older plants with more primitive reproductive mechanisms began to fade away under this unequal contest. They contracted their range into secluded environments. Some, like the giant redwoods, lingered on as relics; many vanished entirely.

The world of the giants was a dying world. These fantastic little seeds skipping and hopping and flying about the woods and valleys brought with them an amazing adaptability. If our whole lives had not been spent in the midst of it, it would astound us. The old, stiff, sky-reaching wooden world had changed into something that glowed here and there with strange colors, put out queer, unheard-of fruits and little intricately carved seed cases, and, most important of all, produced concentrated foods in a way that the land had never seen before, or dreamed of back in the fish-eating, leaf-crunching days of the dinosaurs.

That food came from three sources, all produced by the reproductive system of the flowering plants. There were the tantalizing nectars and pollens intended to draw insects for pollenizing purposes, and which are responsible also for that wonderful jeweled creation, the hummingbird. There were the juicy and enticing fruits to attract larger animals, and in which tough-coated seeds were concealed, as in the tomato, for example. Then, as if this were not enough, there was the food in the actual seed itself, the food intended to nourish the embryo. All over the world, like hot corn in a popper, these incredible elaborations of the flowering plants kept exploding. In a movement that was almost instantaneous, geologically speaking, the angiosperms had taken over the world. Grass was beginning to cover the bare earth until, today, there are over six thousand species. All kinds of vines and bushes squirmed and writhed under new trees with flying seeds.

The explosion was having its effect on animal life also. Specialized groups of insects were arising to feed on the new sources of food and, incidentally and unknowingly, to pollinate the plant. The flowers bloomed and bloomed in ever larger and more spectacular varieties. Some were pale unearthly night flowers intended to lure moths in the evening twilight, some among the orchids even took the shape of female spiders in order to attract wandering males, some flamed redly in the light of noon or twinkled modestly in the meadow grasses. Intricate mechanisms splashed pollen on the breasts of hummingbirds, or stamped it on the bellies of black, grumbling bees droning assiduously from blossom to blossom. Honey ran, insects multiplied, and even the descendants of that toothed and ancient lizard-bird had become strangely altered. Equipped with prodding beaks instead of biting teeth they pecked the seeds and gobbled the insects that were really converted nectar.

Across the planet grasslands were now spreading. A slow continental upthrust which had been a part of the early Age of Flowers had cooled the world's climates. The stalking reptiles and the leather-winged black imps of the seashore cliffs had vanished. Only birds roamed the air now, hot-blooded and high-speed metabolic machines.

The mammals, too, had survived and were venturing into new domains, staring about perhaps a bit bewildered at their sudden eminence now that the thunder lizards were gone. Many of them, beginning as small browsers upon leaves in the forest, began to venture out upon this new sunlit world of the grass. Grass has a high silica content and demands a new type of very tough and resistant tooth enamel, but the seeds taken incidentally in the cropping of the grass are highly nutritious. A new world had opened out for the warm-blooded mammals. Great herbivores like the mammoths, horses and bisons appeared. Skulking about them had arisen savage flesh-feeding carnivores like the now extinct dire wolves and the saber-toothed tiger.

Flesh eaters though these creatures were, they were being sustained on nutritious grasses one step removed. Their fierce energy was being maintained on a high, effective level, through hot days and frosty nights, by the concentrated energy of the angiosperms. That energy, thirty per cent or more of the weight of the entire plant among some of the cereal grasses, was being accumulated and concentrated in the rich proteins and fats of the enormous game herds of the grasslands.

On the edge of the forest, a strange, old-fashioned animal still hesitated. His body was the body of a tree dweller, and though tough and knotty by human standards, he was, in terms of that world into which he gazed, a weakling. His teeth, though strong for chewing on the tough fruits of the forest, or for crunching an occasional unwary bird caught with his prehensile hands, were not the tearing sabers of the great cats. He had a passion for lifting himself up to see about, in his restless, roving curiosity. He would run a little stiffly and uncertainly, perhaps, on his hind legs, but only in those rare moments when he ventured out upon the ground. All this was the legacy of his climbing days; he had a hand with flexible fingers and no fine specialized hoofs upon which to gallop like the wind.

If he had any idea of competing in that new world, he had better forget it; teeth or hooves, he was much too late for either. He was a ne'er-do-well, an in-betweener. Nature had not done well by him. It was as if she had hesitated and never quite made up her mind. Perhaps as a consequence he had a malicious gleam in his eye, the gleam of an outcast who has been left nothing and knows he is going to have to take what he gets. One day a little band of these odd apes—for apes they were—shambled out upon the grass; the human story had begun.

Apes were to become men, in the inscrutable wisdom of nature, because flowers had produced seeds and fruits in such tremendous quantities that a new and totally different store of energy had become available in concentrated form. Impressive as the slow-moving, dim-brained dinosaurs had been, it is doubtful if their age had supported anything like the diversity of life that now rioted across the planet or flashed in and out among the trees. Down on the grass by a streamside, one of those apes with inquisitive fingers turned over a stone and hefted it vaguely. The group clucked together in a throaty tongue and moved off through the tall grass foraging for seeds and insects. The one still held, sniffed, and hefted the stone he had found. He liked the feel of it in his fingers. The attack on the animal world was about to begin.

If one could run the story of that first human group like a speeded-up motion picture through a million years of time, one might see the stone in the hand change to the flint ax and the torch. All that swarming grassland world with its giant bison and trumpeting mammoths would go down in ruin to feed the insatiable and growing numbers of a carnivore who, like the great cats before him, was taking his energy indirectly from the grass. Later he found fire and it altered the tough meats and drained their energy even faster into a stomach ill adapted for the ferocious turn man's habits had taken.

His limbs grew longer, he strode more purposefully over the grass. The stolen energy that would take man across the continents would fail him at last. The great Ice Age herds were destined to vanish. When they did so, another hand like the hand that grasped the stone by the river long ago would pluck a handful of grass seed and hold it contemplatively.

In that moment, the golden towers of man, his swarming millions, his turning wheels, the vast learning of his packed libraries, would glimmer dimly there in the ancestor of wheat, a few seeds held in a muddy hand. Without the gift of flowers and the infinite diversity of their fruits, man and bird, if they had continued to exist at all, would be today unrecognizable. Archaeopteryx, the lizard-bird, might still be snapping at beetles on a sequoia limb; man might still be a nocturnal insectivore gnawing a roach in the dark. The weight of a petal has changed the face of the world and made it ours.

The Real Secret of Piltdown

HOW DID MAN get his brain? Many years ago Charles Darwin's great contemporary, and co-discoverer with him of the principle of natural selection, Alfred Russel Wallace, propounded that simple question. It is a question which has bothered evolutionists ever since, and when Darwin received his copy of an article Wallace had written on this subject he was obviously shaken. It is recorded that he wrote in anguish across the paper, "No!" and underlined the "No" three times heavily in a rising fervor of objection.

Today the question asked by Wallace and never satisfactorily answered by Darwin has returned to haunt us. A skull, a supposedly very ancient skull, long used as one of the most powerful pieces of evidence documenting the Darwinian position upon human evolution, has been proven to be a forgery, a hoax perpetrated by an unscrupulous but learned amateur. In the fall of 1953 the famous Piltdown cranium, known in scientific circles all over the world since its discovery in a gravel pit on the Sussex Downs in 1911, was jocularly dismissed by the world's press as the skull that had "made monkeys out of the anthropologists." Nobody remembered in 1953 that Wallace, the great evolutionist, had protested to a friend in 1913, "The Piltdown skull does not prove much, if anything!"

Why had Wallace made that remark? Why, almost alone among the English scientists of his time, had he chosen to regard with a dubious eye a fossil specimen that seemed to substantiate the theory to which he and Darwin had devoted their lives? He did so for one reason: he did not believe what the Piltdown skull appeared to reveal as to the nature of the process by which the human brain had been evolved. He did not believe in a skull which had a modern brain box attached to an apparently primitive face and given, in the original estimates, an antiquity of something over a million years.

Today we know that the elimination of the Piltdown skull from the growing list of valid human fossils in no way affects the scientific acceptance of the theory of evolution. In fact, only the circumstance that Piltdown had been discovered early, before we had a clear knowledge of the nature of human fossils and the techniques of dating them, made the long survival of this extraordinary hoax possible. Yet in the end it has been the press, absorbed in a piece of clever scientific detection, which has missed the real secret of Piltdown. Darwin saw in the rise of man, with his unique, time-spanning brain, only the undirected play of such natural forces as had created the rest of the living world of plants and animals. Wallace, by contrast, in the case of man, totally abandoned this point of view and turned instead toward a theory of a divinely directed control of the evolutionary process. The issue can be made clear only by a rapid comparison of the views of both men.

As everyone who has studied evolution knows, Darwin propounded the theory that since the reproductive powers of plants and animals potentially far outpace the available food supply, there is in nature a constant struggle for existence on the part of every living thing. Since animals vary individually, the most cleverly adapted will survive and leave offspring which will inherit, and in their turn enhance, the genetic endowment they have received from their ancestors. Because the struggle for life is incessant, this unceasing process promotes endless slow changes in bodily form, as living creatures are subjected to different natural environments, different enemies, and all the vicissitudes against which life has struggled down the ages.

Darwin, however, laid just one stricture on his theory: it could, he maintained, "render each organized being only as perfect or a little more perfect than other inhabitants of the same country." It could allow any animal only a relative superiority, never an absolute perfection—otherwise selection and the struggle for existence would cease to operate. To explain the rise of man through the slow, incremental gains of natural selection, Darwin had to assume a long struggle of man with man and tribe with tribe.

He had to make this assumption because man had far outpaced his animal associates. Since Darwin's theory of the evolutionary process is based upon the practical value of all physical and mental characters in the life struggle, to ignore the human struggle of man with man would have left no explanation as to how humanity by natural selection alone managed to attain an intellectual status so far beyond that of any of the animals with which it had begun its competition for survival.

To most of the thinkers of Darwin's day this seemed a reasonable explanation. It was a time of colonial expansion and ruthless business competition. Peoples of primitive cultures, small societies lost on the world's margins, seemed destined to be destroyed. It was thought that Victorian civilization was the apex of human achievement and that other races with different customs and ways of life must be biologically inferior to Western man. Some of them were even described as only slightly superior to apes. The Darwinians, in a time when there were no satisfactory fossils by which to demonstrate human evolution, were unconsciously minimizing the abyss which yawned between man and ape. In their anxiety to demonstrate our lowly origins they were throwing modern natives into the gap as representing living "missing links" in the chain of human ascent.

It was just at this time that Wallace lifted a voice of lonely protest. The episode is a strange one in the history of science, for Wallace had, independently of Darwin, originally arrived at the same general conclusion as to the nature of the evolutionary process. Nevertheless, only a few years after the publication of Darwin's work, *The Origin of Species*, Wallace had come to entertain a point of view which astounded and troubled Darwin. Wallace, who had had years of experience with natives of the tropical archipelagoes, abandoned the idea that they were of mentally inferior cast. He did more. He committed the Darwinian heresy of maintaining that their mental powers were far in excess of what they really needed to carry on the simple food-gathering techniques by which they survived.

"How, then," Wallace insisted, "was an organ developed so far beyond the needs of its possessor? Natural selection could only have endowed the savage with a brain a little superior to that of an ape, whereas he actually possesses one but little inferior to that of the average member of our learned societies."

At a time when many primitive peoples were erroneously assumed to speak only in grunts or to chatter like monkeys, Wallace maintained his view of the high intellectual powers of natives by insisting that "the capacity of uttering a variety of distinct articulate sounds and of applying to them an almost infinite amount of modulation . . . is not in any way inferior to that of the higher races. An instrument has been developed in advance of the needs of its possessor."

Finally, Wallace challenged the whole Darwinian position on man by insisting that artistic, mathematical, and musical abilities could not be explained on the basis of natural selection and the struggle for existence. Something else, he contended, some unknown spiritual element, must have been at work in the elaboration of the human brain. Why else would men of simple cultures possess the same basic intellectual powers which the Darwinists maintained could be elaborated only by competitive struggle?

"If you had not told me you had made these remarks," Darwin said, "I should have thought they had been added by someone else. I differ grievously from you and am very sorry for it." He did not, however, supply a valid answer to Wallace's queries. Outside of murmuring about the inherited effects of habit—a contention without scientific validity today—Darwin clung to his original position. Slowly Wallace's challenge was forgotten and a great complacency settled down upon the scientific world.

For seventy years after the publication of *The Origin of Species* in 1859, there were only two finds of fossil human skulls which seemed to throw any light upon the Darwin-Wallace controversy. One was the discovery of the small-brained Java Ape Man, the other was the famous Piltdown or "dawn man." Both were originally dated as lying at the very beginning of the Ice Age, and, though these dates were later to be modified, the skulls, for a very long time, were regarded as roughly contemporaneous and very old.

Two more unlike "missing links" could hardly be imagined. Though they were supposed to share a million-year antiquity, the one was indeed quite primitive and small-brained; the other, Piltdown, in spite of what seemed a primitive lower face, was surprisingly modern in brain. Which of these forms told the true story of human development? Was a large brain old? Had ages upon ages of slow, incremental, Darwinian increase produced it? The Piltdown skull seemed to suggest such a development.

Many were flattered to find their anthropoid ancestry seemingly removed to an increasingly remote past. If one looked at the Java Ape Man, one was forced to contemplate an ancestor, not terribly remote in time, who still had a face and a brain which hinted strongly of the ape. Yet, when by geological evidence this "erect walking ape-man" was finally assigned to a middle Ice Age antiquity, there arose the immediate possibility that Wallace could be right in his suspicion that the human brain might have had a surprisingly rapid development. By contrast, the Piltdown remains seemed to suggest a far more ancient and slow-paced evolution of man. The Piltdown hoaxer, in attaching an ape jaw to a human skull fragment, had, perhaps unwittingly, created a creature which supported the Darwinian idea of man, not too unlike the man of today, extending far back into pre–Ice Age times.

Which story was the right one? Until the exposé of Piltdown in 1953, both theories had to be considered possible and the two hopelessly unlike fossils had to be solemnly weighed in the same balance. Today Piltdown is gone. In its place we are confronted with the blunt statement of two modern scientists, M. R. A. Chance and A. P. Mead.

"No adequate explanation," they confess over eighty years after Darwin scrawled his vigorous "No!" upon Wallace's paper, "has been put forward to account for so large a cerebrum as that found in man."[1]

We have been so busy tracing the tangible aspects of evolution in the *forms of animals* that our heads, the little globes which hold the midnight sky and the shining, invisible universes of thought, have been taken about as much for granted as the growth of a yellow pumpkin in the fall.

Now a part of this mystery as it is seen by the anthropologists of today lies in the relation of the brain to time. "If," Wallace had said, "researches in all parts of Europe and Asia fail to bring to light any proofs of man's presence far back in the Age of Mammals, *it will be at least a presumption that he came into existence at a much later date and by a more rapid process of development.*" If human evolution should prove to be comparatively rapid, "explosive" in other words, Wallace felt

---

[1] *Symposia of the Society for Experimental Biology*, VII, Evolution (New York: Academic Press, 1953), p. 395.

that his position would be vindicated, because such a rapid development of the brain would, he thought, imply a divinely directed force at work in man. In the 1870's when he wrote, however, human prehistory was largely an unknown blank. Today we can make a partial answer to Wallace's question. Since the exposure of the Piltdown hoax all of the evidence at our command—and it is considerable—points to man, in his present form, as being one of the youngest and newest of all earth's swarming inhabitants.

The Ice Age extends behind us in time for, at most, a million years. Though this may seem long to one who confines his studies to the written history of man, it is, in reality, a very short period as the student of evolution measures time. It is a period marked more by the extinction of some of the last huge land animals, like the hairy mammoth and the saber-toothed tiger, than it is by the appearance of new forms of life. To this there is only one apparent exception: the rise and spread of man over the Old World land mass.

Most of our knowledge of him—even in his massive-faced, beetle-browed stage—is now confined, since the loss of Piltdown, to the last half of the Ice Age. If we pass backward beyond this point we can find traces of crude tools, stone implements which hint that some earlier form of man was present here and there in Europe, Asia, and particularly Africa in the earlier half of Ice Age time, but to the scientist it is like peering into the mists floating over an unknown landscape. Here and there through the swirling vapor one catches a glimpse of a shambling figure, or a half-wild primordial face stares back at one from some momentary opening in the fog. Then, just as one grasps at a clue, the long gray twilight settles in and the wraiths and the half-heard voices pass away.

Nevertheless, particularly in Africa, a remarkable group of human-like apes have been discovered: creatures with small brains and teeth of a remarkably human cast. Prominent scientists are still debating whether they are on the direct line of ascent to man or are merely near relatives of ours. Some, it is now obvious, existed too late in time to be our true ancestors, though this does not mean that their bodily characters may not tell us what the earliest anthropoids who took the human turn of the road were like.

These apes are not all similar in type or appearance. They are men and yet not men. Some are frailer-bodied, some have great, bone-cracking jaws and massive gorilloid crests atop their skulls. This fact leads us to another of Wallace's remarkable perceptions of long ago. With the rise of the truly human brain, Wallace saw that man had transferred to his machines and tools many of the alterations of parts that in animals take place through evolution of the body. Unwittingly, man had assigned to his machines the selective evolution which in the animal changes the nature of its bodily structure through the ages. Man of today, the atomic manipulator, the aeronaut who flies faster than sound, has precisely the same brain and body as his ancestors of twenty thousand years ago who painted the last Ice Age mammoths on the walls of caves in France.

To put it another way, it is man's ideas that have evolved and changed the world about him. Now, confronted by the lethal radiations of open space and the fantastic speeds of his machines, he has to invent new electronic controls that operate faster than his nerves, and he must shield his naked body against atomic radiation by the use of protective metals. Already he is physically antique in this robot world he has created. All that sustains him is that small globe of gray matter through which spin his ever-changing conceptions of the universe.

Yet, as Wallace, almost a hundred years ago, glimpsed this timeless element in man, he uttered one more prophecy. When we come to trace out history into the past, he contended, sooner or later we will come to a time when the body of man begins to differ and diverge more extravagantly in its appearance. Then, he wrote, we shall know that we stand close to the starting point of the human family. In the twilight before the dawn of the human mind, man will not have been able to protect his body from change and his remains will bear the marks of all the forces that play upon the rest of life. He will be different in his form. He will be, in other words, as variable in body as we know the South African man-apes to be.

Today, with the solution of the Piltdown enigma, we must settle the question of the time involved in human evolution in favor of Wallace, not Darwin; we need not, however, pursue the mystical aspects of Wallace's thought—since other factors yet to be examined may well account for the rise of man. The rapid fading out of archaeological evidence of tools in lower Ice Age times—along with the discovery of man-apes of human aspect but with ape-sized brains, yet possessing a diverse array of bodily characters—suggests that the evolution of the human brain was far more rapid than that conceived of in early Darwinian circles. At that time it was possible to hear the Eskimos spoken of as possible survivals of Miocene men of several million years ago. By contrast to this point of view, man and his rise now appear short in time—explosively short. There is every reason to believe that whatever the nature of the forces involved in the production of the human brain, a long slow competition of human group with human group or race with race would not have resulted in such similar mental potentialities among all peoples everywhere. Something—some other factor—has escaped our scientific attention.

There are certain strange bodily characters which mark man as being more than the product of a dog-eat-dog competition with his fellows. He possesses a peculiar larval nakedness, difficult to explain on survival principles; his periods of helpless infancy and childhood are prolonged; he has aesthetic impulses which, though they vary in intensity from individual to individual, appear in varying manifestations among all peoples. He is totally dependent, in the achievement of human status, upon the careful training he receives in human society.

Unlike a solitary species of animal, he cannot develop alone. He has suffered a major loss of precise instinctive controls of behavior. To make up for this biological lack, society and parents condition the infant, supply his motivations, and promote his long-drawn training at the difficult task of becoming a normal human being. Even today some individuals fail to make this adjustment and have to be excluded from society.

We are now in a position to see the wonder and terror of the human predicament: man is totally dependent on society. Creature of dream, he has created an invisible world of ideas, beliefs, habits, and customs which buttress him about and replace for him the precise instincts of the lower creatures. In this invisible universe he takes refuge, but just as instinct may fail an animal under some shift of environmental conditions, so man's cultural beliefs may prove inadequate to meet a new situation, or, on an individual level, the confused mind may substitute, by some terrible alchemy, cruelty for love.

The profound shock of the leap from animal to human status is echoing still in the depths of our subconscious minds. It is a transition which would seem to have demanded considerable rapidity of adjustment in order for human beings to have survived, and it also involved the growth of prolonged bonds of affection in the subhuman family, because otherwise its naked, helpless offspring would have perished.

It is not beyond the range of possibility that this strange reduction of instincts in man in some manner forced a precipitous brain growth as a compensation—something that had to be hurried for survival purposes. Man's competition, it would thus appear, may have been much less with his own kind than with the dire necessity of building about him a world of ideas to replace his lost animal environment. As we will show later, he is a pedomorph, a creature with an extended childhood.

Modern science would go on to add that many of the characters of man, such as his lack of fur, thin skull, and globular head, suggest mysterious changes in growth rates which preserve, far into human maturity, foetal or infantile characters which hint that the forces creating man drew him fantastically out of the very childhood of his brutal forerunners. Once more the words of Wallace come back to haunt us: "We may safely infer that the savage possesses a brain capable, if cultivated and developed, of performing work of a kind and degree far beyond what he ever requires it to do."

As a modern man, I have sat in concert halls and watched huge audiences floating dazed on the voice of a great singer. Alone in the dark box I have heard far off as if ascending out of some black stairwell the guttural whisperings and bestial coughings out of which that voice arose. Again, I have sat under the slit dome of a mountain observatory and marveled, as the great wheel of the galaxy turned in all its midnight splendor, that the mind in the course of three centuries has been capable of drawing into its strange, nonspatial interior that world of infinite distance and multitudinous dimensions.

Ironically enough, science, which can show us the flints and the broken skulls of our dead fathers, has yet to explain how we have come so far so fast, nor has it any completely satisfactory answer to the question asked by Wallace long ago. Those who would revile us by pointing to an ape at the foot of our family tree grasp little of the awe with which the modern scientist now puzzles over man's lonely and supreme ascent. As one great student of paleoneurology, Dr. Tilly Edinger, recently remarked, "If man has passed through a Pithecanthropus phase, the evolution of his brain has been unique, not only in its result but also in its tempo. . . . Enlargement of the cerebral hemispheres by 50 per cent seems to have taken place, speaking geologically, within an instant, and without having been accompanied by any major increase in body size."

The true secret of Piltdown, though thought by the public to be merely the revelation of an unscrupulous forgery, lies in the fact that it has forced science to reëxamine carefully the history of the most remarkable creation in the world—the human brain.

The Maze

S HORTLY AFTER I had expressed my conclusions about the real secret of Piltdown, I was roundly castigated by a few people who had construed my remarks as an attack upon Darwin and thus an assault upon the theory of evolution itself. A surprising amount of suppressed emotion still lingers about these hundred-year-old controversies, and those who are not historical-minded may be quick to launch themselves, sometimes more valiantly than accurately, into the thick of some forgotten fray. Along an advancing front of science, the man who writes for a nontechnical public runs risks and has curious experiences. Sometimes he is unlucky, as in the case of an acquaintance of mine whose article dealing soberly with the Piltdown skull appeared at the very moment when the hoax was denounced in the press. Sometimes, on the other hand, he may have almost preternatural luck, in that unexpected events may further substantiate a view that he had earlier broached in hesitation and with a minimum of supporting evidence.

After I had expressed myself upon the dangerously controversial subject of the human brain—and I say this avowedly, though so distinguished an authority upon the great apes as Solly Zuckerman has spoken of the "enormous gap" which exists "between the intelligence of Man and that of any other Primate"—two quite astonishing things happened. The first of these it is my intention to chronicle in this chapter; the second event, and the final culmination of the plot, will have to be reserved for the one which follows. The first happening, as it was described in the press, seemed to be a total negation of much that has been expressed in my treatment of the Piltdown story, namely, the recency of man.

The reader may remember that in March of 1956 curious and startling headlines began to appear in the newspapers. In the first excitement it must have seemed to the layman that the whole theory of evolution was about to be overthrown. There were accounts in the press of a ten-million-year-old "human" fossil. Such a discovery seemed, at first thought, to contradict

61

what I had contended was the great youth of man, that is, man as a culture bearer, a user of speech.

The commotion had been touched off by the arrival in New York City of a paleontologist from Switzerland bearing the bones of a small primate long known to science as Oreopithecus. Johannes Hurzeler of Basel presented to a group of scholars gathered at the Wenner-Gren Foundation for Anthropological Research his view that the bones of Oreopithecus showed human rather than anthropoid affinities. Since these bones are estimated to be ten million years older than the earliest known fossil men, his announcement made headlines.

"Fossil Research Questions Darwin Evolution Theory," the *New York Times* announced. The *Herald Tribune* editorialized: "No Missing Link?" Specialists on fossil man were besieged by telephone calls from reporters and by faintly derisive queries from anti-evolutionists whose interest had already been whetted by the Piltdown hoax. Perhaps this new contradiction would mark the final exit of the man-monkey and of the anthropologists along with it.

By the time scientists had begun to respond, the press had passed on to other things, leaving in the mind of the public a confused vision of a sort of "little man" who, so the newspapers said, had been found in a coal mine in Tuscany. Like most such episodes, that of Oreopithecus has a history, and the argument over it is of the same general nature as two similar controversies fought within the memory of men now living.

The incident has served to draw attention to a long-existing debate among anthropologists, which has occasionally waxed acrimonious. The partisans divide basically into two schools: the school of the "little man" and that of the "apeman." The former pursue the figure of man backward until, upon some far wall in time, it appears as a dwarfed, big-headed little shadow; the latter see our earliest ancestor shambling into the light like some great shaggy anthropoid. The argument recalls the ancient dispute between the preformationists, who saw in the human sperm cell a preformed homunculus, or little man, which had only to grow to adult size, and the epigenesists, who judged correctly that each embryo acquires the characteristics of a human being only through development.

Some anthropologists search for human characters—vertical

front teeth, a shortened face, an expanded brain case—early in the human line of descent. They seek, in other words, for something dangerously close to the homunculus of the preformationists. They "prove" evolution by finding, as St. George Jackson Mivart said in 1874, "an ancestral form so like man [that] we have the virtual pre-existence of man's body supposed, in order to account for the actual first appearance of that body as we know it."

The more thoroughgoing evolutionists, in contrast, have looked for forms which contained only the *possibility* of development into man. Such students have generally regarded man as a relatively recent emergent from a group of primates which also gave rise to the modern great apes; in other words, the comparison of man with the anthropoids of today has been based on the assumption that they and we had ancestors in common.

Charles Darwin was not the first to notice our likeness to the monkeys and apes. Such observations extend into antiquity, and by the eighteenth and early nineteenth centuries philosophers were arranging the primates in an order of complexity. As voyagers began to come into contact with primitive peoples, these were often placed on the scale as grades between the anthropoids and civilized European man. The Hottentots of the Cape of Good Hope particularly appealed to the Western mind as candidates for such a place; it was said that their language was only a step above the chatter of apes.

Thus notions of the "missing link" were in existence long before Darwin and long before the appearance of a truly evolutionary philosophy. Darwin himself cautiously refrained from attempting to trace man's precise relationship to the apes. But some of his followers, notably T. H. Huxley, tackled the problem head on. Huxley was provoked to his excursion into man's past by events at the famous meeting of the British Association for the Advancement of Science at Oxford in 1860. He had borne the brunt of the conservatives' attacks on evolution. At this meeting Richard Owen, England's foremost comparative anatomist and a mortal enemy of Darwin and his followers, attempted to maintain man's unique position in the animal world by placing him in a distinct subclass of the mammals for which he proposed the name "Archencephala." This classification was based upon brain characters which Owen maintained did not occur in the lower primates. Huxley, his ire aroused, set out to demonstrate that Owen was wrong, that man was closely related to the other primates. He composed a series of lectures which were published in 1863 under the title *Evidence as to Man's Place in Nature.*

In this work, which more or less set the pattern for much that followed, Huxley thoroughly demolished Owen's position. He took the view that "the surface of the brain of a monkey exhibits a sort of skeleton map of man's, and in the manlike apes the details become more and more filled in, until it is only in minor characters . . . that the chimpanzee's or the orang's brain can be structurally distinguished from man's." Huxley was quite willing to admit that man's own origin was obscure and might go back millions of years to a common ancestor, but he insisted that the modern apes were our closest surviving relatives. If Huxley dwelt too heavily and too emotionally upon anatomical correspondence between ourselves and the great apes, it must be remembered that at the time he wrote the evolutionists were fighting primarily for a principle, against the orthodox "special creationists." Furthermore, it must also be remembered that very few human fossils had been discovered, and these were fragmentary. Our living relatives in the trees could be seen at the zoo, and it was inevitable that they should dominate man's imagination. Serious scholars even came to believe that microcephalic idiots were throwbacks to some remote period of the human past.

By the beginning of the twentieth century the ape origins of modern man seemed pretty well established. The finding of the Pithecanthropus skull cap had bolstered this view. Many felt that from a form something like that of a chimpanzee it was an easy step to the Java man and thence on to Neanderthal and modern man. But at the turn of the century there came a new revolt against the ape.

The attention of anatomists was attracted to a small, tree-living creature in southeast Asia possessing definite characters of a primate. The tarsier (*Tarsius spectrum*), an animal with enormous eyes and about the size of a small kitten, has a brain and other characteristics which ally it to the lower mon-

keys. In 1918, F. Wood Jones, a distinguished English anatomist, had expressed the heretical view, which he has maintained
and developed since, that man arose from a tarsioid rather than
from an anthropoid ancestry.

Wood Jones insists that the human line is very ancient,
going back to a past tens of millions of years old in the Tertiary
Period. He predicts that man's immediate ancestors, if ever
discovered, "will be utterly unlike the slouching, hairy 'ape-men'
of which some have dreamed . . . and will be found in geological strata antedating the heyday of the great apes." The
ancestors of man, he says, were "small, active animals" already
endowed with legs longer than their arms, small jaws without
protruding teeth, and enlarged craniums. They were not swing-
ers in trees: the human hand and foot, he contends, are too
specialized to have been made over rapidly from an arboreal
ancestor's. The present-day tarsiers in the trees, according to
his view, evolved their tree-living specializations later, but our
early tarsioid ancestor walked on the ground.

Wood Jones's proto-man thus sounds like a homunculus.
When he first advocated his views, he found very few followers.
Henry Fairfield Osborn, the late paleontologist, though not a
Wood Jones follower, inclined toward a homuncular dawn
man going back to early Tertiary times many millions of years
ago. "I predict," he said, "that even in Upper Oligocene time
we shall find pro-men, and that they will have pro-human limbs."

Wood Jones and Osborn were vigorously refuted by primatologists who championed the orthodox view that man was a
"made-over ape." They insisted that man's immediate forerunners could not be so ancient as Wood Jones and Osborn said.
"It seems anachronistic," wrote William King Gregory, "to attribute to the very remote Tertiary ancestors of man the long
legs, long thumbs, big brain, short face, small canines, etc., which
are now diagnostic characters." But by the 1940's the "made-
over ape" point of view had moderated. The most important
factor in this change was the discovery in South Africa of the
fossil *Proconsul africanus*—a creature of the early Miocene
(about twenty million years ago) which combined characters
of early Old World monkeys and great apes. William L. Straus,
Jr., of the Johns Hopkins University, voiced a suspicion that

man's immediate ancestors might have been "more monkey-like than anthropoid-like." Straus, who takes a very sane and cautious position on this lengthy controversy over the human ancestry, feels that the anthropoid-ape theory is weakest in its failure to account for anatomical traits which man shares with the monkeys and lemurs. More recently W. C. Osman Hill, the well-known English primatologist, has come to believe that man branched off the primate stock below the great-ape line. He even suggests that Straus's view might be reconciled with Wood Jones's tarsioid hypothesis if some early Oligocene monkey of tarsioid affinities were admitted on the line leading to man—a form, say, like Parapithecus.

Thus, before Hurzeler's recent announcement a slow shift of thought or widening of possible horizons had been under way in the study of human evolution. The theory that man came down late out of the trees has been dropped in some quarters and is less explosively defended in others. There is a greater willingness to reserve judgment and wait upon new evidence. It was in this receptive atmosphere that Hurzeler presented his new study of Oreopithecus.

The fossil has been known since 1872, when it was described by the French paleontologist Paul Gervais, who regarded it as an Old World monkey. Hurzeler, after studying the original fossil and later finds, has become convinced that Oreopithecus is the first manlike form discovered in the Tertiary Period—it is believed to date from the Miocene. He apparently bases this view upon certain technical features of the teeth, including the nonprojecting canines, the vertical bite and the shortened face. It must be noted, however, that only parts of the skull have been found, and its full shape cannot be reconstructed.

Oreopithecus is a lower "monkey," in popular terms. It is not a "man" in the sense that many reporters assumed it to be, in spite of "no tooth gaps, no apelike protruding jaw," and so on. There are both fossil and still living primates which would have no trouble in answering that description, yet I am sure no one would call them men.

So the substance of the story is that Hurzeler has revived interest in a problematical bit of bone we have long been fingering. For the successful reconstruction of the evolution of the horse in the Tertiary Period, paleontologists had thousands of fossil bones to study. Primatologists may therefore be forgiven their fumblings over great gaps of millions of years from which we do not possess a single complete monkey skeleton, let alone the skeleton of a human forerunner. For the whole Tertiary Period, which involves something like sixty to eighty million years, we have to read the story of primate evolution from a few handfuls of broken bones and teeth. Those fossils, moreover, are from places thousands of miles apart on the Old World land mass.

If we were able to follow every step of man's history backward into time, we would see him divested, rag by rag and stitch by stitch, of every vestige of his human garment. That divestment, however, would not occur all at one place. If we accept the evidence of evolution, we must assume that man became man by degrees, that he emerged out of the animal world by the slow accumulation of human characters over long ages—save for that seemingly rapid spurt in brain growth, which has carried him so far from his other relatives.

Our knowledge at present is not sufficient to establish precisely what anatomical traits are peculiarly human. As Straus has very aptly pointed out: "It is this general lack of structural specialization that makes the study of primate phylogeny so difficult." Some traits may have been paralleled in primate lines of evolution which did not lead to man; some traits called human may represent old generalized characters which have survived in man and been lost in some of his modern specialized relatives.

To continue our writing of the story of human evolution we are totally dependent upon finding additional fossils. Until further discoveries accumulate, each student will perhaps inevitably read a little of his own temperament into the record. Some, as Hurzeler has done, will dwell upon short faces, vertical front teeth and little rounded chins. They will catch glimpses of an elfin human figure which mocks us from a remote glade in the forest of time. Others, just as competent, will say that this elusive homuncular elf is a dream spun from our disguised human longing for an ancestor like ourselves. They will say that in the living primate world around us there are lemurs with short faces and vertical teeth, and that there are monkeys which have the genuine faces of elves and the capacious craniums of little men.

In the end we may shake our heads, baffled, and have to admit that many lines of seeming relatives, rather than merely one, lead to man. It is as though we stood at the heart of a maze and no longer remembered how we had come there.

The Dream Animal

IT WILL NOW be seen that in spite of the dramatic press announcements which thundered the end of Darwinism and of missing links, our little homuncular elf proved nothing of the kind. Even if he turned out to be on the main line of evolutionary ascent leading to man—and this is still exceedingly doubtful—he has, at present, nothing to tell us about the human brain. He is small, he is not by any stretch of imagination a man, and if he did indeed become one, the event still lay millions of years in the future. No amount of headlines can turn the little creature from Tuscany into a human being without recourse to evolutionary change. The writers who had seized upon the "little man" as a refutation of Darwin's general thesis had, at best, been merely acclaiming a new "missing link."

We must now examine, however, some recent aspects of the problem to which I have previously given attention: the mystery which enshrouds the rise of the human brain. A most perceptive philosopher once remarked that the truth about man is inside him. This may well prove to be the case, but the difficulty is to get the secret out, if indeed it lies there, and once it is revealed, to be sure that it is read correctly.

Every so often out of the millions of the human population, a six-year-old child or a teen-age youth dies of old age. The cause of this curious disease, known as progeria, or premature aging, is totally unknown. Clinical cases are reported of complete hairlessness, wrinkled and flabby skin, along with senile changes in the heart and blood vessels. Medical science has observed in these rare cases an enormous increase in the velocity of aging, but the mechanism involved remains as yet undiscovered, though the cause may lie somewhere among the ductless glands.

The affliction, rare though it is, reveals a mysterious clock in the body, a clock capable of running fast or slow, shortening life or extending it and, like the more visible portions of our anatomy, being subjected to evolutionary selection. This clock, however, has another even more curious aspect: it may affect

the growth rate of particular organs. In this way certain peculiar animal specializations have appeared, such as the huge
antlers of the extinct Irish elk, or the dagger-like fangs of the
saber-toothed tiger.

Man, too, has a curious specialization of a more abstract and
generalized type, his brain. If this brain, a brain more than
twice as large as that of a much bigger animal—the gorilla—is
to be acquired in infancy, its major growth must take place
with far greater rapidity than in the case of man's nearest living
relatives, the great apes. It must literally spring up like an
overnight mushroom, and this greatly accelerated growth
must take place during the first months after birth. If it took
place in the embryo, man would long since have disappeared
from the planet—it would have been literally impossible for
him to have been born. As it is, the head of the infant is one of
the factors making human birth comparatively difficult. When
we are born, however, our brain size, about 330 cubic centimeters, is only slightly larger than that of a gorilla baby. This is
why human and anthropoid young look so appealingly similar
in their earliest infancy.

A little later, an amazing development takes place in the
human offspring. In the first year of life its brain trebles in size.
It is this peculiar leap, unlike anything else we know in the
animal world, which gives to man his uniquely human qualities. When the leap fails, as in those rare instances where the
brain does not grow, microcephaly, "pinheadedness," is the result,
and the child is then an idiot. Somewhere among the inner
secrets of the body is one which keeps the time for human brain
growth. If we compare our brains with those of other primate
relatives (recognizing, as we do, many similarities of structure)
we are yet unable to perceive at what point in time or under
what evolutionary conditions the actual human forerunner
began to manifest this strange postnatal brain expansion. It has
carried him far beyond the mental span of his surviving relatives. As our previously quoted authority, Dr. Tilly Edinger
of Harvard, has declared, "the brain of *Homo Sapiens* has not
evolved from the brains it is compared with by comparative
anatomy; it developed within the Hominidae, at a late stage
of the evolution of this family whose other species are all
extinct."

We can, in other words, weigh, measure and dissect the brains of any number of existing monkeys. We may learn much in the process, but the key to our human brain clock is not among them. It arose in the germ plasm of the human group alone and we are the last living representatives of that family. As we contemplate, however, the old biological law that, to a certain degree, the history of the development of the individual tends to reproduce the evolutionary history of the group to which it belongs, we cannot help but wonder if this remarkable spurt in brain development may not represent something roughly akin to what happened in the geological past of man—a sudden or explosive increase which was achieved in a relatively short period, geologically speaking. We have already opened this topic in our discussion of the Darwin-Wallace argument. Let us now see what new evidence bears upon the facts we set forth there.

In discussing the significance of the Piltdown hoax and its bearing upon the Darwin-Wallace controversy, I used the accepted orthodox geological estimate of the time involved in that series of fluctuating events which we speak of popularly as the "Ice Age." I pointed out that almost all of what we know about human evolution is confined to this period. Long though one million years may seem compared with our few millennia of written history, it is, in geological terms, in evolutionary terms, a mere minute's tick of the astronomical clock.

Among other forms of life than man, few marked transformations occurred. Rather, the Ice Age was, particularly toward its close, a time of great extinctions. Some of the huge beasts whose intercontinental migrations had laid down the first paths along which man had traveled, vanished totally from the earth. Mammoths, the Temperate Zone elephants, dropped the last of their heavy tusks along the receding fringes of the ice. The long-horned bisons upon whose herds man had nourished himself for many a long century of illiterate wanderings, faded back into the past. The ape whose cultural remnants at the beginning of the first glaciation can scarcely be distinguished from chance bits of stone has, by the ending of the fourth ice, become artist and world rover, penetrator of the five continents, and master of all.

There is nothing quite like this event in all the time that

went before; the end of brute animal dominance upon earth had come at last. For good or ill, the growth of forests or their destruction, the spread of deserts or their elimination, would lie more and more at the whim of that cunning and insatiable creature who slipped so mysteriously out of the green twilight of nature's laboratory a short million years ago.

A million years is a short time as evolution clocks its progress. We assume, of course, that below that point the creature which was to become man was still walking on his hind feet, but there is every reason to think that the bulging cortex which would later measure stars and ice ages was still a dim, impoverished region in a skull box whose capacity was no greater than that of other apes. Still, a million years in the life history of a single active species like man is a long time, and powerful selective forces must have been at work as ice sheets ground their way across vast areas of the temperate zones. But suppose, just suppose for a moment, that this period of the great ice advances did not last a million years—suppose our geological estimates are mistaken. Suppose that this period we have been estimating at one million years should instead have lasted, say, a third of that time.

In that case, what are we to think of the story of man? Into what foreshortened and cramped circumstances is the human drama to be reduced, a drama, moreover, which, besides evolutionary change, involves time for the spread of man into the New World? Such an episode, it is obvious, would involve a complete reëxamination of our thinking upon the subject of human evolution. In 1956 Dr. Cesare Emiliani of the University of Chicago introduced just this startling factor into the dating of the Ice Age. He did it by the application of a new dating process developed in the field of atomic physics.[1]

The method, it should be explained at the outset, is not the carbon-14 technique which has become so widely publicized in the last decade. That method has applications which, at best, can carry us back around thirty to forty thousand years. The new technique elaborated in the University of Chicago laboratories involves oxygen-18. By studying the amount of

[1]"Note on Absolute Chronology of Human Evolution," *Science* 123 (1956), pp. 924–26.

this isotope in the shells of sea creatures it was found that the percentage of oxygen-18 in the limy shell of, say, an oyster would reveal the temperature of the water in which the oyster had lived when its shell was being secreted. This is because oxygen-18 enters chemical reactions differently at different temperatures. For example, as the temperature of the water increases, the oxygen-18 in the shell decreases.

By using marine cores, specimens of undisturbed sediments brought up from the ocean floor, Dr. Emiliani has been able to subject these chalky deposits full of tiny shells to careful oxygen-18 analysis. He has found, as he analyzed the chemical nature of the seas' "long snowfall," that is, the age-long rain of microscopic shells falling gently to the sea bottom, that marked changes in water temperature could be discerned for different periods in the past. As he studied layer after layer of the chalky ooze brought up in sequential order from the depths, he found that the times of maximum ice expansion on the continents coincided with periods of marked cold beyond that of the present, as revealed in the oxygen-18 content of the minute shells from the ocean floor.

Studying Atlantic and Caribbean cores, Emiliani came to the conclusion that the earliest great cold period, most probably coinciding with the onset of the first glaciation in Europe, was probably no earlier than about three hundred thousand years ago. Oxygen-18, of course, indicates periods of relative warmth or cold, not years. The dating triumph was achieved by the well-known carbon-14 technique for the upper levels of the deposits within the forty-thousand-year range, since carbon-14 also occurs in the chalk ooze.

By establishing the beginning of the last ice recession at about twenty thousand years, it was possible, as a result of the undisturbed uniform nature of the sea deposits, to project the datings backward by the combination of the cold graph and the apparent rate at which the deposits had been laid down, as determined from the carbon dates of the more recent levels. The study reveals a considerable degree of regularity in the waxing and waning of the ice sheets at intervals of about fifty to sixty thousand years.

Dr. Emiliani and his co-workers have thus produced an Ice Age chronology startlingly different from orthodox estimates, but one which is being widely and favorably considered. The newer scheme allows about six hundred thousand years for the total of Ice Age time. Actually the modification is more striking than this figure would indicate. Older figures placed the first, or Gunz glaciation, distant from us at the bottom of the Ice Age by almost a million years. The new chronology would place this ice sheet only about three hundred thousand years remote and then allow perhaps three hundred thousand more years, much less accurately computable and quite indefinite, for certain vague preglacial events. These might include our oldest traces of the Australopithecine man-apes and the first dim traces of crude pebble and bone tools, possibly made by some, at least, of these South African anthropoids.

As we have already indicated, most of our collection of human fossils is derived from the last half of the Pleistocene, even by the old chronology. In this new arrangement the bulk of this material is found to be less than two hundred thousand years old. Man, in Dr. Emiliani's own words, had "the apparent ability to evolve rapidly." This is almost an understatement. The new chronology would appear to suggest a spectacular, even more explosive development than I have previously suggested.

Unfortunately the full outlines of this story cannot, as yet, be made out. Our fossils are too scattered and too few. If the Fontechevade cranium from the French third interglacial represents a man essentially like ourselves, as in brain he appears to be, we can date our species as in existence perhaps seventy thousand years ago, though its total diffusion, in terms of area, at that date would be unknown. If the problematical Swanscombe skull—discovered in England—whose face is missing but whose cranial capacity falls within the modern range, should prove, in time, to be also of our own species, "modern" man would have been in existence perhaps one hundred twenty thousand years before the present.

Even if the men of this period should, in the end, prove to have a face somewhat more massive than that of modern man, an essentially modern brain at so early a date can only suggest, in the light of Emiliani's new datings, that the rise of man from a brain level represented in earliest preglacial times by the South African man-apes took place with extreme rapidity.

Either this occurred, or other fossil forms are not on the main line of human ascent at all. This latter theory, if we still try to cling to a slow type of human evolution, would imply that the true origin of our species is lost in some older pre–Ice Age level, and that all the other human fossils represent side lines and blind alleys of development, living fossils already archaic in Pleistocene times.

Some, contending for this view, have pointed out that carbon-14 datings close to the forty-thousand-year mark have recently been recorded in America. This, it has been argued, suggests a remarkably wide and early diffusion for man, if he is really so young as is now suggested. Just lately, however, some of the earliest carbon-14 dates from the Southwest have been challenged. Professor Frederick Zeuner of the University of London has recently (1957) reported that carbon samples subjected to alkaline washing give dates much earlier than they should actually be. Some of the carbon-14 necessary for accurate dating is apparently removed by subjection to this treatment, thus raising the age of the sample. As a consequence, some of the very earliest American dates from the Southwest may be subject to upward revision. There is no doubt that man had reached America in the closing Ice Age, but these earlier dates will be subject to serious scrutiny.

Interestingly enough, the Keilor skull from Australia, once supposed to be a very early third interglacial man of our own species, has now been elevated, on the basis of carbon dates, to definitely postglacial times. Thus, on this remote continent, there is now no reliable evidence of extremely ancient human intrusion. Furthermore, if we turn to the Old World and seek to carry men much like ourselves further back toward the first glaciation, we have to ask why we so rapidly descend into seemingly cultureless or almost cultureless levels. If man approximating ourselves is truly much older than we imagine, it is conceivable that his physical remains might for long escape us. It seems unlikely, however, that a large-brained form, if widely diffused, would have left so little evidence of his activities. It would appear, then, that within the very brief period between about five hundred thousand to one hundred fifty thousand years ago, man acquired the essential features of a modern brain. Admittedly the outlines of this process are dim, but all the evidence at our command points to this process as being surprisingly rapid.

Such rapidity suggests other modes of selection and evolution than those implied in the nineteenth-century literature with its emphasis on intergroup "struggle" which, in turn, would have demanded large populations. We must make plain here, however, that to reject the older Darwinian arguments is not necessarily to reject the principle of natural selection. We may be simply dealing with a situation in which both Darwin and Wallace failed, in different ways, to see what selective forces might be at work in man. Most of the Victorian biologists were heavily concerned with the more visible aspects of the struggle for existence. They saw it in the ruthless, expanding industrialism around them; they tended to see nature as totally "red in tooth and claw."

The anthropologist had yet to subject native societies to careful scrutiny, or to learn that people of different cultures were remarkably like ourselves in their basic mental make-up. They were often regarded as mentally inferior, living fossils pushed to the wall and going under in the struggle with the dominating white. Wallace, as we have already seen, stood somewhat outside this Victorian prejudice, and having himself endured economic want, almost alone among the great biologists of his time, sought for another key to the development of man.

His thoughts led him in a somewhat mystical direction, yet certain of the facts he recorded were valid enough. He wrote early, however, so that natural explanations which could now be offered were, understandably, not available to him at that time. It is impressive that Wallace observed, though he did not understand, what we today call the pedomorphic features of man—his almost hairless body, his helpless childhood, his surprisingly developed brain—which he rightly judged to be in some manner related to the uniqueness of man. His conclusion that the linguistic ability of natives is in no way inferior to that of "higher" races—a commonplace today—was, in its own time, a courageous statement made in considerable contradiction to beliefs widely held even among scientists.

Although there is still much that we do not understand, it is likely that the selective forces working upon the humanization of man lay essentially in the nature of the socio-cultural world itself. Man, in other words, once he had "crossed over" into this new invisible environment, was being as rigorously selected for survival within it as the first fish that waddled up the shore on its fins. I have said that this new world was "invisible." I do so advisedly. It lay, not so much in his surroundings as in man's brain, in his way of looking at the world around him and at the social environment he was beginning to create in his tiny human groupings.

He was becoming something the world had never seen before —a dream animal—living at least partially within a secret universe of his own creation and sharing that secret universe in his head with other, similar heads. Symbolic communication had begun. Man had escaped out of the eternal present of the animal world into a knowledge of past and future. The unseen gods, the powers behind the world of phenomenal appearance, began to stalk through his dreams.

Nature, one might say, through the powers of this mind, grossly superstitious though it might be in its naïve examination of wind and water, was beginning to reach out into the dark behind itself. Nature was beginning to evade its own limitations in the shape of this strange, dreaming and observant brain. It was a weird multiheaded universe, going on, unseen and immaterial save as its thoughts smoldered in the eyes of hunters huddled by night fires, or were translated into pictures upon cave walls, or were expressed in the trappings of myth or ritual. The Eden of the eternal present that the animal world had known for ages was shattered at last. Through the human mind, time and darkness, good and evil, would enter and possess the world.

The Victorian biologists, intent upon the nature of the animal struggle for existence, in some degree misread human society and the kind of social selection toward brain enhancement which would be the product of unceasing struggle, not by ax and spear in the war of nature, but in that world of streaming shadows forever hidden behind the forehead of man. It was a struggle for symbolic communication, for in this new societal world communication meant life. The world of instinct was passing. This emergent creature was not whole, was not made truly human until, in infancy, the dreams of the group, the social constellation amidst which his own orbit was cast, had been implanted in the waiting, receptive substance of his brain.

How did this brain first come? How fast did it come? Probing among rocks and battered skulls, scientists find that the answers are few. There are many living members of the primate order—that order which includes man—who live in groups, but show no signs of becoming men. Their brains bear a family resemblance to our own, but they are not the brains of men. They contain, instead, only the shrewd, wild thoughts that serve to remind us of the solitary door which began to open for us once, and once only, long ago, as the earth swung in some tilted, sunlit orbit far backward on the roads of space.

If one attempts to read the complexities of the story, one is not surprised that man is alone on the planet. Rather, one is amazed and humbled that man was achieved at all. For four things had to happen, and if they had not happened simultaneously, or at least kept pace with each other, the bones of man would lie abortive and forgotten in the sandstones of the past:

1. His brain had almost to treble in size.

2. This had to be effected, not in the womb, but rapidly, after birth.

3. Childhood had to be lengthened to allow this brain, divested of most of its precise instinctive responses, to receive, store, and learn to utilize what it received from others.

4. The family bonds had to survive seasonal mating and become permanent, if this odd new creature was to be prepared for his adult role.

Each one of these major points demanded a multitude of minor biological adjustments, yet all of this—change of growth rate, lengthened age, increased blood supply to the head, moved apparently with rapidity. It is a dizzying spectacle with which we have nothing to compare. The event is complex, it is many-sided, and what touched it off is hidden under the leaf mold of forgotten centuries.

Somewhere in the glacial mists that shroud the past, Nature found a way of speeding the proliferation of brain cells and did it by the ruthless elimination of everything not needed to that end. We lost our hairy covering, our jaws and teeth were reduced in size, our sex life was postponed, our infancy became among the most helpless of any of the animals because everything had to wait upon the development of that fast-growing mushroom which had sprung up in our heads.

Now in man, above all creatures, brain is the really important specialization. As Gavin de Beer, Director of the British Museum of Natural History, has suggested, it appears that if infancy is lengthened, there is a correspondingly lengthier retention of embryonic tissues capable of undergoing change.[2] Here, apparently, is a possible means of stepping up brain growth. The anthropoid ape, because of its shorter life cycle and slow brain growth, does not make use of nearly the amount of primitive neuroblasts—the embryonic and migrating nerve cells—possible in the lengthier, and at the same time paradoxically accelerated development of the human child. The clock in the body, in other words, has placed a limit upon the pace at which the ape brain grows—a limit which, as we have seen, the human ancestors in some manner escaped. This is a simplification of a complicated problem, but it hints at the answer to Wallace's question of long ago as to why man shows such a strange, rich mental life, many of whose artistic aspects can have had little direct value measured in the old utilitarian terms of the selection of all qualities in the struggle for existence.

When these released potentialities for brain growth began, they carried man into a new world where the old laws no longer totally held. With every advance in language, in symbolic thought, the brain paths multiplied. Significantly enough, those which are most heavily involved in the life processes, and are most ancient, mature first. The most recently acquired and less specialized regions of the brain, the "silent areas," mature last. Some neurologists, not without reason, suspect that here may lie other potentialities which only the future of the race may reveal.

Even now, however, the brain of man, with all its individual never-to-be-abandoned richness, is becoming merely a unit in the vast social brain which is potentially immortal, and whose memory is the heaped wisdom of the world's great thinkers. The scientist Haldane, brooding upon the future, has speculated

[2]*Embryos and Ancestors*, rev. ed. (New York, Oxford, 1951), p. 93.

that we will even further prolong our childhood and retard maturity if brain advance continues.

It is unlikely, however, in our present comfortable circumstances, that the pace of human change will ever again speed at the accelerated rate it knew when man strove against extinction. The story of Eden is a greater allegory than man has ever guessed. For it was truly man who, walking memoryless through bars of sunlight and shade in the morning of the world, sat down and passed a wondering hand across his heavy forehead. Time and darkness, knowledge of good and evil, have walked with him ever since. It is the destiny struck by the clock in the body in that brief space between the beginning of the first ice and that of the second. In just that interval a new world of terror and loneliness appears to have been created in the soul of man.

For the first time in four billion years a living creature had contemplated himself and heard with a sudden, unaccountable loneliness, the whisper of the wind in the night reeds. Perhaps he knew, there in the grass by the chill waters, that he had before him an immense journey. Perhaps that same foreboding still troubles the hearts of those who walk out of a crowded room and stare with relief into the abyss of space so long as there is a star to be seen twinkling across those miles of emptiness.

Man of the Future

T HERE ARE DAYS when I may find myself unduly pessimistic about the future of man. Indeed, I will confess that there have been occasions when I swore I would never again make the study of time a profession. My walls are lined with books expounding its mysteries, my hands have been split and raw with grubbing into the quicklime of its waste bins and hidden crevices. I have stared so much at death that I can recognize the lingering personalities in the faces of skulls and feel accompanying affinities and repulsions.

One such skull lies in the lockers of a great metropolitan museum. It is labeled simply: Strandlooper, South Africa. I have never looked longer into any human face than I have upon the features of that skull. I come there often, drawn in spite of myself. It is a face that would lend reality to the fantastic tales of our childhood. There is a hint of Wells's *Time Machine* folk in it—those pathetic, childlike people whom Wells pictures as haunting earth's autumnal cities in the far future of the dying planet.

Yet this skull has not been spirited back to us through future eras by a time machine. It is a thing, instead, of the millennial past. It is a caricature of modern man, not by reason of its primitiveness but, startlingly, because of a modernity outreaching his own. It constitutes, in fact, a mysterious prophecy and warning. For at the very moment in which students of humanity have been sketching their concept of the man of the future, that being has already come, and lived, and passed away.

We men of today are insatiably curious about ourselves and desperately in need of reassurance. Beneath our boisterous self-confidence is fear—a growing fear of the future we are in the process of creating. In such a mood we turn the pages of our favorite magazine and, like as not, come straight upon a description of the man of the future.

The descriptions are never pessimistic; they always, with sublime confidence, involve just one variety of mankind—our own—and they are always subtly flattering. In fact, a distinguished colleague of mine who was adept at this kind of prophecy once allowed a somewhat etherealized version of his own lofty brow to be used as an illustration of what the man of the future was to look like. Even the bald spot didn't matter—all the men of the future were to be bald, anyway.

Occasionally I show this picture to students. They find it highly comforting. Somebody with a lot of brains will save humanity at the proper moment. "It's all right," they say, looking at my friend's picture labeled "Man of the Future." "It's O.K. Somebody's keeping an eye on things. Our heads are getting bigger and our teeth are getting smaller. Look!"

Their voices ring with youthful confidence, the confidence engendered by my persuasive colleagues and myself. At times I glow a little with their reflected enthusiasm. I should like to regain that confidence, that warmth. I should like to but . . .

There's just one thing we haven't quite dared to mention. It's this, and you won't believe it. It's all happened already. Back there in the past, ten thousand years ago. The man of the future, with the big brain, the small teeth.

Where did it get him? Nowhere. *Maybe there isn't any future.* Or, if there is, maybe it's only what you can find in a little heap of bones on a certain South African beach.

Many of you who read this belong to the white race. We like to think about this man of the future as being white. It flatters our ego. But the man of the future in the past I'm talking about was not white. He lived in Africa. His brain was bigger than your brain. His face was straight and small, almost a child's face. He was the end evolutionary product in a direction quite similar to the one anthropologists tell us is the road down which we are traveling.

In the minds of many scholars, a process of "foetalization" is one of the chief mechanisms by which man of today has sloughed off his ferocious appearance of a million years ago, prolonged his childhood, and increased the size of his brain. "Foetalization" or "pedomorphism," as it is termed, means simply the retention, into adult life, of bodily characters which at some earlier stage of evolutionary history were actually only infantile. Such traits were rapidly lost as the animal attained maturity.

If we examine the life history of one of the existing great apes and compare its development with that of man, we observe that the infantile stages of both man and ape are far more similar than the two will be in maturity. At birth, as we have seen, the brain of the gorilla is close to the size of that of the human infant. Both newborn gorilla and human child are much more alike, facially, than they will ever be in adult life because the gorilla infant will, in the course of time, develop an enormously powerful and protrusive muzzle. The sutures of his skull will close early; his brain will grow very little more.

By contrast, human brain growth will first spurt and then grow steadily over an extended youth. Cranial sutures will remain open into adult life. Teeth will be later in their eruption. Furthermore, the great armored skull and the fighting characters of the anthropoid male will be held in abeyance.

Instead, the human child, through a more extended infancy, will approach a maturity marked by the retention of the smooth-browed skull of childhood. His jaws will be tucked inconspicuously under a forehead lacking the huge, muscle-bearing ridges of the ape. In some unknown manner, the ductless glands which stimulate or inhibit growth have, in the course of human evolution, stepped down the pace of development and increased the life span. Our helpless but well-cared-for childhood allows a longer time for brain growth and, as an indirect consequence, human development has slowly been steered away from the ape-like adulthood of our big-jawed forebears.

Modern man retains something of his youthful gaiety and nimble mental habits far into adult life. The great male anthropoids, by contrast, lose the playful friendliness of youth. In the end the massive skull houses a small, savage, and often morose brain. It is doubtful whether our thick-skulled forerunners viewed life very pleasantly in their advancing years.

We of today, then, are pedomorphs—the childlike, yet mature products of a simian line whose years have lengthened and whose adolescence has become long drawn out. We are, for our day and time, civilized. We eat soft food, and an Eskimo child can outbite us. We show signs, in our shortening jaws, of losing our wisdom teeth. Our brain has risen over our eyes and few, even of our professional fighters, show enough trace of a brow ridge to impress a half-grown gorilla. The signs point steadily onward toward a further lightening of the skull box and to additional compression of the jaws.

Imagine this trend continuing in modern man. Imagine our general average cranial capacity rising by two hundred cubic centimeters while the face continued to reduce proportionately. Obviously we would possess a much higher ratio of brain size to face size than now exists. We would, paradoxically, resemble somewhat our children of today. Children acquire facial prominence late in growth under the endocrine stimulus of maturity. Until that stimulus occurs, their faces bear a smaller ratio to the size of the brain case. It was so with these early South Africans.

But no, you may object, this whole process is in some way dependent upon civilization and grows out of it. Man's body and his culture mutually control each other. To that extent we are masters of our physical destiny. This mysterious change that is happening to our bodies is epitomized at just one point today, the point of the highest achieved civilization upon earth—our own.

I believed this statement once, believed it wholeheartedly. Sometimes it is so very logical I believe it still as my colleague's ascetic, earnest, and ennobled face gazes out at me from the screen. It carries the lineaments of my own kind, the race to which I belong. But it is not, I know now, the most foetalized race nor the largest brained. That game had already been played out before written history began—played out in an obscure backwater of the world where sails never came and where the human horde chipped flint as our ancestors had chipped it northward in Europe when the vast ice lay heavy on the land.

These people were not civilized; they were not white. But they meet in every major aspect the physical description of the man of tomorrow. They achieved that status on the raw and primitive diet of a savage. Their delicate and gracefully reduced teeth and fragile jaws are striking testimony to some strange inward hastening of change. Nothing about their environment in the least explains them. They were tomorrow's children surely, born by error into a lion country of spears and sand.

Africa is not a black man's continent in the way we are inclined to think. Like other great land areas it has its uneasy amalgams, its genetically strange variants, its racial deviants whose blood stream is no longer traceable. We know only that the first true men who disturbed the screaming sea birds over Table Bay were a folk that humanity has never looked upon again save as their type has wavered into brief emergence in an occasional mixed descendant. They are related in some dim manner to the modern Kalahari Bushman, but he is dwarfed in brain and body and hastening fast toward eventual extinction. The Bushman's forerunners, by contrast, might have stepped with Weena out of the future eras of the Time Machine.

Widespread along the South African coast, in the lowest strata of ancient cliff shelters, as well as inland in Ice Age gravel and other primeval deposits, lie the bones of these unique people. So remote are they from us in time that the first archaeologists who probed their caves and seashore middens had expected to reveal some distant and primitive human forerunner such as Neanderthal man. Instead their spades uncovered an unknown branch of humanity which, in the words of Sir Arthur Keith, the great English anatomist, "outrivals in brain volume any people of Europe, ancient or modern . . ."

But that is not all. Dr. Drennan of the University of Capetown comments upon one such specimen in anatomical wonder: "It appears ultramodern in many of its features, surpassing the European in almost every direction. That is to say, it is less simian than any modern skull." This ultramodernity Dr. Drennan attributes to the curious foetalization of which I have spoken.

More fascinating than big brain capacity in itself, however, is the relation of the cranium to the base of the skull and to the face. The skull base, that is, the part from the root of the nose to the spinal opening, is buckled and shortened in a way characteristic of the child's skull before the base expands to aid in the creation of the adult face. Thus, on this permanently shortened cranial base, the great brain expands, bulging the forehead heavily above the eyes and leaving the face neatly retracted beneath the brow. There is nothing in this face to suggest the protrusive facial angle of the true Negro. It is, as Dr. Drennan says, "ultramodern," even by Caucasian standards. The bottom of the skull grew, apparently, at a slow and childlike tempo while the pace-setting brain lengthened and broadened to a huge maturity.

When the skull is studied in projection and ratios computed, we find that these fossil South African folk, generally called "Boskop" or "Boskopoids" after the site of first discovery, have the amazing cranium-to-face ratio of almost five to one. In Europeans it is about three to one. This figure is a marked indication of the degree to which face size had been "modernized" and subordinated to brain growth. It is true that Dr. Ronald Singer has recently contended that the "Boskop" people cannot be successfully differentiated from the Bushman because Boskopoid features can be observed in this latter group, but even he would not deny the appearance of the peculiarly pedomorphic and ultrahuman features we have been discussing. At best, he would contend, in contrast to Keith and Drennan, that these characters have emerged in a sporadic fashion throughout the racial history of South Africa. By contrast, the facial structure of existing Caucasians, advanced though we imagine it, has only a mediocre rating.

The teeth vary a little from the usual idea about man of the future, yet they, too, are modern. Our prophecies generally include the speculation that we will, in time, lose our third molar teeth. This seems likely indeed, for the tooth often fails to erupt, crowds, and causes trouble. The Boskop folk had no such difficulty. Their teeth are small, neatly reduced in proportion to their delicate jaws, and free from any sign of the dental ills that trouble us. Here, in a hunter's world that would seem to have demanded at least the stout modern dentition of the Congo Negro, nature had decreed otherwise. These teeth could have nibbled sedately at the Waldorf, nor would the customers have been alarmed.

With the face, however, it would have been otherwise. In its anatomical structure we observe characters which relate these people both with the dwarf modern Bushman and to some ancient Negroid strain distinct from the West Coast blacks. We believe that they had the tightly-kinked "pepper-corn" hair of the Bushman as well as his yellow-brown skin. A branch of the Negro race has thus produced what is actually, so far as we can judge from the anatomical standpoint, one of the most ultrahuman types that ever lived! Had these characters appeared among whites, they would undoubtedly have been used in invidious comparisons with other "lesser" races.

We can, of course, repeat the final, unanswerable question:

What did this tremendous brain mean to the Boskop people? We can marvel over their curious and exotic anatomy. We can wonder at the mysterious powers hidden in the human body, so potent that once unleashed they brought this more than modern being into existence on the very threshold of the Ice Age.

We can debate for days whether that magnificent cranial endowment actually represented a superior brain. We can smile pityingly at his miserable shell heaps, point to the mute stones that were his only tools. We can do this, but in doing it we are mocking our own rude forefathers of a similar day and time. We are forgetting the high artistic sensitivity which flowered in the closing Ice Age of Europe and which, oddly, blossomed here as well, lingering on even among the dwarfed Bushmen of the Kalahari. No, we cannot dismiss the Boskop people on such grounds, for even remarkable potential endowment cannot create high civilization overnight.

What we *can* say is that perhaps the unloosed mechanism ran too fast, that these people may have been ill-equipped physically to compete against the onrush of more ferocious and less foetalized folk. In a certain sense the biological clock had speeded them out of their time and place—a time which ten thousand years later has still not arrived. We may speculate that even mentally they may have lacked something of the elemental savagery of their competitors.

Their evolutionary gallop has led precisely nowhere save to a dwarfed and dying folk—if, with some authorities, we accept the later Bushmen as their descendants. This, then, was the logical end of complete foetalization: a desperate struggle to survive among a welter of more prolific and aggressive stocks. The answer to the one great question is still nowhere, still nothing. But there in the darkened laboratory, after the students have gone, I look once more at the exalted photograph of my friend upon the screen, noting character by character the foetalized refinement by which the artist has attempted to indicate the projected trend of future development—the expanded brain, the delicate face.

I look, and I know I have seen it all before, reading, as I have long grown used to doing, the bones through the living flesh. I have seen this face in another racial guise in another

and forgotten day. And once again I grow aware of that eternal
flickering of forms which we are now too worldly wise to label
progress, and whose meaning forever escapes us.

The man of the future came, and looked out among us once
with wistful, if unsophisticated eyes. He left his bones in the
rubble of an alien land. If we read evolution aright, he may
come again in another million years. Are the evolutionary
forces searching for the right moment of his appearance? Or is
his appearance itself destined always, even in the moment of
emergence, to mark the end of the drama and foretell the extinction of a race?

Perhaps the strange interior clockwork that is here revealed
as so indifferent to environmental surroundings has set, after
all, a limit to the human time it keeps. That is the real question
propounded by my friend's fine face. That is the question that
I sometimes think the Boskop folk have answered. I wish I
could be sure. I wish I knew.

Whatever else these skulls or those of occasional variant
moderns may tell us, one thing they clearly reveal: Those who
contend that because of present human cranial size, and the
limitations of the human pelvis, man's brain is no longer capable of further expansion, are mistaken. Cranial capacities of almost a third more than the modern average have been
occasionally attained among the Boskop people and even in
rare individuals among other, less foetalized races. The secret
does not lie in the size of the brain before birth; rather, as we
have seen, it is contained in that strange spurt which in the first
year of life carries man upward and outward into a social world
from which his fellow beings are excluded. Whether that postnatal expansion is destined to be further enhanced in the long
eras to come there is no telling, nor, perhaps, does it matter
greatly. For in the creation of the social brain, nature, through
man, has eluded the trap which has engulfed in one way or
another every other form of life on the planet. Within the reasonable limits of the brain that now exists, she has placed the
long continuity of civilized memory as it lies packed in the
world's great libraries. The need is not really for more brains,
the need is now for a gentler, a more tolerant people than
those who won for us against the ice, the tiger, and the bear.
The hand that hefted the ax, out of some old blind allegiance

to the past fondles the machine gun as lovingly. It is a habit man will have to break to survive, but the roots go very deep.

I once sat, a prisoner, long ago, and watched a peasant soldier just recently equipped with a submachine gun swing the gun slowly into line with my body. It was a beautiful weapon and his finger toyed hesitantly with the trigger. Suddenly to possess all that power and then to be forbidden to use it must have been almost too much for the man to contain. I remember, also, a protesting female voice nearby—the eternal civilizing voice of women who know that men are fools and children, and irresponsible. Sheepishly the peon slowly dropped the gun muzzle away from my chest. The black eyes over the barrel looked out at me a little wicked, a little desirous of better understanding.

"Thompson, Tome'-son'," he repeated proudly, slapping the barrel. "Tome'-son'." I nodded a little weakly, relaxing with a sigh. After all, we were men together and understood this great subject of destruction. And was I not a citizen of the country that had produced this wonderful mechanism? So I nodded again and said carefully after him, "Thompson, Tome'-son'. *Bueno, si, muy bueno.*" We looked at each other then, smiling a male smile that ran all the way back to the Ice Age. In academic halls since, considering the future of humanity, I have never been quite free of the memory of that soldier's smile. I weigh it mentally against the future whenever one of those delicate forgotten skulls is placed upon my desk.

Little Men and Flying Saucers

TODAY, AS NEVER BEFORE, the sky is menacing. Things seen indifferently last century by the wandering lamplighter now trouble a generation that has grown up to the wail of air-raid sirens and the ominous expectation that the roof may fall at any moment. Even in daytime, reflected light on a floating dandelion seed, or a spider riding a wisp of gossamer in the sun's eye, can bring excited questions from the novice unused to estimating the distance or nature of aerial objects.

Since we now talk, write, and dream endlessly of space rockets, it is no surprise that this thinking yields the obverse of the coin: that the rocket or its equivalent may have come first to us from somewhere "outside." As a youth, I may as well confess, I waited expectantly for it to happen. So deep is the conviction that there must be life out there beyond the dark, one thinks that if they are more advanced than ourselves they may come across space at any moment, perhaps in our generation. Later, contemplating the infinity of time, one wonders if perchance their messages came long ago, hurtling into the swamp muck of the steaming coal forests, the bright projectile clambered over by hissing reptiles, and the delicate instruments running mindlessly down with no report.

Sometimes when young, and fossil hunting in the western Badlands, I had thought it might yet be found, corroding and long dead, in the Tertiary sod that was once green under the rumbling feet of titanotheres. Surely, in the infinite wastes of time, in the lapse of suns and wane of systems, the passage, if it were possible, would have been achieved. But the bright projectile has not been found and now, in sobering middle age, I have long since ceased to look. Moreover, the present theory of the expanding universe has made time, as we know it, no longer infinite. If the entire universe was created in a single explosive instant a few billion years ago, there has not been a sufficient period for all things to occur even behind the star shoals of the outer galaxies. In the light of this fact it is now just conceivable that there may be nowhere in space a mind superior to our own.

If such a mind should exist, there are many reasons why it could not reside in the person of a little man. There is, however, a terrible human fascination about the miniature, and one little man in the hands of the spinner of folk tales can multiply with incredible rapidity. Our unexplainable passion for the small is not quenched at the borders of space, nor, as we shall see, in the spinning rings of the atom. The flying saucer and the much publicized little men from space equate neatly with our own projected dreams.

When I first heard of the little man there was no talk of flying saucers, nor did his owner ascribe to him anything more than an earthly origin. It has been almost a quarter of a century since I encountered him in a bone hunter's camp in the West. A rancher had brought him to us in a box. "I figured you'd maybe know about him," he said. "He'll cost you money, though. There's money in that little man."

"Man?" we said.

"Man," he countered. "What you'd call a pygmy or a dwarf, but smaller than any show dwarf I ever did see. A mummy, too, a little dead mummy. I figure it was some kind of bein' like us, but little. They put him in the place I found him; maybe it was a thousand years ago. You'll likely know."

Our heads met over the box. The last paper was withdrawn. The creature emerged on the man's palm. I've seen a lot of odd things in the years since, and fakes by the score, but that little fellow gave me the creeps. He might have been two feet high in a standing posture—not more. He was mummified in a crouching position, arms folded. The face with closed eyes seemed vaguely evil. I could have sworn I was dreaming.

I touched it. There was a peculiar, fleshy consistency about it, still. It was not a dry mummy. It was more like what you would expect a natural cave mummy to be like. It had no tail. I know because I looked. And to this day the little man sits on there, in my brain, and as plain as yesterday I can see the faint half-smirk of his mouth and the tiny black hands at his knees.

"You can have it for two hundred bucks," said the man. We glanced at each other, sighed, and shook our heads. "We aren't in the market," we said. "We're collecting, not buying, and we're staying with our bones."

"Okay," said the man, and gave us a straight look, closing his box. "I'm going to the carnival down below tonight. There's money in him. There's money in that little man."

I think it may have been just as well for us that we made no purchase. I have never liked the little man, nor the description of the carnival to which he and his owner were going. It may be, I used to think, that I will yet encounter him before I die, in some little colored tent on a country midway. Once, in the years since, I have heard a description that sounded like him in another guise. It involved a fantastic tale of some Paleozoic beings who hunted among the tree ferns when the world was ruled by croaking amphibians. The story did not impress me; I knew him by then for what he was: an anomalous mummified stillbirth with an undeveloped brain.

I never expected to see him emerge again in books on flying saucers, or to see the "little men" multiply and become so common that columnists would take note of them. Nor, though I should have known better, did I expect to live to hear my little man ascribed an extraplanetary origin. There is a story back of him, it is true, but it is a history of this earth, and, of all unlikely things, it involves that great man of science, Charles Darwin, though by a curious, lengthy, and involved route.

Men have been men for so long that they tend not to question the fact. All their experience tells them that their children will precisely resemble themselves; that kittens will become cats and cats will have kittens, and that even caterpillars, though the pattern seems a little odd, will become butterflies, and butterflies will produce caterpillars. It is so habitual an event that we do not stop to ask why this happens, or to consider that this amazing precision in results implies a strange ordering of life in a world we often think is chanceful and meaningless.

A few wise men since the time of the Greeks have found it a source of wonder, but they have been a minority. Most people have shrugged and spoken indifferently of the gods, or contented themselves, as the Christian world did for so long, with the idea of special creation of each species. Nevertheless, the wise ones kept on wondering.

They found, as they began their first groping attempts to classify and arrange the living world, that in spite of the

assumed individual creation of every living species by the supernatural intervention of divine power, a basic similarity of
structure existed among many forms of life. This was a remarkable thing to find among supposedly individual creations.
Offhand one would say that a much greater degree of spontaneous novelty would have been possible. In fact, man once
innocently believed himself part of such a creation. The fabu-
lous animals of the ancient bestiaries, the mermaids, griffins,
and centaurs, not to mention the men whose ears were so large
that their owners slept in them, would have been the natural,
spontaneous products of such uncontrolled, creative whimsy.

But there was the pattern: the ape and the man with their
bone-by-bone correspondence. The very fact that one can add
a plural to the word *reptile* and so suggest anything from a
brontosaurus to a garter snake shows that a pattern exists.
Birds all have feathers, wings, and claws; they are a common
class in spite of their diversities. They have been pulled into
many shapes, but there is still an eternal "birdliness" about
them. They are built on a common plan, just as I share mammalian characters with a small mouse who inhabits my desk
drawer. This is hard to account for in a disordered world, so
that recently, when I came upon this mouse, trapped and terri-
fied in the wastebasket, his similarity to myself rendered me
helpless, and out of sheer embarrassment I connived in his
escape.

Now so long as these remarkable patterns could be observed
only in the living world around us, they occasioned no great
alarm. Even after Cuvier, in 1812, made a magnificent attempt
to reduce the forms of animal life to four basic blueprints or
"archetypes" of divergent character, no one was particularly
disturbed—least of all from the religious point of view. In the
words of one great naturalist, Louis Agassiz, "This plan of
creation . . . has not grown out of the necessary action of
physical laws, but was the free conception of the Almighty Intellect, matured in his thought before it was manifested in
tangible external forms."

It was not long, however, before *pattern*, the divine blue-
print, first recognized in the existing world, was extended by
the geologist across the deeps of time. The animal world of the
past was in the process of discovery. It proved to be a world

without man. Curiously enough, it was soon learned that extinct animals could be fitted into the broad classifications of the existing world. They were mammals or amphibia or reptiles, as the case might be. Though no living eye had beheld them, they seemed to mark the continuation of the divine abstraction, the eternal patterns, across the enormous time gulfs of the past.

The second fact, that man had not been discovered, was a cause for dismay. In the man-centered universe of the time, one can appreciate the anguish of the Reverend Mr. Kirby discovering the Age of Reptiles: "Who can think that a being of unbounded power, wisdom, and goodness, should create a world merely for the habitation of a race of monsters, without a single, rational being in it to serve and glorify him?" This is the wounded outcry of the human ego as it fails to discover its dominance among the beasts of the past. Even more tragically, it learns that the world supposedly made for its enjoyment has existed for untold eons entirely indifferent to its coming. The chill vapors of time and space are beginning to filter under the closed door of the human intellect.

It was in these difficult straits, in the black night of his direst foreboding, that the doctrine of geologic prophecy was evolved by man. For fifty years it would hold time at bay, and in one last great effort its proponents, by clever analogies, would attempt to extend the human drama across the infinite worlds of space; it echoes among us still in the shape of the little men of the flying saucers. No braver *mythos* was ever devised under the cold eye of science.

In an old book from my shelves, Hugh Miller's *The Testimony of the Rocks*, I find this passage: "*Higher still in one of the deposits of the Trias we are startled by what seems to be the impression of a human hand of an uncouth massive shape, but with the thumb apparently set in opposition, as in man, to the other fingers.*"

There is only one way to understand this literature. The biologists of the first half of the nineteenth century had recognized that the unity of animal organization descends into past ages and is observable in forms no living eye has beheld. It was, they believed, an immaterial, a supernatural line of connection. They refused to see in this unity of plan an actual physical relationship. Instead they read the past as a successive series of creations and extinctions upon a divinely modifiable but consistent plan. "Geology," said one writer, "unrolls a prophetic scroll, in which the earlier animated creation points on to the later."

In 1726, before the rise of geological theology, Professor Scheuchzer of Zurich had discovered and described the skeleton of a long extinct amphibian as that of *Homo Diluvii testis*, "Man, witness of the flood." The remains, after being piously termed "a rare relic of the accursed race of the primitive world," were found to be those of an animal, and interest in the fossil ceased. With the development of geological prophecy, however, we find this giant salamander reappearing in the writings of that eminent Scotch philosopher, James McCosh. Admitting the true nature of the relic, McCosh, undaunted, contended in 1857: "Long ages had yet to roll on before the consummation of the vertebrate type; the preparations for man's appearance were not yet completed. *Nevertheless, in this fossil of Scheuchzer's there was a prefiguration of the more perfect type which man's bony framework presents.*" Thus the swinging pick of the geologist at work in the world's bone yards did not, at first, disturb the abstract beauty of the Platonic forms. Instead, the recognition of the past enveloped life with a strange premonitory quality, a sense of prophecy and doom as carefully ordered as the movement on some great stage.

It is in the light of this philosophy that the hand, "massive" and of "uncouth shape," must be interpreted. It foreshadows, out of that slimy concourse of sprawling amphibians and gaping lizards, the eventual emergence of man. Splayed, monstrous, and mud-smeared, it haunts the future. That it is the footprint of some wandering reptilian beast of the coal swamps may be granted, but it is also a vertebrate. Its very body forecasts the times to come.

It would be erroneous, however, to conceive of reptiles as being the major preoccupation of our geological prophets. They scanned the anatomy of fishes, birds, and salamanders, seeking in their skeletons anticipations of the more perfect structure of man. If they found footprints of fossil bipeds it was a "sign" foretelling man. All things led in his direction. Prior to his entrance the stage was merely under preparation.

In this way the blow to the human ego had been softened. The past was only the prologue to the Great Play. Man was at the heart of things after all.

It was a strange half century, as one looks back upon it—that fifty years before the publication of Darwin's *Origin of Species.* It was dominated by a generation that saw the world as a complex symbolic system pointing in the direction of man, who was foreknown and prefigured from the beginning. Man, who comes last, is the end of this strange cycle. With him, in the eyes of many of these thinkers, the process ceases and no further changes in the world of life are to be expected. Since the transcendental "evolutionists" were man-centered, questions involving divergent evolution and adaptation did not come easily to their minds. Working with an immaterial and abstract Platonic concept, it was inevitable that they should seek to extend their doctrine across the deeps of space. Because the pattern was capable of modification, the possibility of the existence of small men, large men, or men of different colors upon other planets did not trouble them, but men they ought to be. There was little comprehension of the fact that man had acquired his particular bodily structure and upright posture through a peculiar set of evolutionary circumstances, not easily to be duplicated.

The theory of the plurality of worlds is a very ancient one; that is, the notion that the lights seen elsewhere in space may be bodies like that which we inhabit. After the rise of the Copernican astronomy and the growing realization that our earth is part of a planetary system revolving around a central sun, it was often contended by philosophers that the other stars seen in space must be similar suns with similar planetary satellites.

Quarrels arose between those who believed God's power infinitely and creatively extended among the stars, and those who regarded it as heresy and dangerous to Christian belief to imply that the Infinite Mind might be concerned with more than the beings of this planet. It was a struggle heightened by an enormous extension of man's vision into the worlds of the infinitely far and the infinitely small, the telescope and the microscope having momentarily stunned the human imagination. Some clung frantically to the little tight-fenced world of the

Middle Ages, refusing to acknowledge what these instruments revealed. Others, with greater willingness to accept the new, tried, nevertheless, to equate what they saw with old beliefs and to elaborate an "astrotheology."

In the fifties of the last century there was a great outburst of interest in the possibility of life on other worlds. The recently discovered life history of our own planet and improvements in astronomical apparatus had all excited great interest on the part of a public wavering in its loyalty between old religious dogmas and the new revelations of science. Speculation, in many instances, was roaming far in advance of actual observation.

"The inhabitants of Jupiter," wrote William Whewell in 1854, "must . . . it would seem, be cartilaginous and glutinous masses. If life be there it does not seem in any way likely, that the living things can be anything higher in the scale of being, than such boneless, watery, pulpy creatures . . ."

This remark is not intended as merely innocent theorizing. In his work, *Plurality of Worlds*, Whewell indicates his definite opposition to the idea that the other planets, or the more remote worlds in other galaxies, are inhabited. At best he is willing to grant the existence of a few gelatinous creatures such as he mentions in the above passage, but that man is to be found elsewhere, he denies. He argues that there are superior and inferior regions of space. Man, preceded by endless eons of lower creatures in time, is yet a superior being. He calls attention to the fact that "the intelligent part of creation is thrust into the compass of a few years, in the course of myriads of ages; why not then into the compass of a few miles, in the expanse of systems?" On this earth a "supernatural interposition" has introduced man; the planet is unique.

Whewell's essay generated a storm of discussion. His was not the popular side of the controversy. Sir David Brewster countered with a volume significantly titled *More Worlds Than One*, in which he bluntly asserts: "The function of one satellite must be the function of all the rest. The function of our Moon, to give light to the Earth, must be the function of the other twenty-two moons of the system; and the function of the Earth, *to support inhabitants*, must be the function of all other planets." He dwells on the "grand combination" of "*infinity of life, with infinity of matter*."

Brewster, moreover, calls attention to the invisible domain revealed by the microscope and argues from this that God has all along been attentive to forms of life of which we had no knowledge. So intriguing became the relativity of size that one author even produced a work whose subtitle bore the query *Are Ultimate Atoms Inhabited Worlds?* Stories like Fitz-James O'Brien's "The Diamond Lens," or Ray Cummings' "The Girl in the Golden Atom," stem from such thought.

Another writer, William Williams, in *The Universe No Desert, the Earth No Monopoly*, strikes more directly at the heart of the argument. He invokes geological prophecy and extends it directly across space: "The archetypal idea of man, revealed in the lower vertebrated animals, proves God's foreknowledge of man's existence; and it equally applies to vertebrates on Jupiter or Neptune as to those on the Earth; and still farther, to the Universe, as these animals were within its precincts."

Williams was not the first nor the last man to utter these sentiments, but he did so with a fierce singleness of purpose. The life plans were immanent, prophetic, and immaterial. They could thus be projected across space. Why, he argues with the same horror that the Reverend Mr. Kirby had exhibited toward the Age of Reptiles, should God "banish his own image to one diminutive enclosure and surround . . . the residue of His immense Person with unintelligent, half-formed, crude monsters?" If man is regarded as a good production here, he must be found in endless duplication throughout the worlds. The pattern in the rocks of this earth is the pattern of the whole.

The shattering of this scheme of geological prophecy was the work of many men, but it was Charles Darwin who brought the event to pass, and who engineered what was to be one of the most dreadful blows that the human ego has ever sustained: the demonstration of man's physical relationship to the world of the lower animals. It is quite apparent, however, that there is an aspect of Darwin's discoveries which has never penetrated to the mind of the general public. It is the fact that once undirected variation and natural selection are introduced as the mechanism controlling the development of plants and animals, the evolution of every world in space becomes a series of unique historical events. The precise accidental duplication of a complex form of life is extremely unlikely to occur in even the same environment, let alone in the different background and atmosphere of a far-off world.

In the modern literature on space travel I have read about cabbage men and bird men; I have investigated the loves of the lizard men and the tree men, but in each case I have labored under no illusion. I have been reading about a man, *Homo sapiens*, that common earthling, clapped into an ill-fitting coat of feathers and retaining all his basic human attributes including an eye for the pretty girl who has just emerged from the space ship. His lechery and miscegenating proclivities have an oddly human ring, and if this is all we are going to find on other planets, I, for one, am going to be content to stay at home. There is quite enough of that sort of thing down here, without encouraging it throughout the starry systems.

The truth is that man is a solitary and peculiar development. I do not mean this in any irreverent or contemptuous sense. I want merely to point out that when Charles Darwin and his colleagues established the community of descent of the living world, and observed the fact of divergent evolutionary adaptation, they destroyed forever the concept of geological prophecy. They did not eliminate the possibility of life on other worlds, but the biological principles which they established have totally removed the likelihood that our descendants, in the next few decades, will be entertaining little men from Mars. I would be much more willing to consider the possibility of sitting down to lunch with a purple polyp, but even this has anatomical comparisons with the life of this planet.

Geologic prophecy was based on two things: first, a belief, as we have seen, in the man-centered nature of the universe, and second, the assumption that since the animals of the past had no physical connection with those of the present, some kind of abstract, immaterial plan in the mind of the Creator linked the forms of the past with those of the present day. The early-nineteenth-century thinkers perceived a genuine relationship, but their attachment to the idea of special creation prevented them from recognizing that the relationship arose out of simple biological "descent with modification."

Man could not be proved preordained or predestined from the beginning simply because he showed certain affinities to

Paleozoic vertebrates. Instead, he was merely one of many descendants of the early vertebrate line. A moose or a mongoose would have had equally good reason to contend that as a modern vertebrate he had been "prefigured from the beginning," and that the universe had been organized with him in mind.

The situation is something like that of walking through a hall of trick mirrors and being pulled out of shape. The mirror of time does that to all things living, and the distortions stay. Nevertheless, there is a pattern of sorts, so that if you have come by the mirror that makes men, and somewhere behind you there is a mirror that makes black cats, you can still see the pattern. You and the cat are related; the shreds of the original shape are in your bones and the shreds of primordial thought patterns move in the eyes of both of you and are understood by both. But somewhere there must be an original pattern; somewhere cat and man and weasel must leap into a single shape. That shape lies inconceivably remote from us now, far back along the time stream. It is historical. In that sense, and in that sense only, the archetype did indeed exist.

Darwin saw clearly that the succession of life on this planet was not a formal pattern imposed from without, or moving exclusively in one direction. Whatever else life might be, it was adjustable and not fixed. It worked its way through difficult environments. It modified and then, if necessary, it modified again, along roads which would never be retraced. Every creature alive is the product of a unique history. The statistical probability of its precise reduplication on another planet is so small as to be meaningless. Life, even cellular life, may exist out yonder in the dark. But high or low in nature, it will not wear the shape of man. That shape is the evolutionary product of a strange, long wandering through the attics of the forest roof, and so great are the chances of failure, that nothing precisely and identically human is likely ever to come that way again.

The picture of the little man of long ago rises before me as I write. As I have said, he was simply a foetal monster, long since scientifically diagnosed and dismissed. The small skull that lent the illusion of maturity to the mummified infant contained a brain which had failed to develop. The describers of two-foot men forget that a normal human brain cannot function with a capacity, at the very minimum, of less than about nine hundred cubic centimeters. A man with a hundred-cubic-centimeter brain will not be a builder of flying saucers; he will be less intelligent than an ape. In any case, he does not exist.

In a universe whose size is beyond human imagining, where our world floats like a dust mote in the void of night, men have grown inconceivably lonely. We scan the time scale and the mechanisms of life itself for portents and signs of the invisible. As the only thinking mammals on the planet—perhaps the only thinking animals in the entire sidereal universe—the burden of consciousness has grown heavy upon us. We watch the stars, but the signs are uncertain. We uncover the bones of the past and seek for our origins. There is a path there, but it appears to wander. The vagaries of the road may have a meaning, however; it is thus we torture ourselves.

Lights come and go in the night sky. Men, troubled at last by the things they build, may toss in their sleep and dream bad dreams, or lie awake while the meteors whisper greenly overhead. But nowhere in all space or on a thousand worlds will there be men to share our loneliness. There may be wisdom; there may be power; somewhere across space great instruments, handled by strange, manipulative organs, may stare vainly at our floating cloud wrack, their owners yearning as we yearn. Nevertheless, in the nature of life and in the principles of evolution we have had our answer. Of men elsewhere, and beyond, there will be none forever.

The Judgment of the Birds

IT IS A commonplace of all religious thought, even the most primitive, that the man seeking visions and insight must go apart from his fellows and live for a time in the wilderness. If he is of the proper sort, he will return with a message. It may not be a message from the god he set out to seek, but even if he has failed in that particular, he will have had a vision or seen a marvel, and these are always worth listening to and thinking about.

The world, I have come to believe, is a very queer place, but we have been part of this queerness for so long that we tend to take it for granted. We rush to and fro like Mad Hatters upon our peculiar errands, all the time imagining our surroundings to be dull and ourselves quite ordinary creatures. Actually, there is nothing in the world to encourage this idea, but such is the mind of man, and this is why he finds it necessary from time to time to send emissaries into the wilderness in the hope of learning of great events, or plans in store for him, that will resuscitate his waning taste for life. His great news services, his world-wide radio network, he knows with a last remnant of healthy distrust will be of no use to him in this matter. No miracle can withstand a radio broadcast, and it is certain that it would be no miracle if it could. One must seek, then, what only the solitary approach can give—a natural revelation.

Let it be understood that I am not the sort of man to whom is entrusted direct knowledge of great events or prophecies. A naturalist, however, spends much of his life alone, and my life is no exception. Even in New York City there are patches of wilderness, and a man by himself is bound to undergo certain experiences falling into the class of which I speak. I set mine down, therefore: a matter of pigeons, a flight of chemicals, and a judgment of birds, in the hope that they will come to the eye of those who have retained a true taste for the marvelous, and who are capable of discerning in the flow of ordinary events the point at which the mundane world gives way to quite another dimension.

New York is not, on the whole, the best place to enjoy the

downright miraculous nature of the planet. There are, I do not doubt, many remarkable stories to be heard there and many strange sights to be seen, but to grasp a marvel fully it must be savored from all aspects. This cannot be done while one is being jostled and hustled along a crowded street. Nevertheless, in any city there are true wildernesses where a man can be alone. It can happen in a hotel room, or on the high roofs at dawn.

One night on the twentieth floor of a midtown hotel I awoke in the dark and grew restless. On an impulse I climbed upon the broad old-fashioned window sill, opened the curtains and peered out. It was the hour just before dawn, the hour when men sigh in their sleep, or, if awake, strive to focus their wavering eyesight upon a world emerging from the shadows. I leaned out sleepily through the open window. I had expected depths, but not the sight I saw.

I found I was looking down from that great height into a series of curious cupolas or lofts that I could just barely make out in the darkness. As I looked, the outlines of these lofts became more distinct because the light was being reflected from the wings of pigeons who, in utter silence, were beginning to float outward upon the city. In and out through the open slits in the cupolas passed the white-winged birds on their mysterious errands. At this hour the city was theirs, and quietly, without the brush of a single wing tip against stone in that high, eerie place, they were taking over the spires of Manhattan. They were pouring upward in a light that was not yet perceptible to human eyes, while far down in the black darkness of the alleys it was still midnight.

As I crouched half asleep across the sill, I had a moment's illusion that the world had changed in the night, as in some immense snowfall, and that if I were to leave, it would have to be as these other inhabitants were doing, by the window. I should have to launch out into that great bottomless void with the simple confidence of young birds reared high up there among the familiar chimney pots and interposed horrors of the abyss.

I leaned farther out. To and fro went the white wings, to and fro. There were no sounds from any of them. They knew man was asleep and this light for a little while was theirs. Or perhaps I had only dreamed about man in this city of wings—which he could surely never have built. Perhaps I, myself, was one of these birds dreaming unpleasantly a moment of old dangers far below as I teetered on a window ledge.

Around and around went the wings. It needed only a little courage, only a little shove from the window ledge to enter that city of light. The muscles of my hands were already making little premonitory lunges. I wanted to enter that city and go away over the roofs in the first dawn. I wanted to enter it so badly that I drew back carefully into the room and opened the hall door. I found my coat on the chair, and it slowly became clear to me that there was a way down through the floors, that I was, after all, only a man.

I dressed then and went back to my own kind, and I have been rather more than usually careful ever since not to look into the city of light. I had seen, just once, man's greatest creation from a strange inverted angle, and it was not really his at all. I will never forget how those wings went round and round, and how, by the merest pressure of the fingers and a feeling for air, one might go away over the roofs. It is a knowledge, however, that is better kept to oneself. I think of it sometimes in such a way that the wings, beginning far down in the black depths of the mind, begin to rise and whirl till all the mind is lit by their spinning, and there is a sense of things passing away, but lightly, as a wing might veer over an obstacle.

To see from an inverted angle, however, is not a gift allotted merely to the human imagination. I have come to suspect that within their degree it is sensed by animals, though perhaps as rarely as among men. The time has to be right; one has to be, by chance or intention, upon the border of two worlds. And sometimes these two borders may shift or interpenetrate and one sees the miraculous.

I once saw this happen to a crow.

This crow lives near my house, and though I have never injured him, he takes good care to stay up in the very highest trees and, in general, to avoid humanity. His world begins at about the limit of my eyesight.

On the particular morning when this episode occurred, the whole countryside was buried in one of the thickest fogs in years. The ceiling was absolutely zero. All planes were grounded, and even a pedestrian could hardly see his outstretched hand before him.

I was groping across a field in the general direction of the railroad station, following a dimly outlined path. Suddenly out of the fog, at about the level of my eyes, and so closely that I flinched, there flashed a pair of immense black wings and a huge beak. The whole bird rushed over my head with a frantic cawing outcry of such hideous terror as I have never heard in a crow's voice before, and never expect to hear again.

He was lost and startled, I thought, as I recovered my poise. He ought not to have flown out in this fog. He'd knock his silly brains out.

All afternoon that great awkward cry rang in my head. Merely being lost in a fog seemed scarcely to account for it—especially in a tough, intelligent old bandit such as I knew that particular crow to be. I even looked once in the mirror to see what it might be about me that had so revolted him that he had cried out in protest to the very stones.

Finally, as I worked my way homeward along the path, the solution came to me. It should have been clear before. The borders of our worlds had shifted. It was the fog that had done it. That crow, and I knew him well, never under normal circumstances flew low near men. He had been lost all right, but it was more than that. He had thought he was high up, and when he encountered me looming gigantically through the fog, he had perceived a ghastly and, to the crow mind, unnatural sight. He had seen a man walking on air, desecrating the very heart of the crow kingdom, a harbinger of the most profound evil a crow mind could conceive of—air-walking men. The encounter, he must have thought, had taken place a hundred feet over the roofs.

He caws now when he sees me leaving for the station in the morning, and I fancy that in that note I catch the uncertainty of a mind that has come to know things are not always what they seem. He has seen a marvel in his heights of air and is no longer as other crows. He has experienced the human world from an unlikely perspective. He and I share a viewpoint in common: our worlds have interpenetrated, and we both have faith in the miraculous.

It is a faith that in my own case has been augmented by two

remarkable sights. As I have hinted previously, I once saw some very odd chemicals fly across a waste so dead it might have been upon the moon, and once, by an even more fantastic piece of luck, I was present when a group of birds passed a judgment upon life.

On the maps of the old voyageurs it is called *Mauvaises Terres*, the evil lands, and, slurred a little with the passage through many minds, it has come down to us anglicized as the Badlands. The soft shuffle of moccasins has passed through its canyons on the grim business of war and flight, but the last of those slight disturbances of immemorial silences died out almost a century ago. The land, if one can call it a land, is a waste as lifeless as that valley in which lie the kings of Egypt. Like the Valley of the Kings, it is a mausoleum, a place of dry bones in what once was a place of life. Now it has silences as deep as those in the moon's airless chasms.

Nothing grows among its pinnacles; there is no shade except under great toadstools of sandstone whose bases have been eaten to the shape of wine glasses by the wind. Everything is flaking, cracking, disintegrating, wearing away in the long, imperceptible weather of time. The ash of ancient volcanic outbursts still sterilizes its soil, and its colors in that waste are the colors that flame in the lonely sunsets on dead planets. Men come there but rarely, and for one purpose only, the collection of bones.

It was a late hour on a cold, wind-bitten autumn day when I climbed a great hill spined like a dinosaur's back and tried to take my bearings. The tumbled waste fell away in waves in all directions. Blue air was darkening into purple along the bases of the hills. I shifted my knapsack, heavy with the petrified bones of long-vanished creatures, and studied my compass. I wanted to be out of there by nightfall, and already the sun was going sullenly down in the west.

It was then that I saw the flight coming on. It was moving like a little close-knit body of black specks that danced and darted and closed again. It was pouring from the north and heading toward me with the undeviating relentlessness of a compass needle. It streamed through the shadows rising out of monstrous gorges. It rushed over towering pinnacles in the red light of the sun, or momentarily sank from sight within their shade. Across that desert of eroding clay and wind-worn stone they came with a faint wild twittering that filled all the air about me as those tiny living bullets hurtled past into the night.

It may not strike you as a marvel. It would not, perhaps, unless you stood in the middle of a dead world at sunset, but that was where I stood. Fifty million years lay under my feet, fifty million years of bellowing monsters moving in a green world now gone so utterly that its very light was travelling on the farther edge of space. The chemicals of all that vanished age lay about me in the ground. Around me still lay the shearing molars of dead titanotheres, the delicate sabers of soft-stepping cats, the hollow sockets that had held the eyes of many a strange, outmoded beast. Those eyes had looked out upon a world as real as ours; dark, savage brains had roamed and roared their challenges into the steaming night.

Now they were still here, or, put it as you will, the chemicals that made them were here about me in the ground. The carbon that had driven them ran blackly in the eroding stone. The stain of iron was in the clays. The iron did not remember the blood it had once moved within, the phosphorus had forgot the savage brain. The little individual moment had ebbed from all those strange combinations of chemicals as it would ebb from our living bodies into the sinks and runnels of oncoming time.

I had lifted up a fistful of that ground. I held it while that wild flight of south-bound warblers hurtled over me into the oncoming dark. There went phosphorus, there went iron, there went carbon, there beat the calcium in those hurrying wings. Alone on a dead planet I watched that incredible miracle speeding past. It ran by some true compass over field and waste land. It cried its individual ecstasies into the air until the gullies rang. It swerved like a single body, it knew itself and, lonely, it bunched close in the racing darkness, its individual entities feeling about them the rising night. And so, crying to each other their identity, they passed away out of my view.

I dropped my fistful of earth. I heard it roll inanimate back into the gully at the base of the hill: iron, carbon, the chemicals of life. Like men from those wild tribes who had haunted these hills before me seeking visions, I made my sign to the great darkness. It was not a mocking sign, and I was not mocked. As I walked into my camp late that night, one man, rousing from his blankets beside the fire, asked sleepily, "What did you see?"

"I think, a miracle," I said softly, but I said it to myself. Behind me that vast waste began to glow under the rising moon.

I have said that I saw a judgment upon life, and that it was not passed by men. Those who stare at birds in cages or who test minds by their closeness to our own may not care for it. It comes from far away out of my past, in a place of pouring waters and green leaves. I shall never see an episode like it again if I live to be a hundred, nor do I think that one man in a million has ever seen it, because man is an intruder into such silences. The light must be right, and the observer must remain unseen. No man sets up such an experiment. What he sees, he sees by chance.

You may put it that I had come over a mountain, that I had slogged through fern and pine needles for half a long day, and that on the edge of a little glade with one long, crooked branch extending across it, I had sat down to rest with my back against a stump. Through accident I was concealed from the glade, although I could see into it perfectly.

The sun was warm there, and the murmurs of forest life blurred softly away into my sleep. When I awoke, dimly aware of some commotion and outcry in the clearing, the light was slanting down through the pines in such a way that the glade was lit like some vast cathedral. I could see the dust motes of wood pollen in the long shaft of light, and there on the extended branch sat an enormous raven with a red and squirming nestling in his beak.

The sound that awoke me was the outraged cries of the nestling's parents, who flew helplessly in circles about the clearing. The sleek black monster was indifferent to them. He gulped, whetted his beak on the dead branch a moment and sat still. Up to that point the little tragedy had followed the usual pattern. But suddenly, out of all that area of woodland, a soft sound of complaint began to rise. Into the glade fluttered small birds of half a dozen varieties drawn by the anguished outcries of the tiny parents.

No one dared to attack the raven. But they cried there in

some instinctive common misery, the bereaved and the unbe-reaved. The glade filled with their soft rustling and their cries. They fluttered as though to point their wings at the murderer. There was a dim intangible ethic he had violated, that they knew. He was a bird of death.

And he, the murderer, the black bird at the heart of life, sat on there, glistening in the common light, formidable, unmov-ing, unperturbed, untouchable.

The sighing died. It was then I saw the judgment. It was the judgment of life against death. I will never see it again so forcefully presented. I will never hear it again in notes so trag-ically prolonged. For in the midst of protest, they forgot the violence. There, in that clearing, the crystal note of a song sparrow lifted hesitantly in the hush. And finally, after painful fluttering, another took the song, and then another, the song passing from one bird to another, doubtfully at first, as though some evil thing were being slowly forgotten. Till suddenly they took heart and sang from many throats joyously together as birds are known to sing. They sang because life is sweet and sunlight beautiful. They sang under the brooding shadow of the raven. In simple truth they had forgotten the raven, for they were the singers of life, and not of death.

I was not of that airy company. My limbs were the heavy limbs of an earthbound creature who could climb mountains, even the mountains of the mind, only by a great effort of will. I knew I had seen a marvel and observed a judgment, but the mind which was my human endowment was sure to question it and to be at me day by day with its heresies until I grew to doubt the meaning of what I had seen. Eventually darkness and subtleties would ring me round once more.

And so it proved until, on the top of a stepladder, I made one more observation upon life. It was cold that autumn eve-ning, and, standing under a suburban street light in a spate of leaves and beginning snow, I was suddenly conscious of some huge and hairy shadows dancing over the pavement. They seemed attached to an odd, globular shape that was magnified above me. There was no mistaking it. I was standing under the shadow of an orb-weaving spider. Gigantically projected against the street, she was about her spinning when everything was going underground. Even her cables were magnified upon the sidewalk and already I was half-entangled in their shadows.

"Good Lord," I thought, "she has found herself a kind of minor sun and is going to upset the course of nature."

I procured a ladder from my yard and climbed up to inspect the situation. There she was, the universe running down around her, warmly arranged among her guy ropes attached to the lamp supports—a great black and yellow embodiment of the life force, not giving up to either frost or stepladders. She ignored me and went on tightening and improving her web.

I stood over her on the ladder, a faint snow touching my cheeks, and surveyed her universe. There were a couple of iridescent green beetle cases turning slowly on a loose strand of web, a fragment of luminescent eye from a moth's wing and a large indeterminable object, perhaps a cicada, that had struggled and been wrapped in silk. There were also little bits and slivers, little red and blue flashes from the scales of anonymous wings that had crashed there.

Some days, I thought, they will be dull and gray and the shine will be out of them; then the dew will polish them again and drops hang on the silk until everything is gleaming and turning in the light. It is like a mind, really, where everything changes but remains, and in the end you have these eaten-out bits of experience like beetle wings.

I stood over her a moment longer, comprehending somewhat reluctantly that her adventure against the great blind forces of winter, her seizure of this warming globe of light, would come to nothing and was hopeless. Nevertheless it brought the birds back into my mind, and that faraway song which had traveled with growing strength around a forest clearing years ago—a kind of heroism, a world where even a spider refuses to lie down and die if a rope can still be spun on to a star. Maybe man himself will fight like this in the end, I thought, slowly realizing that the web and its threatening yellow occupant had been added to some luminous store of experience, shining for a moment in the fogbound reaches of my brain.

The mind, it came to me as I slowly descended the ladder, is a very remarkable thing; it has gotten itself a kind of courage by looking at a spider in a street lamp. Here was something that ought to be passed on to those who will fight our final freezing battle with the void. I thought of setting it down carefully as a message to the future: *In the days of the frost seek a minor sun.*

But as I hesitated, it became plain that something was wrong. The marvel was escaping—a sense of bigness beyond man's power to grasp, the essence of life in its great dealings with the universe. It was better, I decided, for the emissaries returning from the wilderness, even if they were merely descending from a stepladder, to record their marvel, not to define its meaning. In that way it would go echoing on through the minds of men, each grasping at that beyond out of which the miracles emerge, and which, once defined, ceases to satisfy the human need for symbols.

In the end I merely made a mental note: One specimen of Epeira observed building a web in a street light. Late autumn and cold for spiders. Cold for men, too. I shivered and left the lamp glowing there in my mind. The last I saw of Epeira she was hauling steadily on a cable. I stepped carefully over her shadow as I walked away.

The Bird and the Machine

I SUPPOSE THEIR little bones have years ago been lost among the stones and winds of those high glacial pastures. I suppose their feathers blew eventually into the piles of tumbleweed beneath the straggling cattle fences and rotted there in the mountain snows, along with dead steers and all the other things that drift to an end in the corners of the wire. I do not quite know why I should be thinking of birds over the *New York Times* at breakfast, particularly the birds of my youth half a continent away. It is a funny thing what the brain will do with memories and how it will treasure them and finally bring them into odd juxtapositions with other things, as though it wanted to make a design, or get some meaning out of them, whether you want it or not, or even see it.

It used to seem marvelous to me, but I read now that there are machines that can do these things in a small way, machines that can crawl about like animals, and that it may not be long now until they do more things—maybe even make themselves —I saw that piece in the *Times* just now. And then they will, maybe—well, who knows—but you read about it more and more with no one making any protest, and already they can add better than we and reach up and hear things through the dark and finger the guns over the night sky.

This is the new world that I read about at breakfast. This is the world that confronts me in my biological books and journals, until there are times when I sit quietly in my chair and try to hear the little purr of the cogs in my head and the tubes flaring and dying as the messages go through them and the circuits snap shut or open. This is the great age, make no mistake about it; the robot has been born somewhat appropriately along with the atom bomb, and the brain they say now is just another type of more complicated feedback system. The engineers have its basic principles worked out; it's mechanical, you know; nothing to get superstitious about; and man can always improve on nature once he gets the idea. Well, he's got it all right and that's why, I guess, that I sit here in my chair, with the article crunched in my hand, remembering those two birds

113

and that blue mountain sunlight. There is another magazine article on my desk that reads "Machines Are Getting Smarter Every Day." I don't deny it, but I'll still stick with the birds. It's life I believe in, not machines.

Maybe you don't believe there is any difference. A skeleton is all joints and pulleys, I'll admit. And when man was in his simpler stages of machine building in the eighteenth century, he quickly saw the resemblances. "What," wrote Hobbes, "is the heart but a spring, and the nerves but so many strings, and the joints but so many wheels, giving motion to the whole body?" Tinkering about in their shops it was inevitable in the end that men would see the world as a huge machine "subdivided into an infinite number of lesser machines."

The idea took on with a vengeance. Little automatons toured the country—dolls controlled by clockwork. Clocks described as little worlds were taken on tours by their designers. They were made up of moving figures, shifting scenes and other remarkable devices. The life of the cell was unknown. Man, whether he was conceived as possessing a soul or not, moved and jerked about like these tiny puppets. A human being thought of himself in terms of his own tools and implements. He had been fashioned like the puppets he produced and was only a more clever model made by a greater designer.

Then in the nineteenth century, the cell was discovered, and the single machine in its turn was found to be the product of millions of infinitesimal machines—the cells. Now, finally, the cell itself dissolves away into an abstract chemical machine—and that into some intangible, inexpressible flow of energy. The secret seems to lurk all about, the wheels get smaller and smaller, and they turn more rapidly, but when you try to seize it the life is gone—and so, by popular definition, some would say that life was never there in the first place. The wheels and the cogs are the secret and we can make them better in time—machines that will run faster and more accurately than real mice to real cheese.

I have no doubt it can be done, though a mouse harvesting seeds on an autumn thistle is to me a fine sight and more complicated, I think, in his multiform activity, than a machine "mouse" running a maze. Also, I like to think of the possible shape of the future brooding in mice, just as it brooded once in a rather ordinary mousy insectivore who became a man. It leaves a nice fine indeterminate sense of wonder that even an electronic brain hasn't got, because you know perfectly well that if the electronic brain changes, it will be because of something man has done to it. But what man will do to himself he doesn't really know. A certain scale of time and a ghostly intangible thing called change are ticking in him. Powers and potentialities like the oak in the seed, or a red and awful ruin. Either way, it's impressive; and the mouse has it, too. Or those birds, I'll never forget those birds—yet before I measured their significance, I learned the lesson of time first of all. I was young then and left alone in a great desert—part of an expedition that had scattered its men over several hundred miles in order to carry on research more effectively. I learned there that time is a series of planes existing superficially in the same universe. The tempo is a human illusion, a subjective clock ticking in our own kind of protoplasm.

As the long months passed, I began to live on the slower planes and to observe more readily what passed for life there. I sauntered, I passed more and more slowly up and down the canyons in the dry baking heat of midsummer. I slumbered for long hours in the shade of huge brown boulders that had gathered in tilted companies out on the flats. I had forgotten the world of men and the world had forgotten me. Now and then I found a skull in the canyons, and these justified my remaining there. I took a serene cold interest in these discoveries. I had come, like many a naturalist before me, to view life with a wary and subdued attention. I had grown to take pleasure in the divested bone.

I sat once on a high ridge that fell away before me into a waste of sand dunes. I sat through hours of a long afternoon. Finally, as I glanced beside my boot an indistinct configuration caught my eye. It was a coiled rattlesnake, a big one. How long he had sat with me I do not know. I had not frightened him. We were both locked in the sleep-walking tempo of the earlier world, baking in the same high air and sunshine. Perhaps he had been there when I came. He slept on as I left, his coils, so ill discerned by me, dissolving once more among the stones and gravel from which I had barely made him out.

Another time I got on a higher ridge, among some tough little wind-warped pines half covered over with sand in a basin-like depression that caught everything carried by the air up to those heights. There were a few thin bones of birds, some cracked shells of indeterminable age, and the knotty fingers of pine roots bulged out of shape from their long and agonizing grasp upon the crevices of the rock. I lay under the pines in the sparse shade and went to sleep once more.

It grew cold finally, for autumn was in the air by then, and the few things that lived thereabouts were sinking down into an even chillier scale of time. In the moments between sleeping and waking I saw the roots about me and slowly, slowly, a foot in what seemed many centuries, I moved my sleep-stiffened hands over the scaling bark and lifted my numbed face after the vanishing sun. I was a great awkward thing of knots and aching limbs, trapped up there in some long, patient endurance that involved the necessity of putting living fingers into rock and by slow, aching expansion bursting those rocks asunder. I suppose, so thin and slow was the time of my pulse by then, that I might have stayed on to drift still deeper into the lower cadences of the frost, or the crystalline life that glistens pebbles, or shines in a snowflake, or dreams in the meteoric iron between the worlds.

It was a dim descent, but time was present in it. Somewhere far down in that scale the notion struck me that one might come the other way. Not many months thereafter I joined some colleagues heading higher into a remote windy tableland where huge bones were reputed to protrude like boulders from the turf. I had drowsed with reptiles and moved with the century-long pulse of trees; now, lethargically, I was climbing back up some invisible ladder of quickening hours. There had been talk of birds in connection with my duties. Birds are intense, fast-living creatures—reptiles, I suppose one might say, that have escaped out of the heavy sleep of time, transformed fairy creatures dancing over sunlit meadows. It is a youthful fancy, no doubt, but because of something that happened up there among the escarpments of that range, it remains with me a lifelong impression. I can never bear to see a bird imprisoned.

We came into that valley through the trailing mists of a spring night. It was a place that looked as though it might never have known the foot of man, but our scouts had been ahead of us and we knew all about the abandoned cabin of stone that lay far up on one hillside. It had been built in the land rush of the last century and then lost to the cattlemen again as the marginal soils failed to take to the plow.

There were spots like this all over that country. Lost graves marked by unlettered stones and old corroding rim-fire cartridge cases lying where somebody had made a stand among the boulders that rimmed the valley. They are all that remain of the range wars; the men are under the stones now. I could see our cavalcade winding in and out through the mist below us: torches, the reflection of the truck lights on our collecting tins, and the far-off bumping of a loose dinosaur thigh bone in the bottom of a trailer. I stood on a rock a moment looking down and thinking what it cost in money and equipment to capture the past.

We had, in addition, instructions to lay hands on the present. The word had come through to get them alive—birds, reptiles, anything. A zoo somewhere abroad needed restocking. It was one of those reciprocal matters in which science involves itself. Maybe our museum needed a stray ostrich egg and this was the payoff. Anyhow, my job was to help capture some birds and that was why I was there before the trucks.

The cabin had not been occupied for years. We intended to clean it out and live in it, but there were holes in the roof and the birds had come in and were roosting in the rafters. You could depend on it in a place like this where everything blew away, and even a bird needed some place out of the weather and away from coyotes. A cabin going back to nature in a wild place draws them till they come in, listening at the eaves, I imagine, pecking softly among the shingles till they find a hole and then suddenly the place is theirs and man is forgotten.

Sometimes of late years I find myself thinking the most beautiful sight in the world might be the birds taking over New York after the last man has run away to the hills. I will never live to see it, of course, but I know just how it will sound because I've lived up high and I know the sort of watch birds keep on us. I've listened to sparrows tapping tentatively on the outside of air conditioners when they thought no one was listening, and I know how other birds test the vibrations that come up to them through the television aerials.

"Is he gone?" they ask, and the vibrations come up from below, "Not yet, not yet."

Well, to come back, I got the door open softly and I had the spotlight all ready to turn on and blind whatever birds there were so they couldn't see to get out through the roof. I had a short piece of ladder to put against the far wall where there was a shelf on which I expected to make the biggest haul. I had all the information I needed just like any skilled assassin. I pushed the door open, the hinges squeaking only a little. A bird or two stirred—I could hear them—but nothing flew and there was a faint starlight through the holes in the roof.

I padded across the floor, got the ladder up and the light ready, and slithered up the ladder till my head and arms were over the shelf. Everything was dark as pitch except for the starlight at the little place back of the shelf near the eaves. With the light to blind them, they'd never make it. I had them. I reached my arm carefully over in order to be ready to seize whatever was there and I put the flash on the edge of the shelf where it would stand by itself when I turned it on. That way I'd be able to use both hands.

Everything worked perfectly except for one detail—I didn't know what kind of birds were there. I never thought about it at all, and it wouldn't have mattered if I had. My orders were to get something interesting. I snapped on the flash and sure enough there was a great beating and feathers flying, but instead of my having them, they, or rather he, had me. He had my hand, that is, and for a small hawk not much bigger than my fist he was doing all right. I heard him give one short metallic cry when the light went on and my hand descended on the bird beside him; after that he was busy with his claws and his beak was sunk in my thumb. In the struggle I knocked the lamp over on the shelf, and his mate got her sight back and whisked neatly through the hole in the roof and off among the stars outside. It all happened in fifteen seconds and you might think I would have fallen down the ladder, but no, I had a professional assassin's reputation to keep up, and the bird, of course, made the mistake of thinking the hand was the enemy and not the eyes behind it. He chewed my thumb up pretty effectively and lacerated my hand with his claws, but in the end I got him, having two hands to work with.

He was a sparrow hawk and a fine young male in the prime of life. I was sorry not to catch the pair of them, but as I dripped blood and folded his wings carefully, holding him by the back so that he couldn't strike again, I had to admit the two of them might have been more than I could have handled under the circumstances. The little fellow had saved his mate by diverting me, and that was that. He was born to it, and made no outcry now, resting in my hand hopelessly, but peering toward me in the shadows behind the lamp with a fierce, almost indifferent glance. He neither gave nor expected mercy and something out of the high air passed from him to me, stirring a faint embarrassment.

I quit looking into that eye and managed to get my huge carcass with its fist full of prey back down the ladder. I put the bird in a box too small to allow him to injure himself by struggle and walked out to welcome the arriving trucks. It had been a long day, and camp still to make in the darkness. In the morning that bird would be just another episode. He would go back with the bones in the truck to a small cage in a city where he would spend the rest of his life. And a good thing, too. I sucked my aching thumb and spat out some blood. An assassin has to get used to these things. I had a professional reputation to keep up.

In the morning, with the change that comes on suddenly in that high country, the mist that had hovered below us in the valley was gone. The sky was a deep blue, and one could see for miles over the high outcroppings of stone. I was up early and brought the box in which the little hawk was imprisoned out onto the grass where I was building a cage. A wind as cool as a mountain spring ran over the grass and stirred my hair. It was a fine day to be alive. I looked up and all around and at the hole in the cabin roof out of which the other little hawk had fled. There was no sign of her anywhere that I could see.

"Probably in the next county by now," I thought cynically, but before beginning work I decided I'd have a look at my last night's capture.

Secretively, I looked again all around the camp and up and down and opened the box. I got him right out in my hand with his wings folded properly and I was careful not to startle him. He lay limp in my grasp and I could feel his heart pound under the feathers but he only looked beyond me and up.

I saw him look that last look away beyond me into a sky so full of light that I could not follow his gaze. The little breeze flowed over me again, and nearby a mountain aspen shook all its tiny leaves. I suppose I must have had an idea then of what I was going to do, but I never let it come up into consciousness. I just reached over and laid the hawk on the grass.

He lay there a long minute without hope, unmoving, his eyes still fixed on that blue vault above him. It must have been that he was already so far away in heart that he never felt the release from my hand. He never even stood. He just lay with his breast against the grass.

In the next second after that long minute he was gone. Like a flicker of light, he had vanished with my eyes full on him, but without actually seeing even a premonitory wing beat. He was gone straight into that towering emptiness of light and crystal that my eyes could scarcely bear to penetrate. For another long moment there was silence. I could not see him. The light was too intense. Then from far up somewhere a cry came ringing down.

I was young then and had seen little of the world, but when I heard that cry my heart turned over. It was not the cry of the hawk I had captured; for, by shifting my position against the sun, I was now seeing further up. Straight out of the sun's eye, where she must have been soaring restlessly above us for untold hours, hurtled his mate. And from far up, ringing from peak to peak of the summits over us, came a cry of such unutterable and ecstatic joy that it sounds down across the years and tingles among the cups on my quiet breakfast table.

I saw them both now. He was rising fast to meet her. They met in a great soaring gyre that turned to a whirling circle and a dance of wings. Once more, just once, their two voices, joined in a harsh wild medley of question and response, struck and echoed against the pinnacles of the valley. Then they were gone forever somewhere into those upper regions beyond the eyes of men.

I am older now, and sleep less, and have seen most of what there is to see and am not very much impressed any more, I suppose, by anything. "What Next in the Attributes of Machines?" my morning headline runs. "It Might Be the Power to Reproduce Themselves."

I lay the paper down and across my mind a phrase floats insinuatingly: "It does not seem that there is anything in the construction, constituents, or behavior of the human being which it is essentially impossible for science to duplicate and synthesize. On the other hand . . ."

All over the city the cogs in the hard, bright mechanisms have begun to turn. Figures move through computers, names are spelled out, a thoughtful machine selects the fingerprints of a wanted criminal from an array of thousands. In the laboratory an electronic mouse runs swiftly through a maze toward the cheese it can neither taste nor enjoy. On the second run it does better than a living mouse.

"On the other hand . . ." Ah, my mind takes up, on the other hand the machine does not bleed, ache, hang for hours in the empty sky in a torment of hope to learn the fate of another machine, nor does it cry out with joy nor dance in the air with the fierce passion of a bird. Far off, over a distance greater than space, that remote cry from the heart of heaven makes a faint buzzing among my breakfast dishes and passes on and away.

The Secret of Life

I AM MIDDLE-AGED now, but in the autumn I always seek for it again hopefully. On some day when the leaves are red, or fallen, and just after the birds are gone, I put on my hat and an old jacket, and over the protests of my wife that I will catch cold, I start my search. I go carefully down the apartment steps and climb, instead of jump, over the wall. A bit further I reach an unkempt field full of brown stalks and emptied seed pods.

By the time I get to the wood I am carrying all manner of seeds hooked in my coat or piercing my socks or sticking by ingenious devices to my shoestrings. I let them ride. After all, who am I to contend against such ingenuity? It is obvious that nature, or some part of it in the shape of these seeds, has intentions beyond this field and has made plans to travel with me.

We, the seeds and I, climb another wall together and sit down to rest, while I consider the best way to search for the secret of life. The seeds remain very quiet and some slip off into the crevices of the rock. A woolly-bear caterpillar hurries across a ledge, going late to some tremendous transformation, but about this he knows as little as I.

It is not an auspicious beginning. The things alive do not know the secret, and there may be those who would doubt the wisdom of coming out among discarded husks in the dead year to pursue such questions. They might say the proper time is spring, when one can consult the water rats or listen to little chirps under the stones. Of late years, however, I have come to suspect that the mystery may just as well be solved in a carved and intricate seed case out of which the life has flown, as in the seed itself.

In autumn one is not confused by activity and green leaves. The underlying apparatus, the hooks, needles, stalks, wires, suction cups, thin pipes, and iridescent bladders are all exposed in a gigantic dissection. These are the essentials. Do not be deceived simply because the life has flown out of them. It will return, but in the meantime there is an unparalleled opportunity to examine in sharp and beautiful angularity the shape of life without its disturbing muddle of juices and leaves. As I grow older and conserve my efforts, I shall give this season my final and undivided attention. I shall be found puzzling over the saw teeth on the desiccated leg of a dead grasshopper or standing bemused in a brown sea of rusty stems. Somewhere in this discarded machinery may lie the key to the secret. I shall not let it escape through lack of diligence or through fear of the smiles of people in high windows. I am sure now that life is not what it is purported to be and that nature, in the canny words of a Scotch theologue, "is not as natural as it looks." I have learned this in a small suburban field, after a good many years spent in much wilder places upon far less fantastic quests.

The notion that mice can be generated spontaneously from bundles of old clothes is so delightfully whimsical that it is easy to see why men were loath to abandon it. One could accept such accidents in a topsy-turvy universe without trying to decide what transformation of buckles into bones and shoe buttons into eyes had taken place. One could take life as a kind of fantastic magic and not blink too obviously when it appeared, beady-eyed and bustling, under the laundry in the back room.

It was only with the rise of modern biology and the discovery that the trail of life led backward toward infinitesimal beginnings in primordial sloughs, that men began the serious dissection and analysis of the cell. Darwin, in one of his less guarded moments, had spoken hopefully of the possibility that life had emerged from inorganic matter in some "warm little pond." From that day to this biologists have poured, analyzed, minced, and shredded recalcitrant protoplasm in a fruitless attempt to create life from nonliving matter. It seemed inevitable, if we could trace life down through simpler stages, that we must finally arrive at the point where, under the proper chemical conditions, the mysterious borderline that bounds the inanimate must be crossed. It seemed clear that life was a material manifestation. Somewhere, somehow, sometime, in the mysterious chemistry of carbon, the long march toward the talking animal had begun.

A hundred years ago men spoke optimistically about solving the secret, or at the very least they thought the next generation would be in a position to do so. Periodically there were claims that the emergence of life from matter had been observed, but in every case the observer proved to be self-deluded. It became obvious that the secret of life was not to be had by a little casual experimentation, and that life in today's terms appeared to arise only through the medium of preëxisting life. Yet, if science was not to be embarrassed by some kind of mind-matter dualism and a complete and irrational break between life and the world of inorganic matter, the emergence of life had, in some way, to be accounted for. Nevertheless, as the years passed, the secret remained locked in its living jelly, in spite of larger microscopes and more formidable means of dissection. As a matter of fact the mystery was heightened because all this intensified effort revealed that even the supposedly simple amoeba was a complex, self-operating chemical factory. The notion that he was a simple blob, the discovery of whose chemical composition would enable us instantly to set the life process in operation, turned out to be, at best, a monstrous caricature of the truth.

With the failure of these many efforts science was left in the somewhat embarrassing position of having to postulate theories of living origins which it could not demonstrate. After having chided the theologian for his reliance on myth and miracle, science found itself in the unenviable position of having to create a mythology of its own: namely, the assumption that what, after long effort, could not be proved to take place today had, in truth, taken place in the primeval past.

My use of the term *mythology* is perhaps a little harsh. One does occasionally observe, however, a tendency for the beginning zoological textbook to take the unwary reader by a hop, skip, and jump from the little steaming pond or the beneficent chemical crucible of the sea, into the lower world of life with such sureness and rapidity that it is easy to assume that there is no mystery about this matter at all, or, if there is, that it is a very little one.

This attitude has indeed been sharply criticized by the distinguished British biologist Woodger, who remarked some years ago: "Unstable organic compounds and chlorophyll corpuscles do not persist or come into existence in nature on their own account at the present day, and consequently it is necessary to postulate that conditions were once such that this did happen although and in spite of the fact that our knowledge of nature

does not give us any warrant for making such a supposition
. . . It is simple dogmatism—asserting that what you want to
believe did in fact happen."

Yet, unless we are to turn to supernatural explanations or
reinvoke a dualism which is scientifically dubious, we are
forced inevitably toward only two possible explanations of life
upon earth. One of these, although not entirely disproved, is
most certainly out of fashion and surrounded with greater
obstacles to its acceptance than at the time it was formulated. I
refer, of course, to the suggestion of Lord Kelvin and Svante
Arrhenius that life did not arise on this planet, but was wafted
here through the depths of space. Microscopic spores, it was
contended, have great resistance to extremes of cold and might
have come into our atmosphere with meteoric dust, or have
been driven across the earth's orbit by light pressure. In this
view, once the seed was "planted" in soil congenial to its development, it then proceeded to elaborate, evolve, and adjust
until the higher organisms had emerged.

This theory had a certain attraction as a way out of an embarrassing dilemma, but it suffers from the defect of explaining
nothing, even if it should prove true. It does not elucidate the
nature of life. It simply removes the inconvenient problem of
origins to far-off spaces or worlds into which we will never
penetrate. Since life makes use of the chemical compounds of
this earth, it would seem better to proceed, until incontrovert-
ible evidence to the contrary is obtained, on the assumption
that life has actually arisen upon this planet. The now widely
accepted view that the entire universe in its present state is
limited in time, and the apparently lethal nature of unscreened
solar radiation are both obstacles which greatly lessen the
likelihood that life has come to us across the infinite wastes of
space. Once more, therefore, we are forced to examine our
remaining notion that life is not coterminous with matter, but
has arisen from it.

If the single-celled protozoans that riot in roadside pools are
not the simplest forms of life, if, as we know today, these creatures are already highly adapted and really complex, though
minute beings, then where are we to turn in the search for
something simple enough to suggest the greatest missing link
of all—the link between living and dead matter? It is this

problem that keeps me wandering fruitlessly in pastures and weed thickets even though I know this is an old-fashioned naturalist's approach, and that busy men in laboratories have little patience with my scufflings of autumn leaves, or attempts to question beetles in decaying bark. Besides, many of these men are now fascinated by the crystalline viruses and have turned that remarkable instrument, the electron microscope, upon strange molecular "beings" never previously seen by man. Some are satisfied with this glimpse below the cell and find the virus a halfway station on the road to life. Perhaps it is, but as I wander about in the thin mist that is beginning to filter among these decaying stems and ruined spider webs, a kind of disconsolate uncertainty has taken hold of me.

I have come to suspect that this long descent down the ladder of life, beautiful and instructive though it may be, will not lead us to the final secret. In fact I have ceased to believe in the final brew or the ultimate chemical. There is, I know, a kind of heresy, a shocking negation of our confidence in blue-steel microtomes and men in white in making such a statement. I would not be understood to speak ill of scientific effort, for in simple truth I would not be alive today except for the microscopes and the blue steel. It is only that somewhere among these seeds and beetle shells and abandoned grasshopper legs I find something that is not accounted for very clearly in the dissections to the ultimate virus or crystal or protein particle. Even if the secret is contained in these things, in other words, I do not think it will yield to the kind of analysis our science is capable of making.

Imagine, for a moment, that you have drunk from a magician's goblet. Reverse the irreversible stream of time. Go down the dark stairwell out of which the race has ascended. Find yourself at last on the bottommost steps of time, slipping, sliding, and wallowing by scale and fin down into the muck and ooze out of which you arose. Pass by grunts and voiceless hissings below the last tree ferns. Eyeless and earless, float in the primal waters, sense sunlight you cannot see and stretch absorbing tentacles toward vague tastes that float in water. Still, in your formless shiftings, the *you* remains: the sliding particles, the juices, the transformations are working in an exquisitely patterned rhythm which has no other purpose than your preservation—you, the entity, the ameboid being whose substance contains the unfathomable future. Even so does every man come upward from the waters of his birth.

Yet if at any moment the magician bending over you should cry, "Speak! Tell us of that road!" you could not respond. The sensations are yours but not—and this is one of the great mysteries—the power over the body. You cannot describe how the body you inhabit functions, or picture or control the flights and spinnings, the dance of the molecules that compose it, or why they chose to dance into that particular pattern which is you, or, again, why up the long stairway of the eons they dance from one shape to another. It is for this reason that I am no longer interested in final particles. Follow them as you will, pursue them until they become nameless protein crystals replicating on the verge of life. Use all the great powers of the mind and pass backward until you hang with the dire faces of the conquerors in the hydrogen cloud from which the sun was born. You will then have performed the ultimate dissection that our analytic age demands, but the cloud will still veil the secret and, if not the cloud, then the nothingness into which, it now appears, the cloud, in its turn, may be dissolved. The secret, if one may paraphrase a savage vocabulary, lies in the egg of night.

Only along the edges of this field after the frost there are little whispers of it. Once even on a memorable autumn afternoon I discovered a sunning blacksnake brooding among the leaves like the very simulacrum of old night. He slid unhurriedly away, carrying his version of the secret with him in such a glittering menace of scales that I was abashed and could only follow admiringly from a little distance. I observed him well, however, and am sure he carried his share of the common mystery into the stones of my neighbor's wall, and is sleeping endlessly on in the winter darkness with one great coil locked around that glistening head. He is guarding a strange, reptilian darkness which is not night or nothingness, but has, instead, its momentary vision of mouse bones or a bird's egg, in the soft rising and ebbing of the tides of life. The snake has diverted me, however. It was the dissection of a field that was to occupy us—a dissection in search of secrets—a dissection such as a probing and inquisitive age demands.

Every so often one encounters articles in leading magazines with titles such as "The Spark of Life," "The Secret of Life," "New Hormone Key to Life," or other similar optimistic proclamations. Only yesterday, for example, I discovered in the *New York Times* a headline announcing: "Scientist Predicts Creation of Life in Laboratory." The Moscow-date-lined dispatch announced that Academician Olga Lepeshinskaya had predicted that "in the not too distant future, Soviet scientists would create life." "The time is not far off," warns the formidable Madame Olga, "when we shall be able to obtain the vital substance artificially." She said it with such vigor that I had about the same reaction as I do to announcements about atomic bombs. In fact I half started up to latch the door before an invading tide of Russian protoplasm flowed in upon me.

What finally enabled me to regain my shaken confidence was the recollection that these pronouncements have been going on for well over a century. Just now the Russian scientists show a particular tendency to issue such blasts—committed politically, as they are, to an uncompromising materialism and the boastfulness of very young science. Furthermore, Madame Lepeshinskaya's remarks as reported in the press had a curiously old-fashioned flavor about them. The protoplasm she referred to sounded amazingly like the outmoded *Urschleim* or *Autoplasson* of Haeckel—simplified mucoid slimes no longer taken very seriously. American versions—and one must remember they are often journalistic interpretations of scientists' studies rather than direct quotations from the scientists themselves—are more apt to fall into another pattern. Someone has found a new chemical, vitamin, or similar necessary ingredient without which life will not flourish. By the time this reaches the more sensational press, it may have become the "secret of life." The only thing the inexperienced reader may not comprehend is the fact that no one of these items, even the most recently discovered, is *the* secret. Instead, the substance is probably a part, a very small part, of a larger enigma which is well-nigh as inscrutable as it ever was. If anything, the growing list of catalysts, hormones, plasma genes, and other hobgoblins involved in the work of life only serves to underline the enormous complexity of the secret. "To grasp in detail," says the German biologist Von Bertalanffy, "the physico-chemical organization of the simplest cell is far beyond our capacity."

It is not, you understand, disrespect for the laudable and persistent patience of these dedicated scientists happily lost in their maze of pipettes, smells, and gas flames, that has led me into this runaway excursion to the wood. It is rather the loneliness of a man who knows he will not live to see the mystery solved, and who, furthermore, has come to believe that it will not be solved when the first humanly synthesized particle begins—if it ever does—to multiply itself in some unknown solution.

It is really a matter, I suppose, of the kind of questions one asks oneself. Some day we may be able to say with assurance, "We came from such and such a protein particle, possessing the powers of organizing in a manner leading under certain circumstances to that complex entity known as the cell, and from the cell by various steps onward, to multiple cell formation." I mean we may be able to say all this with great surety and elaboration of detail, but it is not the answer to the grasshopper's leg, brown and black and saw-toothed here in my hand, nor the answer to the seeds still clinging tenaciously to my coat, nor to this field, nor to the subtle essences of memory, delight, and wistfulness moving among the thin wires of my brain.

I suppose that in the forty-five years of my existence every atom, every molecule that composes me has changed its position or danced away and beyond to become part of other things. New molecules have come from the grass and the bodies of animals to be part of me a little while, yet in this spinning, light and airy as a midge swarm in a shaft of sunlight, my memories hold, and a loved face of twenty years ago is before me still. Nor is that face, nor all my years, caught cellularly as in some cold precise photographic pattern, some gross, mechanical reproduction of the past. My memory holds the past and yet paradoxically knows, at the same time, that the past is gone and will never come again. It cherishes dead faces and silenced voices, yes, and lost evenings of childhood. In some odd nonspatial way it contains houses and rooms that have been torn timber from timber and brick from brick.

These have a greater permanence in that midge dance which contains them than ever they had in the world of reality. It is for this reason that Academician Olga Lepeshinskaya has not answered the kind of questions one may ask in an open field.

If the day comes when the slime of the laboratory for the first time crawls under man's direction, we shall have great need of humbleness. It will be difficult for us to believe, in our pride of achievement, that the secret of life has slipped through our fingers and eludes us still. We will list all the chemicals and the reactions. The men who have become gods will pose austerely before the popping flashbulbs of news photographers, and there will be few to consider—so deep is the mind-set of an age—whether the desire to link life to matter may not have blinded us to the more remarkable characteristics of both.

As for me, if I am still around on that day, I intend to put on my old hat and climb over the wall as usual. I shall see strange mechanisms lying as they lie here now, in the autumn rain, strange pipes that transported the substance of life, the intricate seed case out of which the life has flown. I shall observe no thing green, no delicate transpirations of leaves, nor subtle comings and goings of vapor. The little sunlit factories of the chloroplasts will have dissolved away into common earth.

Beautiful, angular, and bare the machinery of life will lie exposed, as it now is, to my view. There will be the thin, blue skeleton of a hare tumbled in a little heap, and crouching over it I will marvel, as I marvel now, at the wonderful correlation of parts, the perfect adaptation to purpose, the individually vanished and yet persisting pattern which is now hopping on some other hill. I will wonder, as always, in what manner "particles" pursue such devious plans and symmetries. I will ask once more in what way it is managed, that the simple dust takes on a history and begins to weave these unique and never recurring apparitions in the stream of time. I shall wonder what strange forces at the heart of matter regulate the tiny beating of a rabbit's heart or the dim dream that builds a milkweed pod.

It is said by men who know about these things that the smallest living cell probably contains over a quarter of a million protein molecules engaged in the multitudinous coördinated activities which make up the phenomenon of life. At the instant of death, whether of man or microbe, that ordered, incredible spinning passes away in an almost furious haste of those same particles to get themselves back into the chaotic, unplanned earth.

I do not think, if someone finally twists the key successfully in the tiniest and most humble house of life, that many of these questions will be answered, or that the dark forces which create lights in the deep sea and living batteries in the waters of tropical swamps, or the dread cycles of parasites, or the most noble workings of the human brain, will be much if at all revealed. Rather, I would say that if "dead" matter has reared up this curious landscape of fiddling crickets, song sparrows, and wondering men, it must be plain even to the most devoted materialist that the matter of which he speaks contains amazing, if not dreadful powers, and may not impossibly be, as Hardy has suggested, "but one mask of many worn by the Great Face behind."

THE FIRMAMENT OF TIME

The splendours of the firmament of time
May be eclipsed, but are extinguished not;
Like stars to their appointed height they climb,
And death is a low mist which cannot blot
The brightness it may veil. When lofty thought
Lifts a young heart above its mortal lair,
And love and life contend in it, for what
Shall be its earthly doom, the dead live there
And move like winds of light on dark and stormy
    air.

PERCY BYSSHE SHELLEY

ONE

How the World Became Natural

*That then this Beginning was, is a matter of faith, and so infalli-
ble. When it was, is matter of reason, and therefore various and
perplex'd.*

—JOHN DONNE

MAN IS AT heart a romantic. He believes in thunder, the
destruction of worlds, the voice out of the whirlwind.
Perhaps the fact that he himself is now in possession of powers
wrenched from the atom's heart has enhanced the appeal of
violence in natural events. The human generations are short-
lived. We have difficulty in visualizing the age-long processes
involved in the upheaval of mountain systems, the advance of
continental glaciations or the creation of life. In fact, scarcely
two hundred years have passed since a few wary pioneers began to suspect that the earth might be older than the 4004
years B.C. assigned to it by the theologians. At all events, the
sale of Velikovsky's *Worlds in Collision* a few years ago was a
formidable indication that after the passage of two centuries of
scientific endeavor, man in the mass was still enormously susceptible to the appeal of cataclysmic events, however badly sustained from the scientific point of view. It introduced to our
modern generation, bored long since with the endless small
accretions of scientific truth, the violence and catastrophism in
world events which had so impressed our forefathers.

Man has always had two ways of looking at nature, and these
two divergent approaches to the world can be observed among
modern primitive peoples, as well as being traceable far into
the primitive past. Man has a belief in seen and unseen nature.
He is both pragmatist and mystic. He has been so from the
beginning, and it may well be that the quality of his inquiring
and perceptive intellect will cause him to remain so till the end.

Primitive man, grossly superstitious though he may be, is
also scientist and technologist. He makes tools based upon his
empirical observation of the simple forces around him. Man

would have vanished long ago if he had been content to exist in the wilderness of his own dreams. Instead he compromised. He accepted a world of reality, a natural, everyday, observable world in which he existed, and whose forces he utilized in order to survive. The other aspect of his mind, the mystical part seeking answers to final questions, clothed this visible world in a shimmering haze of magic. Unseen spirits moved in the wood. Today in our sophistication we smile, but we are not satisfied with the appearances of the phenomenal world around us. We wish to pierce beneath to ask the question, "Why does the universe exist?" We have learned a great deal about secondary causes, about the *how* of things. The why, however, eludes us, and as long as this is the case, we will have a yearning for the marvelous, the explosive event in history. Indeed, so restless is man's intellect that were he to penetrate to the secret of the universe tomorrow, the likelihood is that he would grow bored on the day after.

A scientist writing around the turn of this century remarked that all of the past generations of men have lived and died in a world of illusions. The unconscious irony in his observation consists in the fact that this man assumed the progress of science to have been so great that a clear vision of the world without illusion was, by his own time, possible. It is needless to add that he wrote before Einstein, before the spread of Freud's doctrines, at a time when Mendel was just about to be rediscovered, and before advances in the study of radioactivity had made their impact—of both illumination and confusion—upon this century.

Certainly science has moved forward. But when science progresses, it often opens vaster mysteries to our gaze. Moreover, science frequently discovers that it must abandon or modify what it once believed. Sometimes it ends by accepting what it has previously scorned. The simplistic idea that science marches undeviatingly down an ever broadening highway can scarcely be sustained by the historian of ideas. As in other human affairs, there may be prejudice, rigidity, timid evasion and sometimes inability to reorient oneself rapidly to drastic changes in world view.

The student of scientific history soon learns that a given way of looking at things, a kind of unconscious conformity which exists even in a free society, may prevent a new contribution from being followed up, or its implications from being fully grasped. The work of Gregor Mendel, founder of modern genetics, suffered such a fate. Darwin's forerunners endured similar neglect. Semmelweis, the discoverer of the cause of childbed fever, was atrociously abused by his medical colleagues. To rest uneasy consciences, we sometimes ascribe such examples of intolerant behavior to religious prejudice—as though there had been a clean break, with scientists all arrayed under the white banner of truth while the forces of obscurantism parade under the black flag of prejudice.

The truth is better, if less appetizing. Like other members of the human race, scientists are capable of prejudice. They have occasionally persecuted other scientists, and they have not always been able to see that an old theory, given a hairsbreadth twist, might open an entirely new vista to the human reason.

I say this not to defame the profession of learning but to urge the extension of education in scientific history. The study leads both to a better understanding of the process of discovery and to that kind of humbling and contrite wisdom which comes from a long knowledge of human folly in a field supposedly devoid of it. The man who learns how difficult it is to step outside the intellectual climate of his or any age has taken the first step on the road to emancipation, to world citizenship of a high order.

He has learned something of the forces which play upon the supposedly dispassionate mind of the scientist; he has learned how difficult it is to see differently from other men, even when that difference may be incalculably important. It is a study which should bring into the laboratory and the classroom not only greater tolerance for the ideas of others but a clearer realization that even the scientific atmosphere evolves and changes with the society of which it is a part. When the student has become consciously aware of this, he is in a better position to see farther and more dispassionately in the guidance of his own research. A not unimportant by-product of such an awareness may be an extension of his own horizon as a human being.

I have sought to emphasize this point in my beginning discussion because several of my topics will be involved with the intellectual climate of the past. In conclusion I hope to venture some comment upon the world we now call "natural," as if, in some manner, we had tamed it sufficiently to include it under the category of "known and explored," as if it had little in the way of surprises yet in store for us. I shall want to look at this natural world both from the empirical point of view and from one which also takes into account that sense of awe and marvel which is part of man's primitive heritage, and without which man would not be man.

For many of us the Biblical bush still burns, and there is a deep mystery in the heart of a simple seed. If I seem for a time to be telling the story of how man came under the domain of law, how he reluctantly gave up his dreams and found his own footsteps wandering backward until on some far hillside they were transmuted into the footprints of a beast, it is only that we may assess more clearly that strange world into which we have been born—we, compounded of dust, and the light of a star.

Our first effort will be devoted to an examination of that universe which, in the unconsciously prophetic words of Sir Thomas Browne, "God seldom alters or perverts, but like an Excellent Artist, hath so contrived his work, that with the self-same instrument, without a new creation, he may effect his obscurest designs." When the great physician uttered those words in 1635 he was not thinking of evolution, but, as we shall see, he spoke like a blind oracle. The "self-same instrument" effecting design without creation would go unnoticed for two hundred years.

II

It is a moot question in history what brought into existence man's earliest conception of natural law. Certainly what we might call the regularities of nature—the round of the seasons, the passage of day and night—man must have been aware of since the time he began to think at all. There is, however, a difference between this background of observation and the development of the idea that the entire universe lay under some divine system of ordinances which were unalterable.

Whitehead has contended that the medieval insistence on

the rationality of God—something which had its origin in the union of Judeo-Greek philosophy—lies at the root of science. On the other hand, it must be borne in mind that the immediately prescientific era was fascinated by monsters, signs in the heavens—miraculous events which a modern man would regard as unlawful and outside the normal course of nature.

It has also been suggested that the rise of centralized royal authority and the extension of its power in the capitalist state in some manner contributed to the conception of far-ranging law wielded by the Divinity. Like most purely economic explanations of intellectual events, this is doubtless a simplification of a complicated shift in intellectual emphasis. At all events, in the sixteenth and seventeenth centuries, law, natural law, the undeviating law of God, had taken precedence in intellectual circles over the world of the miraculous. There was a rising interest in the Second Book of Revelation; that is, nature. It was assumed that the two books, separately examined, would bear each other out—that the world could be read as though it were part of the Great Book. Since the God of the Old Testament was a God of wrath, it is not surprising that there lingered in the western mind a taste for the violent interpretation of geological events. Across eighteenth-century Europe lay the fallen, transported boulders of what seemed the visible evidence of some vast deluge. In the story of those stones, man's naïve faith in visible catastrophe is countered by the magnificent violence hidden in a raindrop.

III

Time and raindrops! It took enormous effort to discover the potentialities of both those forces. It took centuries before the faint trickling from cottage eaves and gutters caught the ear of some inquiring scholar. Men who could visualize readily the horrors of a universal Flood were deaf to the roar of the invisible Niagara falling into the rain barrel outside their window. They could not hear it because they lived in a time span so short that the only way geologic change could be effected was by the convulsions of earthquakes, or the forty torrential days and nights that brought the Biblical Deluge.

The world of medieval thought was deeply centered upon

itself and upon the traditional myths of Christianity. In spite of sectarian clashes, Christians of the prescientific era saw the earth essentially as the platform of a divine but short-lived drama—a drama so brief that there was little reason to study the stage properties. The full interest centered upon man—his supernatural origins, the drama of his Fall from the deathless Garden, the coming of his Redeemer, and the day of his Judgment.

Outside space was the Empyrean realm beyond time and blemish. Inside were corruption and a falling away from grace which were the consequence of man's sin. The atmosphere was not one to encourage scientific exploration. Men were busied about their souls, not about far voyages either in space or in time. They were contented with the European scene; they were devout and centered inward. It was indeed a centripetally directed society on an earth which itself lay at the center of the universe. Sinful though man had proved to be, he was of enormous importance to himself. The eye of God was constantly and undividedly upon him. The Devil, passing to and fro upon Earth, contended for his soul. If man was not in all ways comfortable, he was at least valuable to divinities, and good and evil strove for the possession of his immortal being.

Then someone found a shell embedded in rock on a mountain top; someone saw the birth of a new star in the inviolable Empyrean heavens, someone watched a little patch of soil carried by a stream into the valley. Another saw a forest buried under ancient clays and wondered. Some heretical idler observed a fish in stone. All these things had doubtless been seen many times before, but human interests were changing. The great voyages that were to open up the physical world had begun. The first telescope was trained upon a star. The first crude microscope was turned upon a drop of ditch water. Because of these small buried events, a world would eventually die, only to be replaced by another—the world in which we now exist.

In the early days of scientific exploration of the universe, the divergence between the world of belief as represented by Biblical tradition and the world of science was not anticipated. It was generally assumed that the investigation of the physical universe would simply reveal more of God's ways toward the care of man and reaffirm Biblical truths. The title of John Ray's well-known book, *The Wisdom of God Manifested in the Works of the Creation*, which first appeared in 1691, suggests this outlook. It did not prevent sharp observations from being made, nor a growing and persistent wonder about fossils, which were then called "formed stones."

Erosion was beginning to be faintly glimpsed as a power at work in nature. "That the height of the mountains doth continually diminish," muses John Ray, "is very likely." Our knowledge of time would have to be greatly altered before anyone would inquire whether persistent forces might be at work which would prevent the total denudation of the land and its eventual disappearance into the engulfing sea.

The change which was to pass over human thinking, however, began in the skies. This astronomical thinking has both its conservative and liberal side.

In the history of science, as in other history, there is rarely a specific place to begin a new story. For our purposes I will select just two events: the discovery of the speed of light, and Newton's formulation of the laws of gravitation. If we were to go farther back in time, we would, of course, have to treat of the opening out of space and the intellectual revolution introduced by the discovery that the earth was not the center of the solar system. Nevertheless, it will suffice for our purposes if we recall that it was a seventeenth-century astronomer, Olaus Roemer, who first deduced the speed of light in 1675. In doing so he opened a doorway upon limitless vistas in space and time. It had previously been supposed that the movement of light was instantaneous. When Roemer discovered a slight lag in the reappearance of one of Jupiter's satellite moons after an eclipse, it became possible, through the calculations of later workers, to estimate the speed of light at 186,000 miles a second. The way now lay open to the light-year—to astronomical time the magnitude of which lay beyond human comprehension. For a period, in fact, time would remain the possession of the astronomers and haunt only the trackless abysses of space. It would not disturb the man in the street.

For that matter, astronomy still reflected much of its ancient attitude toward Empyrean space as inviolable and unchangeable. Newton, a deeply religious man, was of this persuasion.

Kepler had written earlier that the celestial machine is not something like a divine organism, but rather "something like a clockwork in which a single weight drives all the gears." Newton, with his formulation of the laws of gravity, had supplied the single weight. God had been the Creator of the machine, but it could run without his interference. At the most, only an occasional interposition of his power would be needed to set the clock right. In contrast to earlier periods, knowledge of natural forces led to less need for divine intervention in earthly affairs. Newton, however, remained devout in a way that many of his followers of the eighteenth century did not.

The growing interest in mechanics throughout the eighteenth century, the passionate fondness for mechanical devices of all sorts, led to an enthusiastic interest in the *Machina Coelestis*. Miracle was in the process of disappearance save at the moment of creation. The world of science was growing increasingly skeptical as its knowledge increased. Signs in the heavens, wonders in the animal world, were decreasing. The machine reigned. God, who had set the clocks to ticking, was now an anomaly in his own universe. The question of celestial and earthly origins, which Newton had abjured, emerged first in astronomy. It would be Immanuel Kant and the French skeptic Laplace who would introduce cosmic evolution, and who would extend backward into time the laws which Newton had extended across space. The wheels and cogs of the celestial machine for the first time would be pursued backward until they dissolved in spinning vapor. By the midpoint of the century, time was clearly seen as necessary to the development of the cosmos. Thomas Wright had identified as a galactic universe the Milky Way, of which our sun is a minor inhabitant.

The philosopher Kant, drawing inferences of galaxial rotation from Edmund Halley's detection of star movement in 1717, proposed the nebular hypothesis of star and planetary origin in clouds of rotating gas. Kant, at last, was seeking to derive the complex from the simple—"the simplest," as he wrote, "that can succeed the Void."

At the close of the eighteenth century, Laplace, in his *Treatise on Celestial Mechanics*, greatly elaborated the nebular hypothesis. With the concept of historical change and development applied to the heavens, the notion of creation by divine fiat in a universe of short duration began to pass. The way was opening for geologists to pick up the story of the molten earth and carry its development forward into time. Without anyone's knowing the precise way in which the change had been effected, the intellectual climate was altering.

By the 1750's cosmic evolution was openly discussed; geological change, timidly; the evolution of life, in subdued and sporadic whispers. As the idea reached our planet, so to speak, it was greeted with less enthusiasm. It aroused curiosity among the masses but seemed to threaten entrenched religious institutions. It is perhaps not without significance that the chief proponent of cosmic evolutionism arose in radical and free-thinking France, as did two of the first great biological evolutionists.

The devious threads of communication which eventually combined all these ideas of development by physical forces, rather than accepting the theory of creation at the direct hands of a Master Mechanic, are now difficult to trace. Eighteenth-century scientists corresponded, where today they would send a paper or a note to a professional journal. Or they paid visits to each other, or chatted at courts and salons. Moreover, the footnoting of the sources of ideas had not become the traditional practice that it is today.

Whatever the methods used, ideas of development, change, what we would call "historicity," appear in several distinct fields with surprising rapidity, if not simultaneously. In later chapters we shall consider the penetration of this same idea into the life sciences and how it came to extend itself to man. Here, however, I wish merely to examine the way in which Newton's conception of the cosmic machine, the celestial engine, came to extend itself into geology. So far as the life sciences are concerned, we shall see, later on, that the whole idea was both advantageous and, paradoxically, retarding. The flow of ideas from one field into another often takes curious and ambivalent paths. It was so in the case of James Hutton, the founder of historical geology.

IV

The scientific life of James Hutton extended over the last half of the eighteenth century. What Newton achieved and emphasized in astronomy and mathematics, Hutton accomplished in geology. It would be useless here to pursue all of the faint hints and intuitions about geological matters which preceded Hutton's work. They exist, but they do not lessen the fact that it was James Hutton of Edinburgh who, in diligent application of Newton's principles of experimental inquiry and observation, passed from the astronomer's conception of the self-correcting machine of the heavens to the idea that the earth itself constituted a machine which eternally reconstituted and renewed itself.

It has been pointed out that Hutton's doctoral dissertation was on the circulation of the blood in the Microcosm, that is, in man. It is an old idea in western thought, which persisted with unabated force into the eighteenth century, that man the microcosm reproduces in miniature, or is directly influenced by, the events of the Macrocosm, that is, the outside world, the universe. This idea lies at the root of astrology, and persists in a disguised form into modern times. It has been contended that Hutton, as a medical man, applied this idea to the earth, treating it as a living organism with circulation, metabolism, and other correspondences to the organic world. It has been termed Hutton's secret—a secret which happened to yield, in the case of geology, some remarkable insights, because it placed emphasis upon the dynamic qualities of the earth's crust—in short, upon the phenomena of decay and renewal.

It can be maintained, however, that just as Newton and his successors placed emphasis upon the giant celestial machine of the heavens—self-balancing and self-maintaining, set rolling by the hand of the Master Craftsman, God—so Hutton, influenced by this widespread conception, was the first to apply it to another seemingly self-renovating engine under divine care, the earth. Though Hutton, in later years, was not to escape the charge of heresy, the existing documents clearly suggest that he was less a cosmic evolutionist like Laplace than he was a true Newtonian, in that he abjured, or at least evaded, the question of the earth's origins. Rather, he dealt with the planet as a completed mechanism—whether we regard that mechanism as organic in essence or mechanical. He had accepted and read in the rocks, as the astronomers had begun to read in the skies, the message of time. Indeed he states forthrightly of the earth "that we find no vestige of a beginning, no prospect of an end."

Which, then, of these two views of Hutton's achievement is correct? Is Hutton's a machine analogy or an organismic one? The machine analogy, at any rate, bulks large in the interpretation of eighteenth-century thought and descends into our own day. It is only by the hook of the analogy, by the root metaphor, as one philosopher has termed it, that science succeeds in extending its domain.

Occasionally, if not frequently, the analogy is false. Yet so potent is its effect upon a whole generation of scientific thinking that it may lie buried in the lowest stratum of accepted thought, or color unconsciously the thinking of entire generations. While proceeding with what is called "empirical research" and "experiment," the scientist will almost inevitably fit such experiments into an existing comprehensive framework, an integrative formula, until such time as that principle gives way to another. Let us see, in this connection, what ideas Hutton introduced into his examination of a previously neglected subject: the nature of the habitable earth.

V

We have earlier spoken of man's individual life views as colored and influenced by what he can perceive within that lifetime—what, in other words, he can personally observe. The individual is loath to accept explanations of phenomena which come about as the result of forces exceeding the range of his own life span. If it is some type of natural landmark placed before his day, he is apt, rather than consider the effect of the accumulation of small events, to turn to myths incorporating outright violence on a gigantic scale. This, as we have already observed, is the first natural reaction of many laymen unacquainted with the history of geology today. In the seventeenth and eighteenth centuries science attempted its first groping entrance into the vast domain of time. It is not surprising, therefore, that what is now confined to the naïve and scientifically uneducated should have affected the reasoning even of scholars.

The Christian world accepted a surprisingly short time scale of a few thousand years. The calculations of such men as

Bishop Ussher, based upon genealogical charts and other stray
Biblical sources, were not an integral part of the Bible, but
through long association with the volume they had become
so. Moreover, the story of Creation, Eden, and the succeeding
Flood all imply a world controlled and brought into being by
direct supernatural methods which seemed to be devoted solely
to the human drama. These had been the cherished beliefs of
Christendom for over a thousand years. They were graven deep
in the religious consciousness of scientist and layman alike. The
Book of Nature sought by deists and religious liberals was an
embodiment of divine reason and would not contradict the
other great source of direct revelation—the Bible.

The result was that when Hutton, again under the influence
of Newton's mathematical analysis of continuity, postulated
the integration of small events to produce great cumulative
ones in geology, he differed sharply from his associates.
Doubtless these ideas contributed strongly to the charges of
atheism which were hurled at him. Certainly it was not long
before his views were utterly at odds with that school of
thought known as catastrophism, which was destined to obscure his work for a whole generation.

This school of violence is the very antithesis of the Huttonian approach through time, raindrops and aerial erosion.

The catastrophist believed the glacial boulders scattered far
from their point of origin to have been rolled and tossed in the
turbulence of some giant deluge like the Noachian Flood—
visible evidence of wild powers loosed upon the planet at sporadic intervals. Mountain chains were the product of similar
violence. Breaks in the geological record, discontinuities, in
time even abrupt faunal changes, were all assumed to be the
devastating result of world-wide disturbances.

As geological knowledge of the earth's history increased
shortly after Hutton's time, this theory, or modifications of it,
became ascendant in geological circles. It had about it a certain
awe-inspiring Old Testament grandeur. It predicated vast, unknown and perhaps supernatural forces at work. Each cata-
clysm shut one such geological period off almost totally from
another. It was the one great Biblical event multiplied by a
chain of such events extending backward into the past. A series
of shut doors concealed one age from another.

The only continuity, so far as the living world was concerned, lay in an abstract plan, a Platonic ideal in the mind of God, which caused the beings of one age to have an organic phyletic structure related, though only immaterially and with modifications, to the creatures of another.

Catastrophism is one of the prime examples of a scientific world view in transition. Its mysterious geological upheavals and re-creations of life could be paced fast or slow according to the Biblical days of creation as figuratively expressed in the Book of Genesis, or as the tolerance of the individual might incline. Its succession of convulsive movements of the earth's crust accounted for the more dramatic aspects of the European countryside without introducing those limitless and invisibly moving landscapes which seemed, to many Englishmen in the early years of the nineteenth century, to be part of the dreadful culmination of heretical thought as it had afflicted France.

Political and religious considerations aside, however, catastrophism has an appeal of its own, even into our own day. No one likes to watch, listlessly, an hour hand go around the clock. We want the cuckoo bird to erupt violently at intervals from his little box, or a gong to strike. This catastrophism provided. Its time scale was scored and punctuated by violence.

Hutton, on the other hand, presents us with a quite different system. Instead of beginning with ancient catastrophes postulated upon giant tidal waves, he states with the utmost sobriety that "we are to examine the constructions of the present earth, in order to understand the natural operations of times past. The earth," he says, "like the body of an animal, is wasted at the same time that it is repaired. It has a state of growth and augmentation; it has another state, which is that of diminution and decay. This world is thus destroyed in one part, but it is renewed in another." Across Hutton's pages pass a series of small natural operations that over long time periods erode mountains, create valleys, and that, if mountain-building processes did not counteract their effect, would bring whole continents down to sea level.

He saw the bit of soil carried away by a mountain brook or a spring freshet lodge in and nourish a lower valley; he saw the wind endlessly polishing and eroding stones on the high flanks of the world. He saw, with the marvelous all-seeing eye of

Shakespeare, that "water-drops have worn the stones of Troy and blind oblivion swallowed cities up." He knew about the constant passage of water from sea to land and back again. If a leaf fell he knew where it was bound, and multiplied it mentally by ten thousand leaves in ten thousand, thousand autumns. One has the feeling that he sensed, on his remote Scottish farm, when frost split a stone on a winter night. Or when one boulder, poised precariously on a far mountain side, fell after a thousand years. For him and him alone, the water dripping from the cottagers' eaves had become Niagaras falling through unplumbed millennia. "Nature," he wrote simply, "lives in motion." Every particle in the world was hurrying somewhere, or was so destined in the long traverse of time.

In his observation that land was being created while land was being worn away, that there was continental elevation as well as denudation, Hutton shows a great grasp of the earth's interior powers. Though it was impossible for him to be totally correct in small details, he was almost alone in his recognition of the geostrophic cycle.

James Hutton had come upon the secret of the relatively perpetual youth of the planet. Although Hutton was primarily a physical geologist who published little upon fossils, he had, in actuality, set the scene in which, a half-century farther on, the rise of the vertebrates might be better grasped. In fact, he had provided the physical setting for an evolutionary process as lengthy in its implications as his own eroding hills.

Hutton's axiom that the understanding of present forces is the key to the past is now the basis of the natural sciences. Yet he was a flexible man and averse to dogmatic interpretations of his doctrine. "We are not," he added wisely, "to limit Nature with the uniformity of an equable progression." He was aware of violence and occasional spectacular occurrences in nature—he even caught a faint, far glimpse of the European ice age—but he was a child of the century of Enlightenment. "No powers are to be employed that are not natural to the globe," he wrote, disdaining the half-lit supernatural domain of the catastrophists. Such remarks cost his memory ill, later on. For Hutton, who, like Newton, was a devout man, believed in reason because reason itself lay behind nature and had directed its course. He was, in this respect, a typical eighteenth-century deist who accepted the sanctity of undeviating law and avoided the intrusion of the supernatural into the natural realm.

VI

At this point we are faced once more with the question upon which I touched on an earlier page: namely, whether it is correct to interpret Hutton as viewing the earth as a living organism—what we might call the eighteenth-century physician's view—or whether his analogy is not rather that of the celestial machine of Newton. In the answer we will obtain a better glimpse into the preconceptions of the age. I hold for the machine, but it is justifiable to observe that the animal had, in some eyes, also become a machine. The two views are actually unitary.

It is true that Hutton speaks metaphorically of the body of the earth as wasting and being replenished like an animal body. Elsewhere, however, he speaks of the earth as "a machine of peculiar construction" and again he refers to this "beautiful machine."

Besides the influence of the Newtonian celestial machine which so impressed the eighteenth century, there is the powerful example of Newton's experimental method, to which Hutton clung resolutely. Hypotheses were subordinated to experiment. The Newtonian machine was one created by fiat, not one growing like an animal. Hutton seems not to have been greatly affected by evolutionary doctrines save for his acceptance of the long time scale. He believed that the world, like Newton's celestial machine, was run on perfect principles and was self-balancing rather than undergoing what we today would term unreturning, complete historicity. Hutton was also influenced by the general interest in James Watt's steam-engine experiment and is known to have spoken of volcanoes as "safety valves." His world machine sounds at times like a heat engine.

Thus it would appear in the great scientist-physician's memoirs that the world machine was variously conceived. The French experimenters of the seventeenth and eighteenth centuries had been so intrigued, since the days of Descartes, with the idea that animals were pure soulless automata that many cruel and heartless experiments had been performed, as the following contemporary account from La Fontaine attests:

> "They administered beatings to dogs with perfect indifference and made fun of those who pitied the creatures as if they had felt pain. They said that the animals were clocks; that the cries they emitted when struck, were only the noise of a little spring which had been touched, but that the whole body was without feeling. They nailed poor animals up on boards by their four paws to vivisect them and see the circulation of the blood which was a great subject of conversation."

These historical items make plain that the organism-machine analogies were not remote from each other at that time except in so far as man, in contrast to the animal machine, had a soul. When this distinction was no longer scientifically tenable, there would emerge a genuine difference between animal and machine, because organisms form themselves and evolve, while machines do not.

Thus when the Divine Maker was retired from the earthly scene by science, leaving only secondary causes to operate nature for him, men, animals and the celestial and world machines alike were no longer to be quite what they had been in the days of supernatural intrusion, of a tampering by the Unseen. Man's world was finally to be completely natural. Yet at the close of the eighteenth century it was still a world considered to be of divine origin and created for human habitation. Only later would it be found a world without the balance of stabilized perfection. The Microcosm would not repeat the Macrocosm. The celestial clocks would no longer chime in perfect order.

Edmund Halley, as early as 1717, had calculated that the entire solar system must be moving mysteriously toward remote constellations. Man, too, was to become as natural as the wandering stars that lighted his unknown course. He was to learn that his habitation was unfixed. Not only he but his tightly governed universe was soon to be adrift and moorless on the pathways of the night.

TWO

How Death Became Natural

The world is the geologist's great puzzle box.

—LOUIS AGASSIZ

IT IS NECESSARY in surveying the human quest for certainty to consider death before life. I have not done this out of perversity. Rather I have done it because, in the sequence of ideas we have been studying, it is necessary to understand certain aspects of death before we can comprehend the nature of life and its changes.

Man, even primitive man, has tended to take life for granted. Death was the unnatural thing, the result of malice or mistake, the after-message of the gods, or, in the Christian world, the result of the Fall from the Garden. In the development of a scientific approach to life on this planet, therefore, the recognition of death—species death, phylogenetic death—had to precede the rise of serious evolutionary thought. For without the knowledge of extinction in the past, it is impossible to entertain ideas of drastic organic change going on in the present or future.

Moreover, extinction is not something which can be postulated from a philosopher's armchair. It can be ascertained only by careful and precise field observation. Comparative anatomy has to be carried to a sufficient point of accuracy that the existing fauna of the world can be distinguished from the faunas of the past. The deeper our knowledge of the geological record penetrates, the stranger are the forms which can be discerned in the earth's far epochs. Without this historical perspective any suggestion of plant or animal change is bound to be limited and the imagination impoverished. At best such ideas will be confined to what can be observed in the way of change among modern products of the breeder's art. Breeding did, however, promote a kind of incipient evolutionism. Thus one of the very early and anonymous commentators upon selection in England, in discussing domesticated forms, has this to say:

"Amidst these varieties, which have sprung up under our eye, there are not a few which deviate so much from the type of the species, that we seem incapable of assigning a limit to man's power of producing variation; nor when thinking how many similar circumstances accidently occur in nature, is it easy to avoid suspecting that many reputed species may in reality have descended from a common stock."[1]

It was the lack of knowledge of the fossil past which so greatly handicapped the first evolutionists of the eighteenth century. It is just here that the failure of Hutton's views to be received as credible is disappointing. Hutton, while not a student of fossils, was, as we have seen, a student of time. He could read its passage in the rocks and he had been prepared to venture a belief in the enormous antiquity of the earth. There can be no doubt that the conservative English reaction in science after the French Revolution delayed the recognition of Huttonian geology. As an indirect consequence, it may well be that the acceptance of the evolutionary philosophy itself was also delayed by a generation. Time and accompanying geological change are two of the necessary properties without which evolution would be unable to operate. And those two properties bring death as a third factor in their wake.

II

The seventeenth century was, in general, still in the grip of the short Christian time scale, though here and there, toward the close of the century, some hesitant doubts began to be expressed by men like John Ray. It was the general opinion, says Ray, "among Divines and Philosophers that since the first creation there have been no species of animals or vegetables lost, no new ones produced." It was part of the reigning theology of the time that extinction was an impossibility. In fact, this view continued to be reiterated in the eighteenth century and,

[1]Anonymous, "On Systems and Methods In Natural History," *The Quarterly Review* (London), Vol. 41 (1829), p. 307. On the basis of interior evidence I am inclined to suspect that this paper is a youthful and unacknowledged review by Sir Charles Lyell.

by ultraconservative thinkers, into the first decade of the nineteenth century.

Thomas Jefferson, writing in 1782, commented that "such is
the economy of nature that no instance can be produced of
her having permitted any one race of her animals to become
extinct; of her having formed any link in her great work so
weak as to be broken." The well-read Jefferson is, of course,
merely repeating what was the commonly accepted view of his
time. Extinction loomed as something vaguely threatening
and heretical. In fact, for that very reason many refused to accept fossils as representing once-living creatures.

Their reluctance to accept what now seems to us so easily
discernible and commonplace an observation as extinction is
based essentially upon one fact: the benignity of Providence.
"To suppose any species of Creatures to cease cannot consist
with the Divine Providence," writes one seventeenth-century
naturalist, and his comment is frequently reiterated by others.
This point of view is based upon a theory of organic relationships which, though traceable into earlier centuries, reached a
peculiar height of development during the seventeenth and
eighteenth centuries. The belief was not confined to philosophy and theology. It permeated the whole field of letters and
became widely known as the Scala Naturae, or Ladder of
Being.

This conception superficially resembles a line of evolutionary ascent and undoubtedly has played an indirect part in the
promotion of evolutionary notions of a scale of rising complexity in the development of life. It was not, however, a
scheme of evolution. It is based on a gradation which emerged
instantaneously at the moment of creation, and which rises by
imperceptible transitions from the inorganic through the organic world to man, and even beyond him to divine spiritual
natures.

The idea promoted much anatomical work of a devout
character as naturalists attempted to work out the missing or
obscure links in what was regarded as an indissoluble chain
which held together the various parts of creation. It was for
this reason that men viewed with genuine horror the idea that
links could be lost out of the chain of life. The strong belief in
an all-wise Providence which did nothing without intention

caused men to refuse an interpretation of the universe which involved apparently aimless disappearances. If such disappearances were possible, what might not happen to man himself?

Just as there existed the balanced, self-correcting machine of the heavens, and the balanced, self-renewing machine of the earth, so life was similarly linked in the great chain of unalterable law. All was directly under the foreknowing care of the Divine Being. Nevertheless, by the early eighteenth century it began to be whispered among English naturalists "that many Sorts of Shells are wholly lost, or at least out of our Seas." The simpler organisms from marine strata were being identified before the taxonomy of fossil vertebrates had been seriously attempted.

III

It may now be asked, as fossils slowly became accepted as the remains of creatures once living, how men were for so long able to evade the still troubling question of extinction—the existence of death before the Garden. A world view does not dissolve overnight. Rather, like one of Hutton's mountain ranges, it erodes through long centuries. In the case of fossils we must remember how small the European domain of science was in comparison with the vast continental areas which had only recently been opened for examination by the voyagers. Australia, Africa, the Americas, were barely known; their interiors remained unexplored. They provided a providential escape for the devout naturalist who still wished to avoid the dangerous logic implicit in unknown bones and shells.

Since it was beginning to be realized that the separate continents possessed faunas and floras to some degree distinct, the devout could argue that creatures of which there were now no living representatives in Europe might well have been driven out by man, or by changes of climate; in short, that instead of being extinct, they had merely retired to the fastnesses of unknown seas or continents. Jefferson quotes an observer who had heard the mammoth roaring in the Virginia woods. In Europe the bones of Ice Age elephants were ascribed to the living African species imported, so it was claimed, for Roman games. Or the bones were those of Hannibal's war elephants lost on the Alpine passes. The desired point was to make the animals *historic* and thus ascribable still to living species. In America, where great bones lay in profusion, it was rumored that living specimens could be found across the unknown Great Lakes or farther on in the heart of the continent. By the mid-eighteenth century the aroused pursuit of natural curiosities reveals the fascination of a public increasingly alive to foreign rarities and the new mysteries of the living world.

Peter Collinson, the Quaker merchant and naturalist in London, writes testily to John Bartram in the colonies: "If thee know anything of thy own knowledge please to communicate it. The hearsay of others can't be depended on."

Bartram in turn complains peevishly, "The French Indians have been very troublesome, which hath made travelling very dangerous beyond our [territory] where I used to find many curiosities. . . . While we . . . daily expect invasions we have little heart or relish for speculations in Natural History."

Still, in spite of Indians and the depredations of pirates, the letters and the specimens made their slow way to Europe. "The frogs came safe, and lively. I transcribed thy account of them, and had it delivered to the King."

"I received the turtle in good health; and shall be much obliged to him if he will procure me a male and female Bull-frog. Mine are strayed away."

"Your ingenious idea respecting the former existence of certain kinds of animals, now extinct, I confess carries great weight with it and yet, my dear Sir, I cannot implicitly give my assent to it on the whole. With regard to the Unicorn I am rather divided in my judgment, even in respect to their present existence, in the interior region of Africa, of which, we are extremely ignorant."

"That all petrifactions should be attributed to the general deluge, is what I shall never agree," growls another correspondent taking a slight step toward the future.

Reading the old letters, we hear the voices mingle in a mounting symphony. "Every uncommon thing thou finds in any branch of Nature will be acceptable." "The terrapins had bad luck. Some died, others the sailors stole." "I had some doubts, so carefully examined the Ohio Elephant's long teeth with a great number . . . from Asia and from Africa and found they agree with what is called the *Mammot's* Teeth from Siberia. It is all a wonder how they came to America. . . ."

Then Collinson turns to the as yet unknown teeth of the mastodon and the extinct Irish elk. "So here are two animals, the creature to which the great forked or pronged teeth belong [mastodon]. Whether they exist, God Almighty knows,—for no man knows: whether ante-diluvians, or if in being since the flood. *But it is contrary to the common course of Providence to suffer any of his creatures to be annihilated.*"

In Germany one ingenious escapist propounded a theory that the rocks which compose the geological strata of the earth had fallen at various times from the heavens as meteoric or planetary fragments. Thus, he contended, the unknown species of animals whose fossils had now become so troubling to the devout had no necessary connections with this earth at all. They were the remains, instead, of extraterrestrial creatures. It was a disengaging action almost two hundred years ahead of the space age, which would have been fascinated with such a notion.

We are now in position to see that the eighteenth century—powerfully rationalistic and scientifically curious though it was—had difficulty until toward the final decade of the century, in assimilating the idea that species could be utterly extinguished. Deism was the reigning philosophy of the times, and deism repudiated the idea that God immediately interposed his will in nature. Rather, he delegated powers to secondary causes which were self-operating.

We have seen this view expressed in the heavenly and earthly machines together. The deists, however, along with other rationalists, reposed great faith in human reason as reflecting the Divine reason. God, it was thought, could be known through nature and apart from Biblical revelation. The problem which caused such hesitation among the eighteenth-century students of animal life, therefore, was how to explain the apparent irrationality and waste involved in the discovery of extinction. Why would a supremely rational God reject and repudiate his creations? Furthermore, if such repudiation occurred, was there not danger that man himself might be swept from the stage of life?

Secondary forces as controlling the world, man had come to

accept. He had passed beyond the conception of a God super-naturally intervening in mundane affairs. Nevertheless, man had continued to believe that the great machine was rationally orga-nized with human welfare in mind—that it was self-balanced and its aberrations self-correcting. The hint of extinction in the geological past was like a cold wind out of a dark cellar. It chilled men's souls. It brought with it doubts of the rational world men had envisaged on the basis of their own minds. It brought suspi-cions as to the nature of the cozy best-of-all-possible-worlds which had been created specifically for men.

A vast and shadowy history loomed in the rocks. It threat-ened to be a history in which man's entire destiny would lose the significance he had always attached to it. For a few decades the lost links of life might be sought as living beyond the sources of primeval Amazons and Orinocos. In the end man's vision of his world would undergo drastic revision. Out of it would emerge once more, though briefly, a renewed confi-dence in his position in the universe.

IV

We have already had occasion to observe the preference in man to interpret startling features of landscape in terms of catastrophic violence. In the years immediately following Hut-ton's death, the slow alteration of earth's features postulated by him and the Frenchman Lamarck gave way before the wide-spread popularity of the geological doctrine of catastrophism. Linked with it, there appeared a new and highly popular theory in which extinction was finally recognized by science. It was, however, explained in a manner that was pleasing to the public fancy and at least offered a compromise between science and older theological beliefs.

"At certain periods in the development of human knowl-edge," C. D. Broad once remarked, "it may be profitable and even essential for generations of scientists to act on a theory which is philosophically quite ridiculous." This was true in a comparatively short-lived way of catastrophism. It persuaded man to accept both death and progressive change in the uni-verse. It did so by extending such mythological events as the world-shattering Biblical Deluge, and by the creation of a form of geological prophecy which left man still the dominant figure in his universe.

"Half a century ago," wrote the great American botanist Asa Gray, who died in 1880, "the commonly received doctrine was that the earth had been completely depopulated and repopulated over and over; and that the species which now, along with man, occupy the present surface of the earth, belong to an ultimate and independent creation, having an ideal but no genealogical connection with those that preceded. This view . . . has very recently disappeared from science."

This was the philosophy which Sir Charles Lyell was destined to overthrow; this was the view propounded to young Charles Darwin by his geology professors before he went on the voyage of the *Beagle*. This was the concept against which Darwin dueled with Agassiz past midcentury. In what lay the vitality of this weirdly irrational theory, and how did it arise? We have hinted at its appeal to human vanity in a time of growing religious confusion. To examine its roots we must turn again to the closing years of the eighteenth century and the first two decades of the nineteenth. In those years we find a growing amalgamation of new ideas as follows:

1. Cuvier and his associates, working upon the successive vertebrate faunas of the Paris basin, remarked "that none of the large species of quadrupeds whose remains are now found imbedded in regular rocky strata are at all similar to any of the known living species." Furthermore, they denied that the species concerned "are still concealed in the desert and uninhabited parts of the world."

2. Cuvier could observe no graduated changes in his vertebrate fossils. He assumed, therefore, that no intermediate forms existed between one geological level and another. That there was a progression in the complexity of the animal types from age to age as we approach the present, Cuvier had, however, begun to perceive.

3. Cuvier assumed that there were genuine breaks between one age and another brought about by catastrophic floods and alterations in the relation of sea to land. He did not, however, propose such a total break between one age and another as the later catastrophists.

4. There arose among German and French nature philoso-

phers a renewed sense of the unity of plan or biological structure common to large groups of plants and animals—a
morphological advance also marked by the contributions of
Cuvier in France and his disciple, Richard Owen, in England.
The transcendental aspects of this morphology lay in the conception that these major structural plans existed abstractly in
the mind of God, who altered them significantly from age to
age. In the words of Adam Sedgwick, Darwin's old teacher,
"At succeeding epochs, new tribes of beings were called into
existence, not merely as the progeny of those that had appeared
before them, but as new and living proofs of creative interference; and though formed on the same plan, and bearing the
same marks of wise contrivance, oftentimes as unlike those
creatures which preceded them, as if they had been matured in
a different portion of the universe and cast upon the earth by
the collision of another planet."

It was generally conceived that the lower and earlier forms
of life pointed on directly to man, who had been ordained to
appear since the time of the first creation. "It can be shown,"
asserted Louis Agassiz, "that in the great plan of creation . . .
the very commencement exhibits a certain tendency toward the
end . . . The constantly increasing similarity to man of the
creatures successively called into existence, makes the final
purpose obvious . . ." The geological record was being
searched for signs and portents pointing to human emergence
at a later epoch. There is more than a hint of medieval "signs
in the heavens" to be found in these paleontological auguries:
a reptile leaving handlike imprints on some ancient sea beach is
a portent of man's coming; the stride of a bipedal dinosaur
discloses the eventual appearance of bipedal man.

Catastrophism, if we are to examine it in its most mature
form—that of England in the second decade of the nineteenth
century—has, as we have seen, several surprising features. The
deathless Eden of the Biblical first creation has been replaced by
a succession of natural but successive worlds divided from each
other by floods or other violent cataclysms which absolutely ex-
terminate the life of a particular age. Divinity then replaces the
lost fauna with new forms in succeeding eras. Disconformities in
geological strata, breaks in the paleontological record, are taken
as signs of world-wide disaster terminating periods of calm. In

contrast to eighteenth-century concern over the death of species, and anxiety to establish seemingly extinct animals as still in
existence, the natural theologians now revel in violence as exces-
sive as that of the Old Testament. Whole orders of life are swept
out of existence in the great march toward man. The stage
which awaits the coming of the last great drama has to be prepared. Floods destroy the earlier actors. Enormous death demands equally enormous creation, discreetly veiled in the
volcanic mists that hover over this half-supernatural landscape.

From the idea that one lost link in the chain of life might
cause the whole creation to vanish piecemeal, man had passed,
in scarcely more than a generation, to the notion that the entire world was periodically swept clean of living things. The
discoveries of vertebrate paleontology seemed to illuminate, in
solitary lightning flashes, a universe that progressed in leaps
amidst colossal destruction.

Still, there was a pattern amidst the chaos. For while strange
animals arose and perished, it was observed that the great patterns of life, the divine blueprints, one might say, persisted
from one age to another. It was only the individual species or
genera that vanished and each time were re-created in an altered pattern. There was no natural connection, no phylogeny
of descent in a modern sense. There was only the Platonic
ideal of pure substanceless form existing in the mind of God.
The reality of material descent escaped the mind of the
observer.

The public concentrated less upon living animals adjusting
to circumstance than upon this strange spiritual drama which
the natural theologians had read into the rocks. Two steps in the
direction of naturalism had been gained, however. First, the
world of the past was now known to have cherished plants and
animals no longer to be observed among the living. Second,
thanks largely to the pioneer efforts of William Smith of England, who had once caustically remarked to critics that "the
search for a Fossil may be considered at least as rational as the
pursuit of a hare," it was known that fossils could be used to
identify distinct strata.

Smith had recognized the changes in invertebrate fossils
from one geological level to another. Cuvier's observations
upon vertebrate remains had similarly, if more dramatically,

established a type of grand progressive movement among the vertebrates. Life did not return upon its track. The record in the book of stone showed no reversals. Life, in other words, was a historic progression in which the past died totally. But the goal was finalistic—it was man. Even coal forests had been laid down for his use. At times it seemed that the earlier creation existed only as some kind of phylogenetic portent of man.

V

Catastrophism, in essence, may be said to have died of common sense. As a modern historian, Charles Gillispie, has commented, "To imagine the Divine Craftsman as forever fiddling with His materials, forever so dissatisfied with one creation of rocks or animals that He wiped it out in order to try something else, was to invest Him with mankind's attributes instead of the other way about."

Slowly the accumulation of geological information began to lead back toward the pathway pursued earlier by James Hutton and his follower, John Playfair. Sir Charles Lyell, who was born the year of Hutton's death, reapproached the whole subject of uniformitarian geology in his famous *Principles of Geology*, whose first volume was published in 1830. In the intervening half century, much additional information had become available. Lyell was a careful organizer of facts, a man of judicious temperament and an independent thinker. He, like Hutton, had an eye for the common observable workings of sunshine and water drops. In fact, he was one of the first to read in fossil impressions that the raindrops of the past were similar to those of the present, that the eyes of fossil trilobites showed light falling upon the earth many millions of years ago as it falls today. He saw no evidence of world-wide catastrophes. He observed, instead, local disconformities of strata, the rise and fall of coast lines, the slow upthrust of mountain systems. He saw time as illimitable, in the fashion of Hutton.

Moreover, Lyell attempted statistical estimates of the change in molluscan species as one passed from one distinct bed to another. He observed that certain organisms persisted—though in changed proportions. He could not discover the drastic and sudden eliminations of fauna upon which the catastrophists had built their case. As work progressed, more and more of what were termed "passage beds" appeared—strata linking one supposedly separate era with another. Local sequences of this sort began to make clear the essential unity of earth's geological history. By degrees Lyell's more vociferous opponents grew silent. In that silence one thing was clearly apparent. What we might call point-extinction, i.e., extinction of the individual species, had replaced the concept of mass death. Death, in other words, was becoming natural—a product of the struggle for existence.

Lyell observed that the long course of geological change was bound to affect the life upon the planet's surface. He saw that every living creature competed for living space and that every change of season, every shift of shore line, gave advantages to some forms of life and restricted space available to others. Over great periods of time it was inevitable that some species would slowly suffer a reduction in numbers and, by degrees, perish, to be replaced by others.

This incredibly tight and complicated web of life would, Lyell thought, eliminate immediately any newly emerging creatures which might be evolving through natural means. Yet if, as the geological record indicated, species perish, somewhere there must be creation, somewhere there must be a coming in to replace the deficit involved in extinction.

Lyell hesitated. Could there be individual point-emergences as well as vanishings? No one professed to have seen the creation of a complex animal. Was it because so rare a thing was simply not normally observable?

Lyell had inherited from Hutton a distaste for the unseen. He preferred to work with visible, understandable forces susceptible of observation. He did not like catastrophism with its spectacular and unobservable creations any more than he liked its flamboyant geological mechanisms. As early as 1829 he had written privately, "We shall very soon solve the grand problem, whether the various living organic species came into being gradually and singly in insulated spots, or centers of creation, or in various places at once, and all at the same time. The latter cannot, I am already persuaded, be maintained."

Try as he might, Lyell could find no satisfactory explanation for the advances in biological organization which the catastrophists acclaimed. To admit them was to accept miracle—the unknown. To accord them acceptance was equivalent to geological capitulation also—it was, in effect, to say: "I cannot explain your mysterious and advancing creation of life, therefore if life is miraculous, your interpretation of geological forces may just as well be accepted in the same fashion."

At this Lyell balked—not always consistently. He pointed out that the geological record was incomplete. He contended that the discovery of vertebrates in older deposits than they were originally assumed to characterize was a sign of nonprogression, that the serial advances recorded by the progressionist-catastrophist school were largely in error.

Faunas might shift with time and geography, Lyell warned, but this might not involve necessary progression through the vertical realm of geology. Only man Lyell admitted to be young. The extinct forms amidst the great phyla could be accepted without the argument that there was a common, necessary upward trend in creation culminating in man.

Up to the time of the publication of the *Origin of Species*, Lyell was suffering from the lack of a satisfactory explanation of organic change. He had overthrown the extinction-in-mass conception of the catastrophists; he had reintroduced into geology the lengthier time span of Hutton, and Hutton's devotion to purely natural forces. It was, however, difficult to see how those forces applied to the single great mystery—life.

Lyell stood, actually, at the verge of Darwin's discovery. He was, however, within the shadow of another century—the eighteenth. He shared its passion for intellectual order, for the obedient and unmysterious world machine. Hutton had taken life for granted because almost nothing was known in his day of the antediluvian world of fossils. Lyell, by contrast, was confronted by a perverse, unexplainable force that crawled and changed through the strata—life. He made *death* natural, but it could be said that life defeated his efforts to understand it. With all their errors, the catastrophists had been right about one thing. From its early beginnings in the seas, life *had* been journeying and growing in complexity. It is historic.

All the way back into Cambrian time we know that sunlight fell, as it falls now, upon this planet. As Lyell taught, we can tell this by the eyes of fossil sea creatures such as the trilobites. We know that rain fell, as it falls now, upon wet beaches that had never known the step of man. We can read the scampering imprints of the raindrops upon the wet mud that has long since turned to stone. We can view the ripple marks in the sands of vanished coves. In all that time the ways of the inanimate world have not altered; storms and wind, sun and frost, have worked slowly upon the landscape. Mountains have risen and worn down, coast lines have altered. All that world has been the product of blind force and counterforce, the grinding of ice over stone, the pounding of pebbles in the mountain torrents—a workshop of a thousand hammers and shooting sparks in which no conscious hand was ever visible, today or yesterday.

Yet into this world of the machine—this mechanical disturbance surrounded by desert silences—a ghost has come, a ghost whose step must have been as light and imperceptible as the first scurry of a mouse in Cheops' tomb. Musing over the Archean strata, one can hear and see it in the subcellars of the mind itself, a little green in a fulminating spring, some strange objects floundering and helpless in the ooze on the tide line, something beating, beating, like a heart until a mounting thunder goes up through the towering strata, until no drum that ever was can produce its rhythm, until no mind can contain it, until it rises, wet and seaweed-crowned, an apparition from marsh and tide pool, gross with matter, gurgling and inarticulate, ape and man-ape, grisly and fang-scarred, until the thunder is in oneself and is passing—to the ages beyond—to a world unknown, yet forever being born.

"It is carbon," says one, as the music fades within his ear. "It is done with the amino acids," contributes another. "It rots and ebbs into the ground," growls a realist. "It began in the mud," criticizes a dreamer. "It endures pain," cries a sufferer. "It is evil," sighs a man of many disillusionments.

Since the first human eye saw a leaf in Devonian sandstone and a puzzled finger reached to touch it, sadness has lain over the heart of man. By this tenuous thread of living protoplasm, stretching backward into time, we are linked forever to lost beaches whose sands have long since hardened into stone. The stars that caught our blind amphibian stare have shifted far or vanished in their courses, but still that naked, glistening thread winds onward. No one knows the secret of its beginning or its end. Its forms are phantoms. The thread alone is real; the thread is life.

"Nevertheless, there is a goal," we seek to console ourselves. "The thread is there, the thread runs to a goal." But the thread has run a tangled maze. There are strange turns in its history, loops and knots and constrictions. Today the dead beasts decorate the halls of our museums, and that nature of which men spoke so trustingly is known to have created a multitude of forms before the present, played with them, building armor and strange reptilian pleasures, only to let them pass like discarded toys on a playroom floor. Nevertheless, the thread of life ran onward, so that if you look closely you can see the singing reptile in the bird, or some ancient amphibian fondness for the ooze where the child wades in the mud.

One thing alone life does not appear to do; it never brings back the past. Unlike lifeless matter, it is historical. It seems to have had a single point of origin and to be traveling in a totally unique fashion in the time dimension. That life was ever a fixed chain without movement was a human illusion; that it leaped as some mystical abstraction from one giant scene of death to another was also an illusion; that geological prophecy proclaimed the coming of man as Elizabethan astrologers read in the heavens the signs of coming events for kings was an even greater fantasy. Instead, species died irregularly like individual men over the long and scattered waste of eons. And as they died they must, as Lyell foresaw, be replaced in as scattered a fashion as their deaths. But what was the secret? Did a voice speak once in a hundred years in some hidden wood so that a nocturnal flower bloomed, or something new and furry ran away into the dark?

Creation and its mystery could no longer be safely relegated to the past behind us. It might now reveal itself to man at any moment in a farmer's pasture, or a willow thicket. By the comprehension of death man was beginning to glimpse another secret. The common day had turned marvelous. Creation—whether seen or unseen—must be even now about us everywhere in the prosaic world of the present.

THREE

How Life Became Natural

*If we can conceive no end of space, why should we conceive an end
of new creations, whatever our poor little bounds of historical time
might even appear to argue to the contrary.*

—LEIGH HUNT, 1836

GREAT LITERARY GENIUSES often possess an ear or a sensitivity for things in the process of becoming, for ideas which are just about to be born. It is interesting in this connection to compare the remarks of Charles Darwin with certain observations on science made at a much earlier date by the poet Samuel Taylor Coleridge. Darwin, in his autobiography, protests that he saw no evidence that the subject of evolution was "in the air" of his time. "I occasionally sounded out not a few naturalists," he remarks, "and never happened to come across a single one who seemed to doubt about the permanence of species."

By way of contrast, we may note that Coleridge, in a philosophical lecture delivered as early as 1819, makes reference to a belief which "has become quite common even among Christian people, that the human race arose from a state of savagery and then gradually from a monkey came up through various states to be man." Coleridge was not an evolutionist. He is, however, sensitive to a new doctrine, whose presence "in the air" Darwin had failed to discover. He observes in a very shrewd fashion what we have sought to emphasize in previous chapters; namely, the way in which the intellectual climate of a given period may unconsciously retard or limit the theoretical ventures of an exploring scientist. "Whoever is acquainted with the history of philosophy during the last two or three centuries," contended the great poet, "cannot but admit, that there appears to have existed a sort of secret and tacit compact among the learned, not to pass beyond a certain limit in speculative science. The privilege of free thought so highly extolled, has at no time been held valid in actual practice, except within this limit."

Coleridge is here recognizing a fact which escaped the simpler and less philosophically oriented mind of Darwin. The latter had failed to recognize that the silence of his professional colleagues was in some cases just such a restriction as that of which Coleridge speaks. Religious and social pressures had all contributed to making the subject of evolution somewhat taboo and not really in good taste to discuss in public. Time and again one finds naturalists circling all about the subject and then withdrawing timidly from any attempt to derive the final, and what now seems to us the logical, conclusion. Some, in fact, openly contradict themselves in order to stay within the accepted circle of traditional thought.

The entomologist T. Vernon Wollaston is a case in point. In his book *On the Variation of Species*, published in 1856—a volume from which Darwin drew material favorable to the evolutionary point of view—Wollaston dwells sporadically upon organic change. We are led to believe, he says, "that, could the entire living panorama, in all its magnificence and breadth, be spread before our eyes, with its long lost links [of the past and present epochs] replaced, it would be found, from first to last, to be complete and continuous throughout,—a very marvel of perfection, the work of the Master's Hand. . . . From first to last," he contends, "the same truth is re-echoed to our mind, that here all is change." Yet after making these and other suggestive remarks of a similar character, Wollaston disavows in his final chapter the idea of the transmutation of species, though at the same time he can only say that the limitation of a species must be indicated *somewhere*. This timidity, which fits so well Coleridge's observations of the scientific mind, savors more of religious discomfort than genuine scientific conviction. It characterizes a considerable amount of the biological literature of the earlier half of the nineteenth century.

In the end it was an outsider, Robert Chambers, without professional ties, an amateur, as again Coleridge was sharp enough to generalize upon, who would break through the orthodox barrier of conservatism. Finally, it would be another wealthy amateur, Charles Darwin, whose massive treatise would swing world thought into a new channel. The story is a fascinating one. Coleridge's critical observations upon science made forty years before the publication of the *Origin* are an almost exact preview of what happened in the long history of the evolutionary concept.

The period prior to Darwin's enunciation of the full evolutionary doctrine is difficult to define with precision. Just as in the orchestra pit before a great musical performance there is the individual tuning of strings, plucking of stray notes, and discordant thumpings before the conductor, with his baton, brings unity into a composition, so in the period prior to 1859, we can get all manner of tentative retreats and approaches, partial developments, hesitant insights and bold sounds from the wings. When, at last, Darwin picks up the conductor's wand and turns this assemblage of stray notes into a full-throated performance, the world audience is swept off its feet so completely that, hypnotized by Darwin's single figure, it forgets the individual musicians who made his feat possible.

In actuality, life was not made natural in a day, nor in a single generation. Neither, we have observed, were geology and death. Before entering upon the difficult problem of life, therefore, let us now attempt to recall the major shifts which had taken place in scientific thinking upon this subject. It will be remembered that the seventeenth, and to a very considerable extent the eighteenth, century believed in a linked, unbreakable chain of organisms ascending in complexity to man. In spite of the fact that this conception bears a superficial resemblance to the idea of evolution, it is, in reality, a fixed, static, and immovable chain. Nothing changes position, nothing alters, nothing becomes extinct. By degrees, as the rock strata were found to contain organisms unlike those of the present era, two new conceptions arose. One of these came to be called progressionism and is associated with the catastrophic geology which we have previously examined.

This theory assumes that life was first manifested on the planet by simple organisms, but that from the very beginning of time, the stage was being set for man. Each ascending fauna, as we have seen, was successively swept away at the close of a catastrophic geological episode. For reasons unknown to man, but nevertheless mysteriously prophesied in the rocks, this lengthy prologue had preceded the human emergence. Each fauna, however, in spite of its anatomical relationship to the previous order, represented a separate act of creation. A kind of ideal Platonic morphology had been substituted for the notion of a direct physical descent which links the creatures of one age and those of the next.

The second idea which, although originally regarded askance, particularly in England, was being whispered about toward the close of the eighteenth century was that of evolution itself—actual physical descent with bodily alterations from one age to another. The cosmic evolutionism which had begun to enter the speculations of the astronomers had now attracted some inquiring minds among the biologists. These early approaches to the problem were, however, vitiated by a certain casualness and lack of evidence. The age of the world was still being underestimated, and the lack of evidence for any really marked extinction (the notion of evolution actually preceded full-blown progressionism) led to an underestimation of the possibilities contained in the idea. Lamarck, who wrote the most extended treatise upon the subject, never was very sure about the matter of extinction. He succeeded in largely evading the subject by assuming that the missing animals had evolved, without perishing, into other forms. In skirting the problem of extinction, he was a typical child of the eighteenth century. Nevertheless, it must be recognized that Lamarck, besides grasping the reality of the evolutionary process, observed that creatures fitted themselves to the environment they occupied, rather than being made for that specific environment. In this respect alone Lamarck was years ahead of his time, because until geological change has been wedded to organic change, one cannot have a full-fledged evolutionary theory.

It has occasionally been said that Lamarck remained unappreciated because he entertained some ideas which sound ridiculous to the modern ear. Historical hindsight in such matters is rarely unprejudiced. One might as well argue that Newton should have passed unheeded because he wrote lengthily upon theology, or because he manifested paranoid mental tendencies in his declining years. If we were to ignore certain of our scientific forerunners upon such a basis, we would have to dismiss the discoveries of many geniuses of the scientific twilight who entertained advanced notions along with a sincere belief in witches. In time, scientific historians looking back will undoubtedly see our beliefs as shot through and through with the equivalent fantasies of our own age. It is not just to dismiss Lamarck on such a basis, for if we were to catalogue each change in thought that led on to Darwin, we would have to recognize that this much maligned French thinker glimpsed ecological change and adjustment before Darwin. In the process he recognized what Darwin was later to call the "law of divergence" and what the modern world calls "adaptive radiation."

At this moment of recognition, just as Halley had unwittingly set the solar system adrift without a pilot, so Lamarck, though he did not realize it, had destroyed the preordained character of the human emergence. Beside such a momentous observation, the question of whether his evolutionary mechanism was right or wrong lapses into comparative unimportance. The fact that generations of historians have seen this man purely as an advocate of a now rejected explanation of how evolution comes about is the result of two misfortunes: the fact that he was a Frenchman who survived the Revolution, and the additional misfortune that his successor, Charles Darwin, had little interest in, or concern with, the history of the subject. Darwin also cherished the common conservative English attitude toward the thinkers of revolutionary and post-revolutionary France. Something of a conspiracy of silence surrounded Lamarck's name and English naturalists disavowed his theories with almost ritualistic fervor.

Our first step in the effort to understand how life became natural, therefore, is to avoid the commonly held impression that Darwin, by a solitary innovation—natural selection—transformed the western world view. Without detracting in the least from his importance, we may observe that few, if any, scientific discoveries are made in such a fashion. Newton once made the perceptive observation that if he saw far, it was because he stood on the shoulders of giants. Similarly, Charles Darwin was the inheritor of the efforts of his forerunners, but because of a new twist which he gave to those same efforts, the stages in the process leading to Darwin's achievement have passed largely unobserved. The drama of the voyage of the *Beagle*, the isolation of the years at Downe, the great shift in public opinion which began with the final acceptance of geological time through the studies of Lyell—all, in a sense, have obscured rather than illuminated the Darwinian story.

Darwin has been left in solitary grandeur as a kind of psychological father figure to biologists. Let me repeat at this point, since I have already experienced the amount of emotional heat which can still be generated about this man, that I am interested only in the presentation of a succession of some easily verifiable ideas and a view of how they changed in order to bring us to the world we inhabit today. Darwin, some biologists have proclaimed, had nothing to do with his forerunners. Others, similarly aroused, have persisted in the reverse argument that Darwin had certainly made use of the ideas of his forerunners, but that this did not matter in the least because Darwin was the man who brought the public to a recognition of evolution.

Both of these remarks strike one as emotionally oriented. They serve to conceal a certain type of unsophisticated hero worship which still exists among occasional scientists unfamiliar with the history of ideas. What we are interested in at this point is solely how an idea, natural selection, beginning as a conservative eighteenth-century observation, was altered by slow degrees into something which set the world of life adrift in an unfixed wilderness as surely as Halley's sensitivity had set the star streams pouring through unimaginable darkness and distance.

II

Four propositions, it can now be observed, had to be clarified before the theory of organic evolution would prove acceptable to science. *First*, as we have already noted, the great antiquity of the planet had to be grasped. Otherwise life would occupy too narrow a segment of time for change of a slow nature to be possible. In fact, prior to the time of Linnaeus, abrupt spectacular mutations were occasionally discussed in agricultural works—although without reference to evolutionary ideas. The notion that species were completely immutable seems to have come in with a hardening of the religious temper, particularly in the century between about 1750 and 1859.

Second, it was necessary to establish the fact that there had been a true geological succession of forms on the planet. Though again, as we have already observed, this did not lead immediately to the acceptance of the evolutionary hypothesis, it did call attention to a totally forgotten series of worlds stretching into the remote past. Inevitably any rational philosophy would have to account for the inhabitants of former times and, if possible, relate them to the plants and animals of the present. Knowledge of the vertebrate succession began to emerge only after 1800. As a consequence of this gap in our knowledge, paleontology for a time contributed to the growth of the progressionist scheme of mass extinctions and re-creations of life.

Third, the amount of individual variation in the living world and its possible significance in the creation of change had to be understood. Variation began to be noted and speculated about as far back as the seventeenth century, but its role in evolution would not be understood until much later. Nevertheless, the activities of stock breeders would engender some notion of at least limited organic change.

Fourth, the notion of the perpetually balanced world machine, which had been extended to life itself, had to give way to a conception of the organic world as not being in equilibrium at all—or at least being only relatively so—a world whose creations made and transformed themselves throughout eternity. Life had to be seen, in the apt phrase of a later evolutionist, Alfred Russel Wallace, as subject to "indefinite departure" —alteration, in other words, subject to no return.

We have already had occasion to examine what occurred to illuminate the understanding of time and animal succession. It is with the last two of these four propositions, variation and the balanced world machine, that we will now concern ourselves. Before life could be viewed as in any way natural—and it is not my intention to push this word too far—a rational explanation of change through the ages had to be proposed. In a sense, it was Hutton and Lyell's problem of earth change reapplied to the problems of life. It embodied a similar search for natural causes at work in the present day—causes still capable of study and observation. Variation, selection, the struggle for existence, were all known before Darwin. They were seen, however, within the context of a different world view. Their true significance remained obscured or muted in precisely the manner that Coleridge had anticipated in his estimate of the scientific mind.

It was not really new facts that were needed so much as a new way of looking at the world from an old set of data. A few men had tried to accomplish the task even before the close of the eighteenth century. The lack of knowledge of the fossil past, however, made their attempts impoverished and undramatic compared with the catastrophism which relegated their efforts to obscurity. They had failed to supply a satisfactory explanation for evolutionary change. In addition, England was swept by an anti-French wave of conservatism which was the intellectual product first of the Revolution, and then of the following Napoleonic wars. Her science, to a degree, became isolated from that of the Continent.

III

The historian who examines with care the documents of the eighteenth century before the recognition of extinction, and while the scale of nature is still the overriding biological as well as theological concept, will come immediately upon a principle of balance which was believed to prevail throughout the living world. This is what I meant when I said that the eighteenth-century love of order had been extended to the living world. Something of the complexity of the living environment had been observed, something of what the biologist of today would call food chains. To the eighteenth-century mind, however, this world was in a permanent, rather than dynamic, balance. It was a manifestation, in other words, of divine rule and government. Let me explain this subtlety, if I can. To do so, let us examine the work of just one man, John Brückner, a Frenchman. Brückner is the actual forerunner of Thomas Malthus, the man who so profoundly influenced the political and biological thinking of the nineteenth century.

Brückner's book entitled *A Philosophical Survey of the Animal Creation* appeared in English in 1768, over thirty years before Malthus' *Essay on the Human Population*. In it he observes, "Providence seems to have advanced to the utmost verge of possibility in the gift of life conferred upon animated beings." Famine, pestilence and war, the checks upon human population to which Malthus devoted so much attention, occur in Brückner's pages. The struggle for existence is very clear to him: "For it is with the animal as with the vegetable system: the different species can only subsist in proportion to the extent of land they occupy; and wherever the number of individuals exceeds this proportion, they must decline and perish."

Brückner, however, is still obsessed with the short Christian time scale. "It is five thousand years at least," he observes, "that one part of the living substance had waged continual war with the other, yet we do not find that this Law of Nature [i.e., natural selection] has to this day occasioned the extinction of any one species." Here Brückner is quite obviously laboring under the providential belief that extinction is an impossibility. In the next sentence Brückner makes very clear the eighteenth century's recognition of the role of natural selection. "Nay," he says, continuing his discussion of the war of species against species, "it is this which has preserved them in that state of perpetual youth and vigor in which we behold them. . . .

"The effects of the carnivorous race," he goes on to add, "are exactly the same as that of the pruning-hook, with respect to shrubs which are too luxuriant in their growth, or of the hoe to plants that grow too close together. By the diminution of their number, the others arrive at greater perfection." Brückner calls this process "reciprocal attrition." It will take other names before it finally reaches Darwin. Innumerable times it will be referred to by scientific writers as "pruning," "policing," "natural government."

The phrase "natural government" best expresses the eighteenth-century world view of the interlinked web of life. Like Hutton's world machine, its momentary aberrations were self-correcting. Hutton himself, in an unpublished work only recently made available, reveals a clear knowledge of natural selection in this varietally selective sense—proving once more, if proof were needed, that the principle antedates Darwin. Struggle, it was thought, adjusted the quantity of life and eliminated the unfit. Beyond this, selection did not create. Life was held in a static balance. It was not going anywhere.

One other curious phrase in Brückner's work demands

attention, for it, too, is indicative of the time. "If the absolute number of inhabitants is not so great as it might have been, it is nevertheless always approaching toward its plenitude." The word "plenitude" had a very special meaning to the men of this age. It was assumed that God was creative up to the limit of his capacity, that every Platonic form or idea must be expressed in a rational universe. Thus the philosophic notion of plenitude contributed abstractly to what Professor Lovejoy has called a "Malthusian picture of a Nature overcrowded with aspirants for life."

The discovery of extinction did not, at first, disturb the idea of natural selection as a kind of beneficent provision of divine government. It merely operated in the old fashion within each episode of the advancing world of life. Buckland, in 1829, for example, speaks of "the carnivora in each period of the world's history fulfilling their destined office—to check excess in the progress of life and maintain the balance of creation." Portlock, a geologist, referred to the idea as late as 1857 as a "sublime conclusion." The conception of a self-regulating balance of nature had survived the discovery of extinction and been taken up in the new progressionism without undergoing the slightest modification. Selection was still a conservative, not a creative, force in nature. Nevertheless, the age of the world had been lengthened and the concept of the struggle for existence was about to enter a new phase.

When Sir Charles Lyell broke through the strata which separated one age from another and proved that all the separate worlds of the catastrophists were, in reality, one single related and ever altering world, extinction, death, as we have previously seen, became natural. It was no longer veiled behind the mists of smoking volcanoes or hidden in the onrush of tidal waves. Species, it became apparent, died as men died, singly and sporadically. Lyell, like Brückner, continued to see natural selection as a conservative force. He was faced, however, with an increasing handicap. If there was, in actuality, only one ever altering world, then the balance, the self-regulating government of living things, might not exist. If death approached living things in this piecemeal fashion, everything in time might perish.

The only solution lay in the acceptance of the ideas of the

rejected evolutionists or in some form of mysterious point-creation to replace the sporadic disappearance of species. Though Lyell wrote much upon the struggle for existence, in the end he remained content with its negative aspects. He saw the web of life as drawn so tightly that there would be no room for a new form to emerge or evolve. Before it could do so, he believed, it would be overwhelmed by the perfectly adjusted organisms about it. At the same time, Lyell was not insensible to a certain degree of variability within the limits of a species. It was his opinion that this ability to develop minor geographic varieties aided the survival of wide-ranging species. In the main, however, Lyell insisted upon his principle of preoccupancy—that is, the idea that plants and animals long adapted to a specific environment will be able to keep foreign intruders from occupying a new country.

Basically this was once more a static conception. At the time it was made, Darwin was observing the influx of foreign weeds in the pampas, the havoc wrought by the introduction of the fauna of the continents upon oceanic islands. But Darwin was still a young naturalist enjoying his first experience of the world. Back home in England, the century was pressing on.

In 1835 Edward Blyth, a young naturalist of Darwin's own age, wrote a paper on animal varieties. Hidden obscurely in the midst of the paper was Blyth's discussion of what he called his "localizing principle." Blyth, in fact, had described what today we call Darwinian natural selection. Blyth saw his principle as one "intended by Providence to keep up the typical qualities of a species." In this respect Blyth sounds like a pure eighteenth-century exponent of natural government, but by 1837 certain additional thoughts had begun to impress themselves upon the young man.

"A variety of important considerations here crowd upon the mind," Blyth confesses, "foremost of which is the enquiry that, as man, by removing species from their appropriate haunts, superinduces changes on their physical constitution and adaptations, to what extent may not the same take place in wild nature, so that, in a few generations, distinctive characters may be acquired, such as are recognized as indicative of specific diversity. *May not then, a large proportion of what are considered species have descended from a common heritage?*"

These words were written in 1837, a short time after Charles Darwin had returned from the *Beagle* voyage. We know that Darwin was an avid reader of the *Magazine of Natural History*, in which Blyth's papers appeared. We know further that there is interior evidence, in Darwin's two early essays before the *Origin of Species* was attempted, which points strongly to Darwin's early knowledge of Blyth's work. At least one of his letters speaks of his fascination with "some few naturalists" in the *Magazine of Natural History*, and there are other references in the undestroyed portions of his notebook recently published by Sir Gavin de Beer. Just as Coleridge had commented eighteen years before, however, scientific convention kept young Edward Blyth from quite accepting his own speculation. "There is a compact among the learned," Coleridge had said, "not to pass beyond a certain limit in speculative science."

Blyth had not made a world voyage like Darwin. He was overawed by his masters, including Lyell. He had rubbed his eyes and the safe and sane conventional world of the English hedgerows had seemed for an instant to alter into something demonic. In the next moment the vision was gone.

Scientific convention has held for a hundred years that the author of the *Origin* read Malthus' *An Essay on the Principle of Population* in 1838 and received from that work the hint which led to his discovery of natural selection. While Malthus undoubtedly had a wide influence during this period and is a convenient source of reference, it is now unlikely that he was Darwin's main source of inspiration. Darwin opened his first notebook on the species question in 1837, the year young Blyth ventured beyond the unconscious convention of his time. Locked away in Darwin's early essays are some curious similarities of thought which strain credulity to be called coincidence.

John Brückner preceded and was used by Thomas Malthus. Edward Blyth preceded and was used by Charles Darwin. There is one difference. Malthus added almost nothing to the thought of Brückner. Darwin, by contrast, took the immature thought, the nudge from Blyth, and combined it with his own vast and growing experience. He refused to see environments as fixed, and organisms as fixed with them. He argued cogently that no environment is completely static, and that selection therefore is constantly at work in the production of new organisms as time and slowly changing geological conditions alter the existing world. An idea which began as an explanation of "natural government," a stabilizing factor making for providential control of the living world, had changed by slow degrees, even as men openly contradicted what their own eyes professed to see.

The individual variation which all organisms revealed, the hereditary alterations produced by the breeder's art, the unreturning fossils in the rocks, were finally combined in the minds of Charles Darwin and Alfred Russel Wallace with the world of competitive struggle, with that concept of plenitude which was now seen to eternally jostle the living in and out of existence.

It was not natural selection that was born in 1859, as the world believes. Instead it was natural selection without balance. Brückner's "impetuous torrent," which he visualized as beating against its safe restraining dikes, is loose and rolling. The violence in Hutton's raindrop is equaled, if not surpassed, by the violence contained in a microscopic genetic particle. The one, multiplied, carries away a mountain range. The other crosses an ice age and produces, on its far side, a man-ape whose intellectual powers now endanger his own civilization.

The world of geological prophecy has vanished. There is only this vast uneasy river of life spreading into every possible niche, dreaming its way toward every possible form. Since the beginning there have been no breaks in that river. The immaterial blueprints were an illusion generated by physical descent. The lime in our bones, the salt in our blood were not from the direct hand of the Craftsman. They were, instead, part of our heritage from an ancient and forgotten sea.

Yet for all this flood of change, movement and destruction, there is an enormous stability about the morphological plans which are built into the great phyla—the major divisions of life. They have all, or most of them, survived since the first fossil records. They do not vanish. The species alter, one might say, but the *Form*, that greater animal which stretches across the millennia, survives. There is a curious comfort in the discovery. In some parts of the world, if one were to go out into the woods, one would find many versions of oneself, with fur and grimaces, surveying one's activities from behind leaves and thickets. It is almost as though somewhere outside, somewhere beyond the illusions, the several might be one.

IV

Many years ago I was once, by accident, locked in a museum with which I had some association. In the evening twilight I found myself in a lengthy hall containing nothing but Crustacea of all varieties. I used to think they were a rather limited order of life, but as I walked about impatiently in my search for a guard, the sight began to impress, not to say overawe, me.

The last light of sunset, coming through a window, gilded with red a huge Japanese crab on a pedestal at one end of the room. It was one of the stilt-walkers of the nightmare deeps, with a body the size of a human head carried tiptoe on three-foot legs like fire tongs. In the cases beside him there were crabs built and riveted like Sherman tanks, and there were crabs whose claws had been flattened into plates that clapped over their faces and left them shut up inside with little secrets. There were crabs covered with chitinous thorns that would have made them indigestible; there were crabs drawn out and thin, with delicate elongated pinchers like the tools men use to manipulate at a distance in dangerous atomic furnaces.

There were crabs that planted sea growths on their backs and marched about like restless gardens. There were crabs as ragged as waterweed or as smooth as beach pebbles; there were crabs that climbed trees and crabs from beneath the polar ice. But the sea change was on them all. They were one, one great plan that flamed there on its pedestal in the sinister evening light, but they were also many and the touch of Maya, of illusion, lay upon them.

I was shivering a little by the time the guard came to me. Around us in the museum cases was an old pattern, out of the remote sea depths. It was alien to man. I would never underestimate it again. It is not the individual that matters; it is the Plan and the incredible potentialities within it. The forms within the Form are endless and their emergence into time is endless. I leaned there, gazing at that monster from whom the forms seemed flowing, like the last vertebrate on a world whose sun was dying. It was plain that they wanted the planet and meant to have it. One could feel the massed threat of them in this hall.

"It looks alive, Doctor," said the guard at my elbow.

"Davis," I said with relief, "you're a vertebrate. I never appreciated it before, but I do now. You're a vertebrate, and whatever else you are or will be, you'll never be like that thing in there. Never in ten million years. I believe I'm right in congratulating you. Just remember that we're both vertebrates and we've got to stick together. Keep an eye on them now—all of them. I'll spell you in the morning."

Davis did something then that restored my confidence in man. He laughed, and touched my shoulder lightly. I have never heard a crab laugh and I never expect to hear one. It is not in the pattern of the arthropods.

Yet those crabs taught me a lesson really. They reminded me that an order of life is like a diamond of many reflecting surfaces, each with its own pinpoint of light contributing to the total effect. It is a troubling thought, contend some, to be a man and a God-created creature, and at the same time to see animals which mimic our faces in the forest. It is not a good thing to take the center of the stage and to feel at one's back the amused little eyes from the bush. It is not a good thing, someone maintained to me only recently, that animals should stand so close to man.

It depends, I suppose, on the point of view.

On my office wall is a beautiful photograph of a slow loris with round, enormous eyes set in the spectral face of a night-haunter. From a great bundle of fur a small hand protrudes to clasp a branch. Only a specialist would see in that body the far-off simulacrum of our own. Sometimes when I am very tired I can think myself into the picture until I am wrapped securely in a warm coat with a fine black stripe down my spine. And my hands would still grasp a stick as they do today.

At such times a great peace settles on me, and with the office door closed, I can sleep as lemurs sleep tonight, huddled high in the great trees of two continents. Let the storms blow through the streets of cities; the root is safe, the many-faced animal of which we are one flashing and evanescent facet will not pass with us. When the last seared hand has flung the last grenade, an older version of that hand will be stroking a clinging youngster hidden in its fur, high up under some autumn moon. I will think of beginning again, I say to myself then, sleepily. I will think of beginning again, in a different way. . . .

FOUR

How Man Became Natural

*Here below to live is to change, and to be perfect is to have changed
often.*

—CARDINAL NEWMAN

IN MY HOME country, near a small town now almost vanished
from the map, there is a region which remains to this day
uncultivated. It is best seen at nightfall, because it is then that
the full mystery of the place seizes upon the mind. Red granite
boulders hundreds of miles from their point of origin thrust
awkwardly out of the sparse turf. In the last glow from the
west one gets the impression of a waste over which has passed
something inhumanly remote and terrifying—something that
has happened long ago, but which here lies close to the surface. Crows circle above it like disturbed black memories which
rise and fall but never come to rest. It is a barren and disordered landscape, which remembers, and perhaps again antici-
pates, the cold of glacial ice. It has nothing to do with man; its
gravels, its red afterglow, are remnants of another era, in which
man was of no consequence.

One instinctively feels that if anything were to graze here it
would be mammoth—great sullen, shapeless hulks in the dead
light. No one crosses these fields. An invisible barrier confronts
one at every turn. Man literally ends here. Beyond lies something morosely violent, of which we have no knowledge, or of
which it might be better said that we have a traumatic eagerness to forget. For lurking in this domain is still the nature that
created us: the nature of ringing ice fields, of choked forests, of
unseasonable thunders. It is the unpredictable nature of the
time before the gods, before man had laid hold upon any
powers with his mind. It is a season of helplessness that stirs
our submerged memories and that causes us to turn back at
twilight to the safe road and the lights of town. Behind us
whisper the ancient, uncontrollable winds. An animal cries
harshly from the dark field we refuse to enter. But enter it we

must, though the effort lifts a long-vanished ruff of hair along our nape. This is man's place of birth, this region of inarticulate terror—and this is the story of how he came through the clouds of forgetfulness to find himself in a world which has vanished.

All across Europe, from the British Isles through the North German plain to the Alps and well beyond into Russia, lie the transported boulders of an event so massive in its effects and so strange in its mechanical explanation that European man for a long time failed to digest its meaning. The same marks of that vanished episode—the closest thing to a real cataclysmic event that the planet has ever produced—stretch from sea to sea across temperate America. It is a curious coincidence in history that the moraines of the great continental ice fields should lie across those regions where men believed most implicitly in world deluges and the reality of incalculable violence. The co-incidence is not without meaning. Western Christian man, both in America and Europe, saw—where huge boulders lay scattered in windrows—the signs of the passage of a giant tidal wave.

European man was surrounded by sea. He knew its force. In no other manner could he account for those isolated and far-flung monuments of the past. His religious mythology made him even more ready to accept this version of events. "It is interesting to realize," remarked Philip Lake in 1930, on the hundredth anniversary of the publication of Lyell's *Principles of Geology*, "how strong was the case for a great debacle until the glacial deposits of northern Europe had been recognized as glacial." Ironically enough, man had begun to peer into the dark field of his origin, in just that region which offered the greatest hints of disturbance, of mystery, of an inviolate line across which it was impossible to pass.

"If species have changed by degrees," wrote Cuvier in 1815, "we ought to find traces of this gradual modification. We should be able to discover some intermediate forms; and yet no discovery has ever been made." After reviewing the then existing knowledge of fossil elephants and other creatures nearly allied to those of the present, Cuvier says of the coming of man that it "must necessarily have been posterior . . . to the revolutions which covered up these bones. No argument for the antiquity of the human race in those countries can be founded upon these fossil bones or upon the rocks by which they are covered." Humanity, Cuvier asserts, "did not exist in the countries in which the fossil bones of animals have been discovered."

In Cuvier's time, and for many years thereafter, during the reign of catastrophic geology, man's thoughts about his past would falter inevitably before those boulder-strewn wastes in which the human story seemed to end abruptly. Although Cuvier's single reservation was lost in later, more enthusiastically religious volumes, it can be noted that the master anatomist did have a moment of hesitation. "I do not presume," he said, "to conclude that man did not exist at all before these epochs. He may have then inhabited some narrow region, whence he went forth to re-people the earth after the cessation of these terrible revolutions. Perhaps even the places which he then inhabited may have sunk into the abyss."

As the pre-Darwinian years began to pass, men returned again and again to the contemplation of what seemed the shallow time-depths of the human species in Europe. Industrial activities here and there resulted in the disturbance of ancient strata. A few bones and crude implements faintly suggested that beneath historic Europe, the Europe of the Romans and Greeks, might lie an unknown, shadowy world in which the white race had cracked marrow bones by campfires as primitive as those of red Indians. It was not a nice thought to intrude into civilized minds. Most people rejected it. "Man," insisted Darwin's geology teacher, Sedgwick, "has been but a few years' dweller on the earth. He was called into being within a few thousand years of the days in which we live by a provident contriving power."

If any association of the bones of Cuvier's mammoths with men was brought forward, it was suggested that the human remains and the fossils had been accidentally mixed in later times. The first paleolithic archaeologists were apt to find themselves doubted by their more conservative colleagues. No one had ever seen such crude tools before. Were they really of human manufacture? Even Darwin, in his earlier years, had been dubious. To make things worse, workmen faked finds to gain shillings or francs from overeager investigators. Thus fraud confused the issue.

But the boulders still stretched in mysterious lines across Europe. Hutton's inspired guess about ice action had long since been forgotten. Even his intellectual descendant, Charles Lyell, speculated that the stones were transported by icebergs in greater seas. As Sir James Geikie, the great Scotch geologist, was later to observe, "Even a cautious thinker like Lyell saw less difficulty in sinking the whole of Central Europe under the sea and covering the waters with floating icebergs, than in conceiving that the Swiss glaciers were once large enough to reach to the Jura. Men shut their eyes to the meaning of the unquestionable fact that, while there was absolutely no evidence for a marine submergence, the former track of the glaciers could be followed mile after mile . . ." The episode is again an apt illustration of how difficult it is even for trained scientists to break out of a prevailing mode of thought.

II

In 1837, however, two remarkable discoveries occurred. They did not change the intellectual climate of the age overnight. Nevertheless they resulted, eventually, in a drastic revision of the ruling conception of human antiquity and of the nature of those boulder-strewn fields whose wastes had long tantalized and intimidated the inquiring human spirit.

The first of these events was the discovery of the fossil remains of great apes and other primates in the Tertiary beds of northern India. It had long been assumed that man's nearest anatomical relatives were totally contemporaneous with himself, and were, in catastrophist terms, part of the living creation which included man.

In 1830, Hugh Falconer, a young Scotch doctor with a deep interest in natural history, had gone out to India as an assistant surgeon in the service of the East India Company. He rapidly acquired a reputation as a scientist and in 1832 became superintendent of the Botanic Gardens at Suharunpoor, which lies about twenty-five miles from the Siwalik hills, an eroded Tertiary outlyer of the Himalayas. Here Falconer, with the assistance of army friends, began the paleontological explorations which were to lead to the discovery of one of the finest Tertiary fossil beds in the world. Falconer, with his friend Captain

Proby Cautley, established the age of the deposits and brought to light a subtropical mammalian fauna which excelled in richness any other Cenozoic fossil beds then known.

Although Cautley and Falconer made the first discovery of a fossil primate from these beds, they deferred publication because of the fragmentary nature of their discovery. In the meantime, their friends W. E. Baker and H. M. Durand, of the army engineers, located and reported upon a much larger species, equivalent in size to the orang. The specimen, they indicated, revealed "the existence of a gigantic species of quadrumanous animal contemporaneously with the Pachydermata of the sub-Himalayas and thus supplies . . . proof of the existence, in a fossil state, of the type of organization most nearly resembling that of man."

Here at last was more than a new fossil for the taxonomist. Here, in Falconer's words, was a mixture of the old and new together, "affording another illustration of constancy in the order of nature, of an identity of condition in the earth with what it exhibits now." Most scientists were not yet evolutionists, but the barrier to the past which they had encountered in Europe had been most dramatically pierced on the far-off flanks of the Himalayas. There "was clear evidence, physical and organic," Falconer reminisced later, "that the present order of things had set in from a very remote period in India. Every condition was suited to the requirements of man. The wide spread of the plains of India showed no signs of the unstratified superficial Gravels, Sands and Clays, which for a long time were confidently adduced as evidence that a great Diluvial wave had suddenly passed over Europe and other continents, overwhelming terrestrial life."

It was not in Europe, nor in North America, that the earliest relics of the human race were to be sought. Man in those regions, argued Falconer, was a creature of yesterday. It was in the great alluvial valleys of the tropics and subtropics that his earlier history would be found.

The second discovery which was drastically to alter our notions of the environment of prehistoric man was made, or rather reintroduced into science, in the same year that Falconer and his friends were busying themselves with the Siwalik primates. Louis Agassiz, whose remarkable researches upon the fossil fishes of Europe had brought him international acclaim, turned, at the age of thirty-three, to a study of glaciation. Before his time, save for the premonitory glimpses of a few men, including Hutton, who did not pursue the problem, the European and American ice erratics were accounted for in one of two ways: as the product of turbulent flood waters of oceanic proportions, or as the result of transportal by floating ice drifting over submerged areas.

Agassiz studied the effects of ice action in modern glaciers and observed the marks of scouring and engraving which no water action could duplicate. He was the first to recognize the enormous extent of the continental ice fields in both Europe and America. He grasped equally well the fact that such a widespread phenomenon must involve causes at work throughout the whole northern hemisphere.

The British, in particular, were loath to give up the conception of sea-borne ice, but in the end Agassiz carried the day and became, in the process, the founder of glacial geology. Flood-swept Europe became ice-locked Europe, though, as we have intimated, it was growing apparent that man—at least late glacial man—had managed to subsist under rude conditions in caves and rock shelters south of the ice desert. There had been no diluvial wave to carry him off.

The crude tools associated with the bones of extinct animals, which had been so long scorned, were read. There actually *was* a way through the doorway of the past. Paradoxically, Agassiz lived on to oppose Darwin and reiterate his belief in world-wide extinctions. Falconer, too, though more temperate and a warm friend of Darwin's, had reservations about the reality of evolution.

Nevertheless, the two men had opened new vistas. We have already examined the mechanism which Darwin had provided as a natural explanation of organic change. We will now want to observe the manner in which the Darwinian circle applied this conception to human evolution. The creation of natural man encountered unexpected obstacles, not all of which were provided by resentful theologians. Darwin and his followers, carrying over concepts which had proved useful in the other realms of biology, were not always judicious in their examination of man. Their biological triumph had been so complete, however, that they, in turn, had created an intellectual climate which accepted their views unquestioningly.

III

We have already had occasion to observe that to the catastrophist school of thought man was not the incidental product of variation in a ground-dwelling ape, but rather a creature foreordained and foreseen from the beginning of geologic time. When evolution began to be taken seriously in England, and it seemed inevitable that it would be extended to man, the latter's peculiarly human characteristics began to be scrutinized with care. The biologist sought to "animalize" man, even if he had to humanize the animal world in order to demonstrate a genetic connection.

The advocate of man's divinity, on the other hand, sought to identify human traits of which there was no trace in the surrounding organic world and thus to confront the Darwinists with a break, or discontinuity, in their system. This was apt to be particularly nettling, because Darwin had placed such emphasis upon slow, almost invisible change over vast time periods. It needs scarcely to be said that both sides were forced to argue almost entirely on theoretical grounds, for although human antiquity had been greatly extended after the full understanding of the ice age, only one extinct specimen of man had been doubtfully recognized, while none of the really subhuman links in the phylogeny of man were yet known. Human evolution, therefore, was for long debated upon largely abstract grounds. Indeed, the very lack of human fossils was occasionally pointed out as an argument against attaching man to the rest of the organic world. To many thinkers he remained a divine interposition in the universe.

I have said that the biologist of that day, lacking enough fossils to substantiate his points, sought to thrust living men and living animals closer together than the facts sometimes warranted. This was not a conscious attempt at deception, but again was the product of a particular intellectual climate. There was, for example, a tendency to see human evolution in terms of the eighteenth-century scale of nature as based on living forms. Thus one could begin with one of the existing great apes, the gorilla, orang or chimpanzee, as representing the earliest "human" stage, and from this pass to the existing human races, which were arranged in a vertical series with white Victorian man representing the summit of evolutionary ascent.

The fact that the existing apes are creatures contemporaneous with ourselves, who have evolved down another evolutionary pathway which is also highly specialized, was lost from sight. Even the contemporary races of man came to be regarded as living fossils. Today this whole approach to the human problem has been abandoned, but it was peculiarly attractive to many nineteenth-century thinkers. The past, so to speak, had been quietly transported into the present, and evolutionary roles—not always attractive ones—assigned to living actors without their consent. As a matter of fact, a curious twofold interpretation of the human psyche has descended from the Darwinian epoch into modern science. Darwin himself seems to have hesitated in his book, *The Descent of Man*, between a conception of man the warrior—the product of ruthless, competitive forces—and man the weak-bodied, unarmed primate who, until his intellectual powers were strengthened, might not have been able to survive on the ground in competition with the great carnivores.

Today it is frequently recalled, because of fossil discoveries made there in the last decade, that Darwin once suggested Africa as the possible original home of man. What is less often remembered is the fact that Darwin, in the same volume, suggested that man, even if far more helpless and defenseless than any existing savages, would not have been exposed to any special danger had he inhabited some warm continent or large island such as Australia, New Guinea or Borneo. The creation of a sort of Garden of Eden for early man seems to have been forced upon Darwin by the criticism of the Duke of Argyll, who pointed out that physically man is one of the most helpless creatures in the world.

The Duke raised the question, in the light of Darwin's and Huxley's emphasis on the struggle for existence, how, by their own philosophy, the human body could have diverged in the direction of greater helplessness in the early phase of its evolution. This criticism forced upon Darwin the wary evasion of his island Eden. "We do not know," he admitted, "whether man is descended from some small species, like the chimpanzee, or from one as powerful as the gorilla. Therefore we cannot say whether man has become larger and stronger or smaller and weaker than his ancestors." He confessed, however, that an animal as powerful as the gorilla might not have become social.

In spite of this hesitation as to the precise nature of early man, it is evident that the existing great apes played a large part in Darwinian thinking and that some gorilloid characters came to be projected upon the first Neanderthal fossils. Such characters seemed appropriate in the light of the Darwinian interpretation of the struggle for existence.

IV

Just a little over a hundred years ago two German quarrymen were digging in a small cave along the gorge of the Neander near Düsseldorf, Germany. In a dark interior chamber which was filled to a depth of four or five feet by earth which had been swept into the cave at some time in the long past, the workmen stumbled upon some strange bones. As the deposit of clay and stones was removed, the men came first upon a skull lying near the entrance of the grotto. Later, stretched along the floor of the cave, other skeletal remains appeared. With no idea that they had unearthed a human skeleton destined to become the scientific sensation of its time, these rough and unskilled quarrymen picked up some of the larger bones and tossed them out with the debris of their excavation.

Weeks later, what could be found of the broken and dispersed skeleton was placed in the hands of a local physician, Dr. Fuhlrott of Elberfeld. This individual, an enlightened collector and one of those early scientific pioneers of whom the medical profession affords so many examples, immediately dispatched the bones to a skilled anatomist, Professor Hermann Schaafhausen of Bonn University. Thus, three years before Charles Darwin gave to the world his theory of evolution, science found placed in its hands an important clue to the prehistoric past of the human race.

Today we know that this low-browed, thick-walled skull

vault, which its describer promptly characterized as "due to a natural conformation *hitherto not known to exist even in the most barbarous races*," belongs to an era far more remote than even Schaafhausen dreamed. This skull is man's first relic of the ice age—of a long-vanished world where men and women had endured glacial cold and had struggled for existence armed only with crude spears and sharpened flints.

When this dead man had been interred in the little cave by the gorge of the Neander, all the enormous triumphs of humanity, the piled wealth of our great cities, still lay tens of thousands of years in the future. On the glacial uplands of Europe, winds had howled and dust had swirled endlessly over naked grasslands. Only in sheltered gullies and occasional caves flickered the fires of the sparse human population, which lived by hunting mammoth through the phantom streets of future Paris and Berlin.

By Darwinian standards, these creatures were an odd and unimagined link with the past. Their skulls, in spite of jutting brow ridges and massive chinless faces, had brains as large as or larger than our own. Huxley, the swashbuckling evolutionist, hesitated over their meaning with the reluctance of a choirboy. Darwin saw them as armed with gorilloid fangs, and an artist pictured them with the grasping feet of apes. A distinguished anatomist spoke of them as "the quintessence of brute-benightedness."

These were the men who had scouted for game within a few miles of that great blue barrier which had once lain like a sea over all of northern Europe. These were the men whose skin and eye color we do not know, but who fought a battle for naked survival against cold and the prowling beasts thrust southward by the long inexorable advance of the continental ice field. Generation by generation life fell back before it. Ice grumbled across the North German plain; in the Alps the glaciers of the high altitudes moved downward, coalesced and blocked Italy away from the north. The vast northern sheet reduced the corridor between eastern Europe and the west to a narrow and freezing channel between two great ice masses.

Here then was the "ice island." On the south lay the barrier of the Mediterranean Sea; on the north was a blue-white silence which could be matched today only in the Antarctic; a place of winter unbroken save by the crack of ice and the fall of avalanches. Almost where the borders of the Iron Curtain run today, ran the ice curtain. Below it men moved, tiny and over-awed, surrounded and cut off from Asia and the world farther east. And still the ice came on, and still, warmed by the Atlantic waters and the Mediterranean, the little groups of naked hunters clung to life.

Season followed season. Men grubbed for roots and nuts and berries in the short, cool summers, and followed remorse-lessly upon the trail of animals heavy with young. The spear and cutting flints were their only weapons; the bow was as far away in the future as gunpowder from the bow. All killing was personal. If a thrust failed, men stood to the charge of big game with nothing but a wooden spear and a flint knife. In the camps were children to be fed—big-browed children, the sides of whose foreheads were beginning to roof out with the heavy bone of their strange fathers.

Somewhere, perhaps around the forty-thousand-year mark, the ice hesitated. A little more and the men of West Europe might have vanished. They huddled in the caves of France and along the warm Mediterranean shore, little bands composed of a few score people, tough and enduring. Illiterate, they had no knowledge of any other world. This was their life, the sound of ice in the mountains, the endless search for game.

Thousands of years had passed while the ice ground south-ward, thousands more while the great sheet thrust out explor-ing tongues into the western valleys, paused, and slowly, very slowly, began a slight retreat. All this time western man in scant numbers had lived alone and without contact. We have information which suggests that outside of Europe a branch of the old big-browed strain was modifying in the direction of today's human type. Here in West Europe, however, the caves of this period yield up only the massive-muzzled creature who seems, in the eyes of modern science, to be in some degree the creation of the ice.

We know two things. We know that wherever small popula-tions of any type of living thing are marooned as Neanderthal man was marooned, a process of genetic drift sets in. That is, mutations, little changes in the germ plasm, may run with comparative quickness over a whole group. If the new physical characters promote survival, so much the better. In any case, little groups of primitive men are always intimately related. The novelist Thomas Hardy expressed this in his poem about family likeness "leaping from place to place over oblivion." Even today, in a few regions in the mountain South, one can observe a physical likeness which is the product of local isolation over several generations. The people are interrelated; a local type has arisen and is revealed in a vague physical similarity. In West Europe such isolation continued for thousands of years. In addition, it may well be that something in that desperate struggle with the ice enhanced the value of these primitive characters.

Now the classic western Neanderthal type has been carefully analyzed. Such notions as those of the Darwinian years that our cave men had threatening gorilloid fangs or clasping feet have long since been abandoned. Nevertheless, these people who were living in the earlier part of the last ice advance in caves in France, Italy, Spain and Belgium are distinctly different from ourselves. Their skulls, which housed a brain in some instances larger than the modern average, were low, long and broad. Where the modern skull vault is high, Neanderthal man, by contrast, had a low but very wide skull. The eye orbits were large and overhung by a pronounced bar of bone suggestive of that which can be observed in a modern chimpanzee or gorilla. The face is massive, the jaw lacking the pointed chin of today.

The chest was barrel-like and the stature short—just slightly over five feet. The forearm was short and powerful, as was the lower leg. The over-all picture is that of a very powerful but economically built man—a creature selected for survival under conditions demanding great hardihood and physical strength sustained often on a minimum of food.

Nevertheless, the Neanderthal people reveal no such quick descent to beastdom as the nineteenth-century writers presumed with their lurid pictures of a creature "in the highest degree hideous and ferocious." Indeed, we have clear evidence that they buried their dead with offerings. There is evidence also that these men were capable of altruistic care of adults. The big-brained primitives who seem to characterize the upper reaches of the ice age in the end forced scholars to reassess the time involved in the human transformation and to extend it. It has become more and more evident with the passing years that the place of the human emergence, whether it be Darwin's fanged gorilla of unsocial nature or his Eden-like creature of comparative helplessness, lies much farther back in the time scale than the nineteenth century realized. The nature of the original animal-man is still a matter of some debate.

V

The likelihood of man's origin on some idyllic Bornean island has been abandoned long since. However man managed his transformation, it was achieved amidst the great mammals of the Old World land mass, and most probably in Africa. Here the discoveries of the last few decades have revealed small-bodied, upright-walking man-apes whose brains appear little, if at all, larger than those of the existing great apes. Whether they could speak is not known, but it now appears that they were capable of at least some crude tool-making capacities.

As might be expected, these bipedal apes are quite different from our arboreal cousins, the great apes of today. Their teeth are heavy-molared, but they lack completely the huge canine teeth with which the nineteenth century, using gorilla models, endowed our ancestors. In spite of a powerful jaw musculature the creatures are light-bodied and apparently pygmoid, compared with modern man.

In spite, also, of exaggerated guesses when the first massive jaws were recovered, the several forms of these creatures in no case indicate man's descent from a giant primate—quite the reverse, in fact. Instead, we seem to be confronted with a short-faced, big-jawed ape of quite moderate body dimensions and a brain qualitatively, if not quantitatively, superior to that of any existing anthropoid.

It is unlikely that all of the several species known became men. Some, indeed, are late enough in time that primitive men were probably already in existence. Surveyed as a whole, however, they suggest an early half-human or near-human level, revealing clearly that the creature who became man was a ground-dwelling, well-adapted ape long before his skull and brain underwent their final transformation.

Interestingly, the hesitation exhibited by Darwin over the

psychical nature of the human forerunner continues into these remote time depths of the lower ice age. Raymond Dart, one of the pioneer discoverers of these ape-men, regards them as successful carnivores and killers of big game. He sees them as brutal and aggressive primates, capable of killing their own offspring, and carrying in their genetic structure much of the sadism and cruelty still manifested by modern man. They are club-swingers par excellence, already ably balanced on their two feet, and the terror of everything around them.

One cannot help but feel, however, that Dart tends to minimize the social nature which even early man must have had to survive and care for young who needed ever more time for adaptation to group life. He paints a picture too starkly overshadowed by struggle to be quite believable. It would seem that into his paleontological studies has crept a touch of disillusionment and distaste which has been projected backward upon that wild era in which the human predicament began. This judgment is, of course, a subjective one. It reveals that the hesitations of Darwin's day, in spite of increased knowledge, follow man back into the past. Perhaps it is a matter of temperament on the part of the observer. Perhaps man has always been both saint and sinner—even in his raw beginnings.

As we press farther back in time, however, back until the long, desolate years of the fourth ice age lie somewhere far in the remote future, we come upon something almost unbelievable. We come upon man, near-man, "the bridge to man," estimated as over seven hundred thousand years removed from us, in the Olduvai Gorge in Tanganyika. He is a step beyond Dart's man-apes, according to Dr. L. S. B. Leakey, his discoverer. Like them, he is small, huge-jawed, with a sagittal crest like that of a gorilla, but a true shaper of tools. Unlike Dart, Leakey describes his specimen as a semivegetarian eater of nuts, small rodents and lizards. Perhaps massive jaws and molar teeth, just as in the case of the living gorilla, were developed for other food than flesh.

The thing which appears the strangest of any news to come down from that far epoch, however, is the report from London that the youthful ape-man's body had apparently been carefully protected from scavenging hyenas until rising lake waters buried him among his tools.

Three years ago, in a symposium on the one hundredth anniversary of the discovery of Neanderthal man, I made these remarks: "When we consider this creature of 'brute benightedness' and 'gorilloid ferocity,' as most of those who peered into that dark skull vault chose to interpret what they saw there, let us remember what was finally revealed at the little French cave near La Chapelle-aux-Saints in 1908. Here, across millennia of time, we can observe a very moving spectacle. For these men whose brains were locked in a skull foreshadowing the ape, these men whom scientists had contended to possess no thoughts beyond those of the brute, had laid down their dead in grief.

"Massive flint-hardened hands had shaped a sepulcher and placed flat stones to guard the dead man's head. A haunch of meat had been left to aid the dead man's journey. Worked flints, a little treasure of the human dawn, had been poured lovingly into the grave. And down the untold centuries the message had come without words: 'We too were human, we too suffered, we too believed that the grave is not the end. We too, whose faces affright you now, knew human agony and human love.'

"It is important to consider," I said then, "that across fifty thousand years nothing has changed or altered in that act. It is the human gesture by which we know a man, though he looks out upon us under a brow reminiscent of the ape."

If the London story is correct, an aspect of that act has now been made distant from us by almost a million years. The creature who made it could only be identified by specialists as human. He is far more distant from Neanderthal man than the latter is from us.

Man, bone by bone, flint by flint, has been traced backward into the night of time more successfully than even Darwin dreamed. He has been traced to a creature with an almost gorilloid head on the light, fast body of a still completely upright, plains-dwelling creature. In the end he partakes both of Darwinian toughness, resilience, and something else, a humanity —if this story is true—that runs well nigh as deep as time itself.

Man has, in scientific terms, become natural, but the nature of his "naturalness" escapes him. Perhaps his human freedom has left him the difficult choice of determining what it is in his nature to be. Perhaps the two sides of the dark question Darwin speculated upon were only an evolutionary version of man's ancient warfare with himself—a drama as great in its hidden fashion as the story of the Garden and the Fall.

FIVE

How Human Is Man?

Be not under any Brutal Metempsychosis while thou livest and walkest about erectly under the scheme of Man.
—SIR THOMAS BROWNE

O VER A HUNDRED years ago a Scandinavian philosopher, Sören Kierkegaard, made a profound observation about the future. Kierkegaard's remark is of such great, though hidden, importance to our subject that I shall begin by quoting his words. "He who fights the future," remarked the philosopher, "has a dangerous enemy. The future is not, it borrows its strength from the man himself, and when it has tricked him out of this, then it appears outside of him as the enemy he must meet."

We in the western world have rushed eagerly to embrace the future—and in so doing we have provided that future with a strength it has derived from us and our endeavors. Now, stunned, puzzled and dismayed, we try to withdraw from the embrace, not of a necessary tomorrow, but of that future which we have invited and of which, at last, we have grown perceptibly afraid. In a sudden horror we discover that the years now rushing upon us have drained our moral resources and have taken shape out of our own impotence. At this moment, if we possess even a modicum of reflective insight, we will give heed to Kierkegaard's concluding wisdom: "Through the eternal," he enjoins us, "we can conquer the future."

The advice is cryptic; the hour late. Moreover, what have we to do with the eternal? Our age, we know, is littered with the wrecks of war, of outworn philosophies, of broken faiths. We profess little but the new and study only change.

Three hundred years have passed since Galileo, with the telescope, opened the enormous vista of the night. In those three centuries the phenomenal world, previously explored with the unaided senses, has undergone tremendous alteration in our minds. A misty light so remote as to be scarcely sensed by the unaided eye has become a galaxy. Under the microscope the previously unseen has become a cosmos of both beautiful and repugnant life, while the tissues of the body have been resolved into a cellular hierarchy whose constituents mysteriously produce the human personality.

Similarly, the time dimension, by the use of other sensory extensions and the close calculations made possible by our improved knowledge of the elements, has been plumbed as never before, and even its dead, forgotten life has been made to yield remarkable secrets. The great stage, in other words, the world stage where the Elizabethans saw us strutting and mouthing our parts, has the skeletons of dead actors under the floor boards, and the dusty scenery of forgotten dramas lies abandoned in the wings. The idea necessarily comes home to us then with a sudden chill: What if we are not playing on the center stage? What if the Great Spectacle has no terminus and no meaning? What if there is no audience beyond the footlights, and the play, in spite of bold villains and posturing heroes, is a shabby repeat performance in an echoing vacuity? Man is a perceptive animal. He hates above all else to appear ridiculous. His explorations of reality in the course of just three hundred years have so enlarged his vision and reduced his ego that his tongue sometimes fumbles for the proper lines to speak, and he plays his part uncertainly, with one dubious eye cast upon the dark beyond the stage lights. He is beginning to feel alone and to hear nothing but echoes reverberating back.

It will do no harm then, if in this moment of hesitation we survey the history of our dilemma. Man's efforts to understand his predicament can be compassed in the simple mechanics of the theatre. We have examined the time allowed the play, the nature of the stage, and what appears to be the nature of the plot. All else is purely incidental to this drama, and it may well be that we can see our history in no other terms, being mentally structured to look within as well as without, and to be influenced within by what we consider the nature of the "without" to be. It is for this reason that the "without," and our modes of apprehending it, assume so pressing an importance. Nor is it fully possible to understand the human drama, the drama of the great stage, without a historical knowledge of how the characters have interpreted their parts in the play, and in doing so perhaps affected the nature of the plot itself.

This, in brief, epitomizes the role of the human mind in history. It has looked through many spectacles in the last several centuries, and each time the world has appeared real, and the plot has been played accordingly. Strange colorings have been given to reality and the colors have come mostly from within. As science extends itself, the colors, and through them the nature of reality, continue to change. The "within" and "without" are in some strange fashion intermingled. Perhaps, in a sense, the great play is actually a great magic, and we, the players, are a part of the illusion, making and transforming the plot as we go.

If the play has its magical aspect, however, there is an increasing malignancy about it. A great Russian novelist ventures to remark mildly that the human heart, rather than the state, is the final abode of goodness. He is immediately denounced by his colleagues as a heretic. In the West, psychological studies are made of human "rigidity," and although there is a dispassionate scientific air about them the suggestion lingers that the "normal" man should conform; that the deviant is pathological. The television networks seek the lowest denominator which will entrance their mass audience. There is a muted intimation that we can do without the kind of intellectual individualists who used to declaim along the edges of the American wilderness and who have left the world some highly explosive literature in the shape of *Walden* and *Moby Dick*. It is obvious that the whole of western ethic, whether Russian or American, is undergoing change, and that the change is increasingly toward conformity in exterior observance and, at the same time, toward confusion and uncertainty in deep personal relations. In our examination of this phenomenon there will emerge for us the meaning of Kierkegaard's faith in the eternal as the only way of achieving victory against the corrosive power of the human future.

II

If we examine the living universe around us which was before man and may be after him, we find two ways in which that universe, with its inhabitants, differs from the world of man: first, it is essentially a stable universe; second, its inhabitants are intensely concentrated upon their environment. They respond to it totally, and without it, or rather when they relax from it, they merely sleep. They reflect their environment but they do not alter it. In Browning's words, "It has them, not they it."

Life, as manifested through its instincts, demands a security guarantee from nature that is largely forthcoming. All the release mechanisms, the instinctive shorthand methods by which nature provides for organisms too simple to comprehend their environment, are based upon this guarantee. The inorganic world could, and does, exist in a kind of chaos, but before life can peep forth, even as a flower, or a stick insect, or a beetle, it has to have some kind of unofficial assurance of nature's stability, just as we have read that stability of forces in the ripples impressed in stone, or the rain marks on a long-vanished beach, or the unchanging laws of light in the eye of a four-hundred-million-year-old trilobite.

The nineteenth century was amazed when it discovered these things, but wasps and migratory birds were not. They had an old contract, an old promise, never broken till man began to interfere with things, that nature, in degree, is steadfast and continuous. Her laws do not deviate, nor the seasons come and go too violently. There is change, but throughout the past life alters with the slow pace of geological epochs. Calcium, iron, phosphorus, could exist in the jumbled world of the inorganic without the certainties that life demands. Taken up into a living system, however, *being* that system, they must, in a sense, have knowledge of the future. Tomorrow's rain may be important, or tomorrow's wind or sun. Life, in contrast to the inorganic, is historic in a new way. It reflects the past, but must also expect something of the future. It has nature's promise—a guarantee that has not been broken in three billion years—that the universe has this queer rationality and "expectedness" about it. "Whatever interrupts the even flow and luxurious monotony of organic life," wrote Santayana, "is odious to the primeval animal."

This is a true observation, because on the more simple levels of life, monotony is a necessity for survival. The life in pond and thicket is not equipped for the storms that shake the human world. Its small domain is frequently confined to a splinter of sunlight, or the hole under a root. What life does under such circumstances, how it meets the precarious future (for even here the future can be precarious), is written into its substance by the obscure mechanisms of nature. The snail recoils into his house, the dissembling caterpillar who does not know he dissembles, thrusts stiffly, like a budding twig, from his branch. The enemy is known, the contingency prepared for. But still the dreaming comes from below, from somewhere in the molecular substance. It is as if nature in a thousand forms played games against herself, but the games were each one known, the rules ancient and observed.

It is with the coming of man that a vast hole seems to open in nature, a vast black whirlpool spinning faster and faster, consuming flesh, stones, soil, minerals, sucking down the lightning, wrenching power from the atom, until the ancient sounds of nature are drowned in the cacophony of something which is no longer nature, something instead which is loose and knocking at the world's heart, something demonic and no longer planned—escaped, it may be—spewed out of nature, contending in a final giant's game against its master.

Yet the coming of man was quiet enough. Even after he arrived, even after his strange retarded youth had given him the brain which opened up to him the dimensions of time and space, he walked softly. If, as was true, he had sloughed instinct away for a new interior world of consciousness, he did something which at the same time revealed his continued need for the stability which had preserved his ancestors. Scarcely had he stepped across the border of the old instinctive world when he began to create the world of custom. He was using reason, his new attribute, to remake, in another fashion, a substitute for the lost instinctive world of nature. He was, in fact, creating another nature, a new source of stability for his conflicting erratic reason. Custom became fixed: order, the new order imposed by cultural discipline, became the "nature" of human society. Custom directed the vagaries of the will. Among the fixed institutional bonds of society man found once more the security of the animal. He moved in a patient renewed orbit with the seasons. His life was directed, the gods had ordained it so. In some parts of the world this long twilight, half in and half out of nature, has persisted into the present. Viewed over a wide domain of history this cultural edifice, though somewhat less stable than the natural world, has yet appeared a fair substitute—a structure, like nature, reasonably secure. But the security in the end was to prove an illusion. It was in the West that the whirlpool began to spin. Ironically, it began in the search for the earthly Paradise.

The medieval world was limited in time. It was a stage upon which the great drama of the human Fall and Redemption was being played out. Since the position in time of the medieval culture fell late in this drama, man's gaze was not centered scientifically upon the events of an earth destined soon to vanish. The ranks of society, even objects themselves, were Platonic reflections from eternity. They were as unalterable as the divine Empyrean world behind them. Life was directed and fixed from above. So far as the Christian world of the West was concerned, man was locked in an unchanging social structure well nigh as firm as nature. The earth was the center of divine attention. The ingenuity of intellectual men was turned almost exclusively upon theological problems.

As the medieval culture began to wane toward its close, men turned their curiosity upon the world around them. The era of the great voyages, of the breaking through barriers, had begun. Indeed, there is evidence that among the motivations of those same voyagers, dreams of the recovery of the earthly Paradise were legion. The legendary Garden of Eden was thought to be still in existence. There were stories that in this or that far land, behind cloud banks or over mountains, the abandoned Garden still survived. There were speculations that through one of those four great rivers which were supposed to flow from the Garden, the way back might still be found. Perhaps the angel with the sword might still be waiting at the weed-grown gateway, warning men away; nevertheless, the idea of that haven lingered wistfully in the minds of captains in whom the beliefs of the Middle Ages had not quite perished.

There was, however, another, a more symbolic road into the Garden. It was first glimpsed and the way to its discovery charted by Francis Bacon. With that act, though he did not intend it to be so, the philosopher had opened the doorway of the modern world. The paradise he sought, the dreams he dreamed, are now intermingled with the countdown on the latest model of the ICBM, or the radioactive cloud drifting downwind from a megaton explosion. Three centuries earlier, however, science had been Lord Bacon's road to the earthly Paradise. "Surely," he wrote in the *Novum Organum*, "it would be disgraceful if, while the regions of the material globe, that is, of the earth, of the sea, and of the stars—have been in our times laid widely open and revealed, the intellectual globe should remain shut up within the narrow limits of the old discoveries."

Instead, Bacon chafed for another world than that of the restless voyagers. "I am now therefore to speak touching Hope," he rallied his audience, who believed, many of them, in a declining and decaying world. Much, if not all, that man lost in his ejection from the earthly Paradise might, Bacon thought, be regained by application, so long as the human intellect remained unimpaired. "Trial should be made," he contends in one famous passage, "whether the commerce between the mind of men and the nature of things . . . might by any means be restored to its perfect and original condition, or if that may not be, yet reduced to a better condition than that in which it now is." To the task of raising up the new science he devoted himself as the bell ringer who "called the wits together."

Bacon was not blind to the dangers in his new philosophy. "Through the premature hurry of the understanding," he cautioned, "great dangers may be apprehended . . . against which we ought even now to prepare." Out of the same fountain, he saw clearly, could pour the instruments of beneficence or death.

Bacon's warning went unheeded. The struggle between those forces he envisaged continues into the modern world. We have now reached the point where we must look deep into the whirlpool of the modern age. Whirlpool or flight, as Max Picard has called it, it is all one. The stability of nature on the planet—that old and simple promise to the living, which is written in every sedimentary rock—is threatened by nature's own product, man.

Not long ago a young man—I hope not a forerunner of the coming race on the planet—remarked to me with the colossal insensitivity of the new asphalt animal, "Why can't we just eventually kill off everything and live here by ourselves with more room? We'll be able to synthesize food pretty soon." It was his solution to the problem of overpopulation.

I had no response to make, for I saw suddenly that this man was in the world of the flight. For him there was no eternal, nature did not exist save as something to be crushed, and that second order of stability, the cultural world, was, for him, also ceasing to exist. If he meant what he said, pity had vanished, life was not sacred, and custom was a purely useless impediment from the past. There floated into my mind the penetrating statement of a modern critic and novelist, Wright Morris. "It is not fear of the bomb that paralyzes us," he writes, "not fear that man has no future. Rather, it is the *nature* of the future, not its extinction, that produces such foreboding in the artist. It is a numbing apprehension that such future as man has may dispense with art, with man as we now know him, and such as art has made him. The survival of men who are strangers to the nature of this conception is a more appalling thought than the extinction of the species."

There before me stood the new race in embryo. It was I who fled. There was no means of communication sufficient to call across the roaring cataract that lay between us, and down which this youth was already figuratively passing toward some doom I did not wish to see. Man's second rock of certitude, his cultural world, that had gotten him out of bed in the morning for many thousand years, that had taught him manners, how to love, and to see beauty, and how, when the time came, to die—this cultural world was now dissolving even as it grew. The roar of jet aircraft, the ugly ostentation of badly designed automobiles, the clatter of the supermarkets could not lend stability nor reality to the world we face.

Before us is Bacon's road to Paradise after three hundred years. In the medieval world, man had felt God both as exterior lord above the stars, and as immanent in the human heart. He was both outside and within, the true hound of Heaven. All this alters as we enter modern times. Bacon's world to explore opens to infinity, but it is the world of the outside. Man's whole attention is shifted outward. Even if he looks within, it is largely with the eye of science, to examine what can be learned of the personality, or what excuses for its behavior can be found in the darker, ill-lit caverns of the brain.

The western scientific achievement, great though it is, has not concerned itself enough with the creation of better human beings, nor with self-discipline. It has concentrated instead upon things, and assumed that the good life would follow. Therefore it hungers for infinity. Outward in that infinity lies the Garden the sixteenth-century voyagers did not find. We no longer call it the Garden. We are sophisticated men. We call it, vaguely, "progress," because that word in itself implies the endless movement of pursuit. We have abandoned the past without realizing that without the past the pursued future has no meaning, that it leads, as Morris has anticipated, to the world of artless, dehumanized man.

III

Some time ago there was encountered, in the litter of a vacant lot in a small American town, a fallen sign. This sign was intended to commemorate the names of local heroes who had fallen in the Second World War. But that war was over, and another had come in Korea. Probably the population of that entire town had turned over in the meantime. Tom and Joe and Isaac were events of the past, and the past of the modern world is short. The names of yesterday's heroes lay with yesterday's torn newspaper. They had served their purpose and were now forgotten.

This incident may serve to reveal the nature of what has happened, or seems to be happening, to our culture, to that world which science was to beautify and embellish. I do not say that science is responsible except in the sense that men are responsible, but men increasingly are the victims of what they themselves have created. To the student of human culture, the rise of science and its dominating role in our society presents a unique phenomenon.

Nothing like it occurs in antiquity, for in antiquity nature represented the divine. It was an object of worship. It contained mysteries. It was the mother. Today the phrase has disappeared. It is nature we shape, nature, without the softening application of the word mother, which under our control and guidance hurls the missile on its path. There has been no age in history like this one, and men are increasingly brushed aside who speak of the possibility of another road into the future.

Some time ago, in a magazine of considerable circulation, I spoke about the role of love in human society, and about pressing human problems which I felt, rightly or wrongly, would not be solved by the penetration of space. The response amazed me, in some instances, by its virulence. I was denounced for interfering with the colonization of other planets, and for corruption of the young. Most pathetically of all, several people wrote me letters in which they tried to prove, largely for their own satisfaction, that love did not exist, that parents abused and murdered their children and children their parents. They concentrated upon sparse incidents of pathological violence, and averted their eyes from the normal.

It was all too plain that these individuals were seeking rationalizations behind which they might hide from their own responsibilities. They were in the whirlpool, that much was evident. But so are we all. In 1914 the London *Times* editorialized confidently that no civilized nation would bomb open cities from the air. Today there is not a civilized nation on the face of the globe that does not take this aspect of warfare for granted. Technology demands it. In Kierkegaard's deadly future man strives, or rather ceases to strive, against himself.

But crime, moral deficiencies, inadequate ethical standards, we are prone to accept as part of the life of man. Why, in this respect, should we be regarded as unique? True, we have had Buchenwald and the Arctic slave camps, but the Romans had their circuses. It is just here, however, that the uniqueness enters in. After the passage of three hundred years from Bacon and his followers—three hundred years on the road to the earthly Paradise—there is a rising poison in the air. It crosses frontiers and follows the winds across the planet. It is manmade; no treaty of the powers has yet halted it.

Yet it is only a symbol, a token of that vast maelstrom which has caught up states and stone-age peoples equally with the modern world. It is the technological revolution, and it has brought three things to man which it has been impossible for him to do to himself previously.

First, it has brought a social environment altering so rapidly with technological change that personal adjustments to it are frequently not viable. The individual either becomes anxious and confused or, what is worse, develops a superficial philosophy intended to carry him over the surface of life with the least possible expenditure of himself. Never before in history has it been literally possible to have been born in one age and to die in another. Many of us are now living in an age quite different from the one into which we were born. The experience is not confined to a ride in a buggy, followed in later years by a ride in a Cadillac. Of far greater significance are the social patterns and ethical adjustments which have followed fast upon the alterations in living habits introduced by machines.

Second, much of man's attention is directed exteriorly upon the machines which now occupy most of his waking hours. He has less time alone than any man before him. In dictator-controlled countries he is harangued and stirred by propaganda projected upon him by machines to which he is forced to listen. In America he sits quiescent before the flickering screen in the living room while horsemen gallop across an American wilderness long vanished in the past. In the presence of so compelling an instrument, there is little opportunity in the evenings to explore his own thoughts or to participate in family living in the way that the man from the early part of the century remembers. For too many men, the exterior world with its mass-produced daydreams has become the conqueror. Where are the eager listeners who used to throng the lecture halls; where are the workingmen's intellectual clubs? This world has vanished into the whirlpool.

Third, this outward projection of attention, along with the rise of a science whose powers and creations seem awe-inspiringly remote, as if above both man and nature, has come dangerously close to bringing into existence a type of man who is not human. He no longer thinks in the old terms; he has ceased to have a conscience. He is an instrument of power. Because his mind is directed outward upon this power torn from nature, he does not realize that the moment such power is brought into the human domain it partakes of human freedom. It is no longer safely *within* nature; it has become violent, sharing in human ambivalence and moral uncertainty.

At the same time that this has occurred, the scientific worker has frequently denied personal responsibility for the way his discoveries are used. The scientist points to the evils of the statesmen's use of power. The statesmen shrug and remind the scientist that they are encumbered with monstrous forces that science has unleashed upon a totally unprepared public. But there are few men on either side of the Iron Curtain able to believe themselves in any sense personally responsible for this situation. Individual conscience lies too close to home, and is archaic. It is better, we subconsciously tell ourselves, to speak of inevitable forces beyond human control. When we reason thus, however, we lend powers to the whirlpool; we bring nearer the future which Kierkegaard saw, not as the *necessary* future, but one just as inevitable as man has made it.

IV

We have now glimpsed, however briefly and inadequately, the fact that modern man is being swept along in a stream of things, giving rise to other things, at such a pace that no substantial ethic, no inward stability, has been achieved. Such stability as survives, such human courtesies as remain, are the remnants of an older Christian order. Daily they are attenuated. In the name of mass man, in the name of unionism, for example, we have seen violence done and rudeness justified. I will not argue the justice or injustice of particular strikes. I can only remark that the violence to which I refer has been the stupid, meaningless violence of the rootless, nonhuman members of the age that is close to us. It is the asphalt man who defiantly votes the convicted labor boss back into office and who says: "He gets me a bigger pay check. What do I care what he does?" This is a growing aspect of modern society that runs from teen-age gangs to the corporation boards of amusement industries that deliberately plan the further debauchment of public taste.

It is, unfortunately, the "ethic" of groups, not of society. It cannot replace personal ethic or a sense of personal responsibility for society at large. It is, in reality, group selfishness, not ethics. In the words of Max Picard, "Spirit has been divided, fragmented; here is a spirit belonging to this and to that sociological group, each group having its own peculiar little spirit, exactly what one needs in the Flight, where, in order to flee more easily, one breaks the whole up into parts; and as always happens when one separates the part from the whole . . . one magnifies the tiny part, making it ridiculously important, so that no one may notice that the tiny part is not the whole."

All over the world this fragmentation is taking place. Small nationalisms, as in Cyprus or Algiers, murder in the name of freedom. In America, child gangs battle in the streets. The group ethic as distinct from personal ethic is faceless and obscure. It is whatever its leaders choose it to mean; it destroys the innocent and justifies the act in terms of the future. In Russia this has been done on a colossal scale. The future is no more than the running of the whirlpool. It is not divinely ordained. It has been wrought by man in ignorance and folly. That folly has two faces; one is our secularized conception of progress; the other is the mass loss of personal ethic as distinguished from group ethic.

It would be idle to deny that progress has its root in the Christian ethic, or that history, viewed as progression toward a goal, as unique rather than cyclic, is also a product of Christian thinking. There is a sense in which one can say that man entered into history through Christianity, for as Berdyaev somewhere observes, it is this religion, par excellence, which took God out of nature and elevated man above nature. The struggle for the realization of the human soul, the attempt to lift it beyond its base origins, became, in the earlier Christian centuries, the major preoccupation of the Church.

When science developed, in the hands of Bacon and his followers, the struggle for progress ceased to be an interior struggle directed toward the good life in the soul of the individual. Instead, the enormous success of the experimental method focused attention upon the power which man could exert over nature. Now he found, through Bacon's road back to the Garden, that he could share once more in that fruit of the legendary tree. With the rise of industrial science, "progress" became the watchword of the age, but it was a secularized progress. It was the increasing whirlpool of goods, cannon, bodies and yielded-up souls that an outward concentration upon the mastery of material nature was sure to bring.

Let us admit at once that the interpretation of secular

progress is two-sided. If this were not so, men would more easily recognize their dilemma. Science has brought remedies for physical pain and disease; it has opened out the far fields of the universe. Gross superstition and petty dogmatism have withered under its glance. It has supplied us with fruits unseen in nature, and given an opportunity, has told us dramatically of the paradise that might be ours if we could struggle free of ancient prejudices that still beset us. No man can afford to ignore this aspect of science, no man can evade those haunting visions.

It is the roar of the whirlpool, nonetheless, that breaks now most constantly upon our listening ears, increasingly instructing us upon the most important aspect of progress—that which in secularizing the concept we have forgotten. Its sound marks the dangerous near-dissolution of man's second nature, custom. Ideas, heresies, run like wildfire and death over the crackling static of the air. They no longer pick their way slowly through the experience of generations. Tax burdens multiply and reach upward year by year as man pays for his engines of death and underwrites ever more wearily the cost of the "progress" to which this road has led him. There is no retreat. The great green forest that once surrounded us Americans and behind which we could seek refuge has been consumed.

And thus, though more symbolically, has it been everywhere for man. We have re-entered nature, not like a Greek shepherd on a hillside hearing joyfully the returning pipes of Pan, but rather as an evil and precocious animal who slinks home in the night with a few stolen powers. The serenity of the gods is not disturbed. They know well on whose head the final lightning will fall.

Progress secularized, progress which pursues only the next invention, progress which pulls thought out of the mind and replaces it with idle slogans, is not progress at all. It is a beckoning mirage in a desert over which stagger the generations of men. Because man, each individual man among us, possesses his own soul and by that light must live or perish, there is no way by which Utopias—or the lost Garden itself—can be brought out of the future and presented to man. Neither can he go forward to such a destiny. Since in the world of time every man lives but one life, it is in himself that he must search for the secret of the Garden. With the fading of religious emphasis and the growth of the torrent, modern man is confused. The tumult without has obscured those voices that still cry desperately to man from somewhere within his consciousness.

V

One hundred years ago last autumn, Charles Darwin published the *Origin of Species*. Epic of science though it is, it was a great blow to man. Earlier, man had seen his world displaced from the center of space; he had seen the Empyrean heaven vanish to be replaced by a void filled only with the wandering dust of worlds; he had seen earthly time lengthen until man's duration within it was only a small whisper on the sidereal clock. Finally, now, he was to be taught that his trail ran backward until in some lost era it faded into the night-world of the beast. Because it is easier to look backward than to look forward, since the past is written in the rocks, this observation, too, was added to the whirlpool.

"I am an animal," man considered. It was a fair judgment, an outside judgment. Man went into the torrent along with the steel of the first ironclads and a new slogan, "the survival of the fittest." There would be one more human retreat when, in the twentieth century, human values themselves would fall under scrutiny and be judged relative, shifting and uncertain. It is the way of the torrent—everything touched by it begins to circle without direction. All is relative, there is nothing fixed, and of guilt there can, of course, remain but little. Moral responsibility has difficulty in existing consistently beside the new scientific determinism.

I remarked at the beginning of this discussion that the "within," man's subjective nature, and the things that come to him from without often bear a striking relationship. Man cannot be long studied as an object without his cleverly altering his inner defenses. He thus becomes a very difficult creature with which to deal. Let me illustrate this concretely.

Not long ago, hoping to find relief from the duties of my office, I sought refuge with my books on a campus bench. In a little while there sidled up to me a red-faced derelict whose parboiled features spoke eloquently of his particular weakness. I was resolved to resist all blandishments for a handout.

"Mac," he said, "I'm out of a job. I need help."

I remained stonily indifferent.

"Sir," he repeated.

"Uh huh," I said, unwillingly looking up from my book.

"I'm an alcoholic."

"Oh," I said. There didn't seem to be anything else to say. You can't berate a man for what he's already confessed to you.

"I'm an alcoholic," he repeated. "You know what that means? I'm a sick man. Not giving me alcohol is ill-treating a sick man. I'm a sick man. I'm an alcoholic. I have to have a drink. I'm telling you honest. It's a disease. I'm an alcoholic. I can't help myself."

"Okay," I said, "you're an alcoholic." Grudgingly I contributed a quarter to his disease and his increasing degradation. But the words stayed in my head. "I can't help myself." Let us face it. In one disastrous sense, he was probably right. At least at this point. But where had the point been reached, and when had he developed this clever neo-modern, post-Freudian panhandling lingo? From what judicious purloining of psychiatric or social-work literature and lectures had come these useful phrases?

And he had chosen his subject well. At a glance he had seen from my book and spectacles that I was susceptible to this approach. I was immersed in the modern dilemma. I could have listened, gazing into his mottled face, without an emotion if he had spoken of home and mother. But he was an alcoholic. He knew it and he guessed that I might be a scientist. He had to be helped from the outside. It was not a moral problem. He was ill.

I settled uncomfortably into my book once more, but the phrase stayed with me, "I can't help myself." The clever reversal. The outside judgment turned back and put to dubious, unethical use by the man inside.

"I can't help myself." It is the final exteriorization of man's moral predicament, of his loss of authority over himself. It is the phrase that, above all others, tortures the social scientist. In it is truth, but in it also is a dreadful, contrived folly. It is society, a genuinely sick society, saying to its social scientists, as it says to its engineers and doctors: "Help me. I'm rotten with hate and ignorance that I won't give up, but you are the doctor; fix me." This, says society, is *our* duty. We are social scientists. Individuals, poor blighted specimens, cannot assume such responsibilities. "True, true," we mutter as we read the case histories. "Life is dreadful, and yet—"

Man on the inside is quick to accept scientific judgments and make use of them. He is conditioned to do this. This new judgment is an easy one; it deadens man's concern for himself. It makes the way into the whirlpool easier. In spite of our boasted vigor we wait for the next age to be brought to us by Madison Avenue and General Motors. We do not prepare to go there by means of the good inner life. We wait, and in the meantime it slowly becomes easier to mistake longer cars or brighter lights for progress. And yet—

Forty thousand years ago in the bleak uplands of southwestern Asia, a man, a Neanderthal man, once labeled by the Darwinian proponents of struggle as a ferocious ancestral beast—a man whose face might cause you some slight uneasiness if he sat beside you—a man of this sort existed with a fearful body handicap in that ice-age world. He had lost an arm. But still he lived and was cared for. Somebody, some group of human things, in a hard, violent and stony world, loved this maimed creature enough to cherish him.

And looking so, across the centuries and the millennia, toward the animal men of the past, one can see a faint light, like a patch of sunlight moving over the dark shadows on a forest floor. It shifts and widens, it winks out, it comes again, but it persists. It is the human spirit, the human soul, however transient, however faulty men may claim it to be. In its coming man had no part. It merely came, that curious light, and man, the animal, sought to be something that no animal had been before. Cruel he might be, vengeful he might be, but there had entered into his nature a curious wistful gentleness and courage. It seemed to have little to do with survival, for such men died over and over. They did not value life compared to what they saw in themselves—that strange inner light which has come from no man knows where, and which was not made by us. It has followed us all the way from the age of ice, from the dark borders of the ancient forest into which our footprints vanish. It is in this that Kierkegaard glimpsed the eternal, the way of the heart, the way of love which is not of today, but is of the whole journey and may lead us at last to the end. Through this, he thought, the future may be conquered. Certainly it is true. For man may grow until he towers to the skies, but without this light he is nothing, and his place is nothing. Even as we try to deny the light, we know that it has made us, and what we are without it remains meaningless.

We have come a long road up from the darkness, and it well may be—so brief, even so, is the human story—that viewed in the light of history, we are still uncouth barbarians. We are potential love animals, wrenching and floundering in our larval envelopes, trying to fling off the bestial past. Like children or savages, we have delighted ourselves with technics. We have thought they alone might free us. As I remarked before, once launched on this road, there is no retreat. The whirlpool can be conquered, but only by placing it in proper perspective. As it grows, we must learn to cultivate that which must never be permitted to enter the maelstrom—ourselves. We must never accept utility as the sole reason for education. If all knowledge is of the outside, if none is turned inward, if self-awareness fades into the blind acquiescence of the mass man, then the personal responsibility by which democracy lives will fade also.

Schoolrooms are not and should not be the place where man learns only scientific techniques. They are the place where selfhood, what has been called "the supreme instrument of knowledge," is created. Only such deep inner knowledge truly expands horizons and makes use of technology, not for power, but for human happiness. As the capacity for self-awareness is intensified, so will return that sense of personal responsibility which has been well-nigh lost in the eager yearning for aggrandizement of the asphalt man. The group may abstractly desire an ethic, theologians may preach an ethic, but no group ethic ever could, or should, replace the personal ethic of individual, responsible men. Yet it is just this which the Marxist countries are seeking to destroy; and we, in a vague, good-natured indifference, are furthering its destruction by our concentration upon material enjoyment and our expressed contempt for the man who thinks, to our mind, unnecessarily.

Let it be admitted that the world's problems are many and wearing, and that the whirlpool runs fast. If we are to build a stable cultural structure above that which threatens to engulf us by changing our lives more rapidly than we can adjust our habits, it will only be by flinging over the torrent a structure as taut and flexible as a spider's web, a human society deeply self-conscious and undeceived by the waters that race beneath it, a society more literate, more appreciative of human worth than any society that has previously existed. That is the sole prescription, not for survival—which is meaningless—but for a society worthy to survive. It should be, in the end, a society more interested in the cultivation of noble minds than in change.

There is a story about one of our great atomic physicists— a story for whose authenticity I cannot vouch, and therefore I will not mention his name. I hope, however, with all my heart that it is true. If it is not, then it ought to be, for it illustrates well what I mean by a growing self-awareness, a sense of responsibility about the universe.

This man, one of the chief architects of the atomic bomb, so the story runs, was out wandering in the woods one day with a friend when he came upon a small tortoise. Overcome with pleasurable excitement, he took up the tortoise and started home, thinking to surprise his children with it. After a few steps he paused and surveyed the tortoise doubtfully.

"What's the matter?" asked his friend.

Without responding, the great scientist slowly retraced his steps as precisely as possible, and gently set the turtle down upon the exact spot from which he had taken him up.

Then he turned solemnly to his friend. "It just struck me," he said, "that perhaps, for one man, I have tampered enough with the universe." He turned, and left the turtle to wander on its way.

The man who made that remark was one of the best of the modern men, and what he had devised had gone down into the whirlpool. "I have tampered enough," he said. It was not a denial of science. It was a final recognition that science is not enough for man. It is not the road back to the waiting Garden, for that road lies through the heart of man. Only when man has recognized this fact will science become what it was for Bacon, something to speak of as "touching upon Hope." Only then will man be truly human.

SIX

How Natural Is "Natural"?

The very design of imagination is to domesticate us in another, a celestial nature.

—RALPH WALDO EMERSON

IN THE MORE obscure scientific circles which I frequent there is a legend circulating about a late distinguished physicist who, in his declining years, persisted in wearing enormous padded boots much too large for him. He had developed, it seems, what to his fellows was a wholly irrational fear of falling through the interstices of that largely empty molecular space which common men in their folly speak of as the world. A stroll across his living-room floor had become, for him, something as dizzily horrendous as the activities of a window washer on the Empire State Building. Indeed, with equal reason he could have passed a ghostly hand through his own ribs.

The quivering network of his nerves, the awe-inspiring movement of his thought had become a vague cloud of electrons interspersed with the light-year distances that obtain between us and the farther galaxies. This was the natural world which he had helped to create, and in which, at last, he had found himself a lonely and imprisoned occupant. All around him the ignorant rushed on their way over the illusion of substantial floors, leaping, though they did not see it, from particle to particle, over a bottomless abyss. There was even a question as to the reality of the particles which bore them up. It did not, however, keep insubstantial newspapers from being sold, or insubstantial love from being made.

Not long ago I became aware of another world perhaps equally natural and real, which man is beginning to forget. My thinking began in New England under a boat dock. The lake I speak of has been pre-empted and civilized by man. All day long in the vacation season high-speed motorboats, driven with the reckless abandon common to the young Apollos of our society, speed back and forth, carrying loads of equally attractive girls. The shores echo to the roar of powerful motors and the delighted screams of young Americans with uncounted horsepower surging under their hands. In truth, as I sat there under the boat dock, I had some desire to swim or to canoe in the older ways of the great forest which once lay about this region. Either notion would have been folly. I would have been gaily chopped to ribbons by teen-age youngsters whose eyes were always immutably fixed on the far horizons of space, or upon the dials which indicated the speed of their passing. There was another world, I was to discover, along the lake shallows and under the boat dock, where the motors could not come.

As I sat there one sunny morning when the water was peculiarly translucent, I saw a dark shadow moving swiftly over the bottom. It was the first sign of life I had seen in this lake, whose shores seemed to yield little but washed-in beer cans. By and by the gliding shadow ceased to scurry from stone to stone over the bottom. Unexpectedly, it headed almost directly for me. A furry nose with gray whiskers broke the surface. Below the whiskers green water foliage trailed out in an inverted V as long as his body. A muskrat still lived in the lake. He was bringing in his breakfast.

I sat very still in the strips of sunlight under the pier. To my surprise the muskrat came almost to my feet with his little breakfast of greens. He was young, and it rapidly became obvious to me that he was laboring under an illusion of his own, and that he thought animals and men were still living in the Garden of Eden. He gave me a friendly glance from time to time as he nibbled his greens. Once, even, he went out into the lake again and returned to my feet with more greens. He had not, it seemed, heard very much about men. I shuddered. Only the evening before I had heard a man describe with triumphant enthusiasm how he had killed a rat in the garden because the creature had dared to nibble his petunias. He had even showed me the murder weapon, a sharp-edged brick.

On this pleasant shore a war existed and would go on until nothing remained but man. Yet this creature with the gray, appealing face wanted very little: a strip of shore to coast up and down, sunlight and moonlight, some weeds from the deep water. He was an edge-of-the-world dweller, caught between a vanishing forest and a deep lake pre-empted by unpredictable machines full of chopping blades. He eyed me nearsightedly, a green leaf poised in his mouth. Plainly he had come with some poorly instructed memory about the lion and the lamb.

"You had better run away now," I said softly, making no movement in the shafts of light. "You are in the wrong universe and must not make this mistake again. I am really a very terrible and cunning beast. I can throw stones." With this I dropped a little pebble at his feet.

He looked at me half blindly, with eyes much better adjusted to the wavering shadows of his lake bottom than to sight in the open air. He made almost as if to take the pebble up into his forepaws. Then a thought seemed to cross his mind—a thought perhaps telepathically received, as Freud once hinted, in the dark world below and before man, a whisper of ancient disaster heard in the depths of a burrow. Perhaps after all this was not Eden. His nose twitched carefully; he edged toward the water.

As he vanished in an oncoming wave, there went with him a natural world, distinct from the world of girls and motorboats, distinct from the world of the professor holding to reality by some great snowshoe effort in his study. My muskrat's shoreline universe was edged with the dark wall of hills on one side and the waspish drone of motors farther out, but it was a world of sunlight he had taken down into the water weeds. It hovered there, waiting for my disappearance. I walked away, obscurely pleased that darkness had not gained on life by any act of mine. In so many worlds, I thought, how natural is "natural" —and is there anything we can call a natural world at all?

II

Nature, contended John Donne in the seventeenth century, is the common law by which God governs us. Donne was already aware of the new science and impressed by glimpses of those vast abstractions which man was beginning to build across the gulfs of his ignorance. Donne makes, however, a reservation which rings strangely in the modern ear. If nature is the common law, he said, then Miracle is God's Prerogative.

By the nineteenth century, this spider web of common law

had been flung across the deeps of space and time. "In astronomy," meditates Emerson, "vast distance, but we never go into a foreign system. In geology, vast duration, but we are never strangers. Our metaphysic should be able to follow the flying force through all its transformations."

Now admittedly there is a way in which all these worlds are real and sufficiently natural. We can say, if we like, that the muskrat's world is naïve and limited, a fraction, a bare fraction, of the world of life: a view from a little pile of wet stones on a nameless shore. The view of the motor speedsters in essence is similar and no less naïve. All would give way to the priority of that desperate professor, striving like a tired swimmer to hold himself aloft against the soft and fluid nothingness beneath his feet. In terms of the modern temper, the physicist has penetrated the deepest into life. He has come to that place of whirling sparks which are themselves phantoms. He is close upon the void where science ends and the manifestation of God's Prerogative begins. "He can be no creature," argued Donne, "who is present at the first creation."

Yet there is a way in which the intelligence of man in this era of science and the machine can be viewed as having taken the wrong turning. There is a dislocation of our vision which is, perhaps, the product of the kind of creatures we are, or at least conceive ourselves to be. As we mentioned earlier, man, as a two-handed manipulator of the world about him, has projected himself outward upon his surroundings in a way impossible to other creatures. He has done this since the first half-human man-ape hefted a stone in his hand. He has always sought mastery over the materials of his environment, and in our day he has pierced so deeply through the screen of appearances that the age-old distinctions between matter and energy have been dimmed to the point of disappearance. The creations of his clever intellect ride in the skies and the sea's depths; he has hurled a great fragment of metal at the moon, which he once feared. He holds the heat of suns within his hands and threatens with it both the lives and happiness of his unborn descendants.

Man, in the words of one astute biologist, is "caught in a physiological trap and faced with the problem of escaping from his own ingenuity." Pascal, with intuitive sensitivity, saw this at the very dawn of the modern era in science. "There is nothing which we cannot make natural," he wrote, and then, prophetically, comes the full weight of his judgment upon man, "there is nothing natural which we do not destroy." *Homo faber*, the toolmaker, is not enough. There must be another road and another kind of man lurking in the mind of this odd creature, but whether the attraction of that path is as strong as the age-old primate addiction to taking things apart remains to be seen.

We who are engaged in the life of thought are likely to assume that the key to an understanding of the world is knowledge, both of the past and of the future—that if we had that knowledge we would also have wisdom. It is not my intention here to decry learning. It is only to say that we must come to understand that learning is endless and that nowhere does it lead us behind the existent world. It may reduce the prejudices of ignorance, set our bones, build our cities. In itself it will never make us ethical men. Yet because ours, we conceive, is an age of progress, and because we know more about time and history than any men before us, we fallaciously equate ethical advance with scientific progress in a point-to-point relationship. Thus as society improves physically, we assume the improvement of the individual, and are all the more horrified at those mass movements of terror which have so typified the first half of this century.

On the morning of which I want to speak, I was surfeited with the smell of mortality and tired of the years I had spent in archaeological dustbins. I rode out of a camp and across a mountain. I would never have believed, before it happened, that one could ride into the past on horseback. It is true I rode with a purpose, but that purpose was to settle an argument within myself.

It was time, I thought, to face up to what was in my mind— to the dust and the broken teeth and the spilled chemicals of life seeping away into the sand. It was time I admitted that life was of the earth, earthy, and could be turned into a piece of wretched tar. It was time I consented to the proposition that man had as little to do with his fate as a seed blown against a grating. It was time I looked upon the world without spectacles and saw love and pride and beauty dissolve into effervescing juices. I could be an empiricist with the best of them. I would be deceived by no more music. I had entered a black cloud of merciless thought, but the horse, as it chanced, worked his own way over that mountain.

I could hear the sudden ring of his hooves as we came cautiously treading over a tilted table of granite, past the winds that blow on the high places of the world. There were stones there so polished that they shone from the long ages that the storms had rushed across them. We crossed the divide then, picking our way in places scoured by ancient ice action, through boulder fields where nothing moved, and yet where one could feel time like an enemy hidden behind each stone.

If there was life on those heights, it was the thin life of mountain spiders who caught nothing in their webs, or of small gray birds that slipped soundlessly among the stones. The wind in the pass caught me head on and blew whatever thoughts I had into a raveling stream behind me, until they were all gone and there was only myself and the horse, moving in an eternal dangerous present, free of the encumbrances of the past.

We crossed a wind that smelled of ice from still higher snow-fields, we cantered with a breeze that came from somewhere among cedars, we passed a gust like Hell's breath that had risen straight up from the desert floor. They were winds and they did not stay with us. Presently we descended out of their domain, and it was curious to see, as we dropped farther through gloomy woods and canyons, how the cleansed and scoured mind I had brought over the mountain began, like the water in those rumbling gorges, to talk in a variety of voices, to debate, to argue, to push at stones, or curve subtly around obstacles. Sometimes I wonder whether we are only endlessly repeating in our heads an argument that is going on in the world's foundations among crashing stones and recalcitrant roots.

"Fall, fall, fall," cried the roaring water and the grinding pebbles in the torrent. "Let go, come with us, come home to the place without light." But the roots clung and climbed and the trees pushed up, impeding the water, and forests filled even the wind with their sighing and grasped after the sun. It is so in the mind. One can hear the rattle of falling stones in the night, and the thoughts like trees holding their place. Sometimes one can shut the noise away by turning over on the other ear, sometimes the sounds are as dreadful as a storm in the mountains, and one lies awake, holding, like the roots that wait for daylight. It was after such a night that I came over the mountain, but it was the descent on the other side that suddenly struck me as a journey into the eons of the past.

I came down across stones dotted with pink and gray lichens —a barren land dreaming life's last dreams in the thin air of a cold and future world.

I passed a meadow and a meadow mouse in a little shower of petals struck from mountain flowers. I dismissed it—it was almost my own time—a pleasant golden hour in the age of mammals, lost before the human coming. I rode heavily toward an old age far backward in the reptilian dark.

I was below timber line and sinking deeper and deeper into the pine woods, whose fallen needles lay thick and springy over the ungrassed slopes. The brown needles and the fallen cones, the stiff, endless green forests were a mark that placed me in the Age of Dinosaurs. I moved in silence now, waiting a sign. I saw it finally, a green lizard on a stone. We were far back, far back. He bobbed his head uncertainly at me, and I reined in with the nostalgic intent, for a moment, to call him father, but I saw soon enough that I was a ghost who troubled him and that he would wish me, though he had not the voice to speak, to ride on. A man who comes down the road of time should not expect to converse—even with his own kin. I made a brief, uncertain sign of recognition, to which he did not respond, and passed by. Things grew more lonely. I was coming out upon the barren ridges of an old sea beach that rose along the desert floor. Life was small and grubby now. The hot, warning scarlet of peculiar desert ants occasionally flashed among the stones. I had lost all trace of myself and thought regretfully of the lizard who might have directed me.

A turned-up stone yielded only a scorpion who curled his tail in a kind of evil malice. I surveyed him reproachfully. He was old enough to know the secret of my origin, but once more an ancient, bitter animus drawn from that poisoned soil possessed him and he raised his tail. I turned away. An enormous emptiness by degrees possessed me. I was back almost,

in a different way, to the thin air over the mountain, to the end
of all things in the cold starlight of space.

I passed some indefinable bones and shells in the salt-crusted
wall of a dry arroyo. As I reined up, only sand dunes rose like
waves before me and if life was there it was no longer visible.
It was like coming down to the end—to the place of fires where
we began. I turned about then and let my gaze go up, tier after
tier, height after height, from crawling desert bush to towering
pine on the great slopes far above me.

In the same way animal life had gone up that road from these
dry, envenomed things to the deer nuzzling a fawn in the meadows far above. I had come down the whole way into a place
where one could lift sand and ask in a hollow, dust-shrouded
whisper, "Life, what is it? Why am I here? Why am I here?"

And my mind went up that figurative ladder of the ages, bone
by bone, skull by skull, seeking an answer. There was none, except that in all that downrush of wild energy that I had passed in
the canyons there was this other strange organized stream that
marched upward, gaining a foothold here, tossing there a pine
cone a little farther upward into a crevice in the rock.

And again one asked, not of the past this time, but of the
future, there where the winds howled through open space and
the last lichens clung to the naked rock, "Why did we live?"
There was no answer I could hear. The living river flowed out
of nowhere into nothing. No one knew its source or its departing. It was an apparition. If one did not see it there was no
way to prove that it was real.

No way, that is, except within the mind itself. And the mind,
in some strange manner so involved with time, moving against
the cutting edge of it like the wind I had faced on the mountain, has yet its own small skull-borne image of eternity. It is
not alone that I can reach out and receive within my head a
handsbreadth replica of the far fields of the universe. It is not
because I can touch a trilobite and know the fall of light in
ages before my birth. Rather, it lies in the fact that the human
mind can transcend time, even though trapped, to all appearances, within that medium. As from some remote place, I see
myself as child and young man, watch with a certain dispassionate objectivity the violence and tears of a remote youth
who was once I, shaping his character, for good or ill, toward

the creature he is today. Shrinking, I see him teeter at the edge of abysses he never saw. With pain I acknowledge acts undone that might have saved and led him into some serene and noble pathway. I move about him like a ghost, that vanished youth. I exhort, I plead. He does not hear me. Indeed, he too is already a ghost. He has become me. I am what I am. Yet the point is, we are not wholly given over to time—if we were, such acts, such leaps through that gray medium, would be impossible. Perhaps God himself may rove in similar pain up the dark roads of his universe. Only how would it be, I wonder, to contain at once both the beginning and the end, and to hear, in helplessness perhaps, the fall of worlds in the night?

This is what the mind of man is just beginning to achieve—a little microcosm, a replica of whatever it is that, from some unimaginable "outside," contains the universe and all the fractured bits of seeing which the world's creatures see. It is not necessary to ride over a mountain range to experience historical infinity. It can descend upon one in the lecture room.

I find it is really in daylight that the sensation I am about to describe is apt to come most clearly upon me, and for some reason I associate it extensively with crowds. It is not, you understand, an hallucination. It is a reality. It is, I can only say with difficulty, a chink torn in a dimension life was never intended to look through. It connotes a sense beyond the eye, though the twenty years' impressions are visual. Man, it is said, is a time-binding animal, but he was never intended for this. Here is the way it comes.

I mount the lecturer's rostrum to address a class. Like any work-worn professor fond of his subject, I fumble among my skulls and papers, shuffle to the blackboard and back again, begin the patient translation of three billion years of time into chalk scrawls and uncertain words ventured timidly to a sea of young, impatient faces. Time does not frighten them, I think enviously. They have, most of them, never lain awake and grasped the sides of a cot, staring upward into the dark while the slow clock strokes begin.

"Doctor." A voice diverts me. I stare out nearsightedly over the class. A hand from the back row gesticulates. "Doctor, do you believe there is a direction to evolution? Do you believe, Doctor . . . Doctor, do you believe? . . ." Instead of the words, I hear a faint piping, and see an eager scholar's face squeezed and dissolving on the body of a chest-thumping ape. "Doctor, is there a direction?"

I see it then—the trunk that stretches monstrously behind him. It winds out of the door, down dark and obscure corridors to the cellar, and vanishes into the floor. It writhes, it crawls, it barks and snuffles and roars, and the odor of the swamp exhales from it. That pale young scholar's face is the last bloom on a curious animal extrusion through time. And who among us, under the cold persuasion of the archaeological eye, can perceive which of his many shapes is real, or if, perhaps, the entire shape in time is not a greater and more curious animal than its single appearance?

I too am aware of the trunk that stretches loathsomely back of me along the floor. I too am a many-visaged thing that has climbed upward out of the dark of endless leaf falls, and has slunk, furred, through the glitter of blue glacial nights. I, the professor, trembling absurdly on the platform with my book and spectacles, am the single philosophical animal. I am the unfolding worm, and mud fish, the weird tree of Igdrasil shaping itself endlessly out of darkness toward the light.

I have said this is not an illusion. It is when one sees in this manner, or a sense of strangeness halts one on a busy street to verify the appearance of one's fellows, that one knows a terrible new sense has opened a faint crack into the Absolute. It is in this way alone that one comes to grips with a great mystery, that life and time bear some curious relationship to each other that is not shared by inanimate things.

It is in the brain that this world opens. To our descendants it may become a commonplace, but me, and others like me, it has made a castaway. I have no refuge in time, as others do who troop homeward at nightfall. As a result, I am one of those who linger furtively over coffee in the kitchen at bedtime or haunt the all-night restaurants. Nevertheless, I shall say without regret: there are hazards in all professions.

III

It may seem at this point that I have gone considerably round about in my examination of the natural world. I have done so in the attempt to indicate that the spider web of law which has been flung, as Emerson indicated, across the deeps of time and space and between each member of the living world, has brought us some quite remarkable, but at the same time disquieting, knowledge. In rapid summary, man has passed from a natural world of appearances invisibly controlled by the caprice of spirits to an astronomical universe visualized by Newton, through the law of gravitation, as operating with the regularity of a clock.

Newton, who remained devout, assumed that God, at the time of the creation of the solar system, had set everything to operating in its proper orbit. He recognized, however, certain irregularities of planetary movement which, in time, would lead to a disruption of his perfect astronomical machine. It was here, as a seventeenth-century scholar, that he felt no objection to the notion that God interfered at periodic intervals to correct the deviations of the machine.

A century later Laplace had succeeded in dispensing with this last vestige of divine intervention. Hutton had similarly dealt with supernaturalism in earth-building, and Darwin, in the nineteenth century, had gone far toward producing a similar mechanistic explanation of life. The machine that began in the heavens had finally been installed in the human heart and brain. "We can make everything natural," Pascal had truly said, and surely the more naïve forms of worship of the unseen are vanishing.

Yet strangely, with the discovery of evolutionary, as opposed to purely durational, time, there emerges into this safe-and-sane mechanical universe something quite unanticipated by the eighteenth-century rationalists—a kind of emergent, if not miraculous, novelty.

I know that the word "miraculous" is regarded dubiously in scientific circles because of past quarrels with theologians. The word has been defined, however, as an event transcending the known laws of nature. Since, as we have seen, the laws of nature have a way of being altered from one generation of scientists to the next, a little taste for the miraculous in this broad sense will do us no harm. We forget that nature itself is one vast miracle transcending the reality of night and nothingness. We forget that each one of us in his personal life repeats that miracle.

Whatever may be the power behind those dancing motes to which the physicist has penetrated, it makes the light of the muskrat's world as it makes the world of the great poet. It makes, in fact, all of the innumerable and private worlds which exist in the heads of men. There is a sense in which we can say that the planet, with its strange freight of life, is always just passing from the unnatural to the natural, from that Unseen which man has always reverenced to the small reality of the day. If all life were to be swept from the world, leaving only its chemical constituents, no visitor from another star would be able to establish the reality of such a phantom. The dust would lie without visible protest, as it does now in the moon's airless craters, or in the road before our door.

Yet this is the same dust which, dead, quiescent and unmoving, when taken up in the process known as life, hears music and responds to it, weeps bitterly over time and loss, or is oppressed by the looming future that is, on any materialist terms, the veriest shadow of nothing. How natural was man, we may ask, until he came? What forces dictated that a walking ape should watch the red shift of light beyond the island universes or listen by carefully devised antennae to the pulse of unseen stars? Who, whimsically, conceived that the plot of the world should begin in a mud puddle and end—where, and with whom? Men argue learnedly over whether life is chemical chance or antichance, but they seem to forget that the life *in* chemicals may be the greatest chance of all, the most mysterious and unexplainable property in matter.

"The special value of science," a perceptive philosopher once wrote, "lies not in what it makes of the world, but in what it makes of the knower." Some years ago, while camping in a vast eroded area in the West, I came upon one of those unlikely sights which illuminate such truths.

I suppose that nothing living had moved among those great stones for centuries. They lay toppled against each other like fallen dolmens. The huge stones were beasts, I used to think, of a kind man ordinarily lived too fast to understand. They seemed inanimate because the tempo of the life in them was slow. They lived ages in one place and moved only when man was not looking. Sometimes at night I would hear a low rumble as one drew itself into a new position and subsided again.

Sometimes I found their tracks ground deeply into the hillsides.

It was with considerable surprise that while traversing this barren valley I came, one afternoon, upon what I can only describe as a very remarkable sight. Some distance away, so far that for a little space I could make nothing of the spectacle, my eyes were attracted by a dun-colored object about the size of a football, which periodically bounded up from the desert floor. Wonderingly, I drew closer and observed that something rope-like which glittered in the sun appeared to be dangling from the ball-shaped object. Whatever the object was, it appeared to be bouncing faster and more desperately as I approached. My surroundings were such that this hysterical dance of what at first glance appeared to be a common stone was quite unnerving, as though suddenly all the natural objects in the valley were about to break into a jig. Going closer, I penetrated the mystery.

The sun was sparkling on the scales of a huge blacksnake which was partially looped about the body of a hen pheasant. Desperately the bird tried to rise, and just as desperately the big snake coiled and clung, though each time the bird, falling several feet, was pounding the snake's body in the gravel. I gazed at the scene in astonishment. Here in this silent waste, like an emanation from nowhere, two bitter and desperate vapors, two little whirlwinds of contending energy, were beating each other to death because their plans—something, I suspected, about whether a clutch of eggs was to turn into a thing with wings or scales—this problem, I say, of the onrushing nonexistent future, had catapulted serpent against bird.

The bird was too big for the snake to have had it in mind as prey. Most probably, he had been intent on stealing the pheasant's eggs and had been set upon and pecked. Somehow in the ensuing scuffle he had flung a loop over the bird's back and partially blocked her wings. She could not take off, and the snake would not let go. The snake was taking a heavy battering among the stones, but the high-speed metabolism and tremendous flight exertion of the mother bird were rapidly exhausting her. I stood a moment and saw the bloodshot glaze deepen in her eyes. I suppose I could have waited there to see what would happen when she could not fly; I suppose it might have been worth scientifically recording. But I could not stand that ceaseless, bloody pounding in the gravel. I thought of the eggs somewhere about, and whether they were to elongate and writhe into an armor of scales, or eventually to go whistling into the wind with their wild mother.

So I, the mammal, in my way supple, and less bound by instinct, arbitrated the matter. I unwound the serpent from the bird and let him hiss and wrap his battered coils around my arm. The bird, her wings flung out, rocked on her legs and gasped repeatedly. I moved away in order not to drive her further from her nest. Thus the serpent and I, two terrible and feared beings, passed quickly out of view.

Over the next ridge, where he could do no more damage, I let the snake, whose anger had subsided, slowly uncoil and slither from my arm. He flowed away into a little patch of bunch grass—aloof, forgetting, unaware of the journey he had made upon my wrist, which throbbed from his expert constriction. The bird had contended for birds against the oncoming future; the serpent writhing into the bunch grass had contended just as desperately for serpents. And I, the apparition in that valley—for what had I contended?—I who contained the serpent and the bird and who read the past long written in their bodies.

Slowly, as I sauntered dwarfed among overhanging pinnacles, as the great slabs which were the visible remnants of past ages laid their enormous shadows rhythmically as life and death across my face, the answer came to me. Man could contain more than himself. Among these many appearances that flew, or swam in the waters, or wavered momentarily into being, man alone possessed that unique ability.

The Renaissance thinkers were right when they said that man, the Microcosm, contains the Macrocosm. I had touched the lives of creatures other than myself and had seen their shapes waver and blow like smoke through the corridors of time. I had watched, with sudden concentrated attention, myself, this brain, unrolling from the seed like a genie from a bottle, and casting my eyes forward, I had seen it vanish again into the formless alchemies of the earth.

For what then had I contended, weighing the serpent with

the bird in that wild valley? I had struggled, I am now convinced, for a greater, more comprehensive version of myself.

IV

I am a man who has spent a great deal of his life on his knees, though not in prayer. I do not say this last pridefully, but with the feeling that the posture, if not the thought behind it, may have had some final salutary effect. I am a naturalist and a fossil hunter, and I have crawled most of the way through life. I have crawled downward into holes without a bottom, and upward, wedged into crevices where the wind and the birds scream at you until the sound of a falling pebble is enough to make the sick heart lurch. In man, I know now, there is no such thing as wisdom. I have learned this with my face against the ground. It is a very difficult thing for a man to grasp today, because of his power; yet in his brain there is really only a sort of universal marsh, spotted at intervals by quaking green islands representing the elusive stability of modern science—islands frequently gone as soon as glimpsed.

It is our custom to deny this; we are men of precision, measurement and logic; we abhor the unexplainable and reject it. This, too, is a green island. We wish our lives to be one continuous growth in knowledge; indeed, we expect them to be. Yet well over a hundred years ago Kierkegaard observed that maturity consists in the discovery that "there comes a critical moment where everything is reversed, after which the point becomes to understand more and more that there is something which cannot be understood."

When I separated the serpent from the bird and released them in that wild upland, it was not for knowledge; not for anything I had learned in science. Instead, I contained, to put it simply, the serpent and the bird. I would always contain them. I was no longer one of the contending vapors; I had embraced them in my own substance and, in some insubstantial way, reconciled them, as I had sought reconciliation with the muskrat on the shore. I had transcended feather and scale and gone beyond them into another sphere of reality. I was trying to give birth to a different self whose only expression lies again in the deeply religious words of Pascal, "You would not seek me had you not found me."

I had not known what I sought, but I was aware at last that something had found me. I no longer believed that nature was either natural or unnatural, only that nature now appears natural to man. But the nature that appears natural to man is another version of the muskrat's world under the boat dock, or the elusive sparks over which the physicist made his trembling passage. They were appearances, specialized insights, but unreal because in the constantly onrushing future they were swept away.

What had become of the natural world of that gorilla-headed little ape from which we sprang—that dim African corner with its chewed fish bones and giant ice-age pigs? It was gone more utterly than my muskrat's tiny domain, yet it had given birth to an unimaginable thing—ourselves—something overreaching the observable laws of that far epoch. Man since the beginning seems to be awaiting an event the nature of which he does not know. "With reference to the near past," Thoreau once shrewdly commented, "we all occupy the region of common sense, but in the prospect of the future we are, by instinct, transcendentalists." This is the way of the man who makes nature "natural." He stands at the point where the miraculous comes into being, and after the event he calls it "natural." The imagination of man, in its highest manifestations, stands close to the doorway of the infinite, to the world beyond the nature that we know. Perhaps, after all, in this respect man constitutes the exertion of that act which Donne three centuries ago called God's Prerogative.

Man's quest for certainty is, in the last analysis, a quest for meaning. But the meaning lies buried within himself rather than in the void he has vainly searched for portents since antiquity. Perhaps the first act in its unfolding was taken by a raw beast with a fearsome head who dreamed some difficult and unimaginable thing denied his fellows. Perhaps the flashes of beauty and insight which trouble us so deeply are no less prophetic of what the race might achieve. All that prevents us is doubt—the power to make everything natural without the accompanying gift to see, beyond the natural, to that inexpressible realm in which the words "natural" and "supernatural" cease to have meaning.

Man, at last, is face to face with himself in natural guise. "What we make natural, we destroy," said Pascal. He knew, with superlative insight, man's complete necessity to transcend the worldly image that this word connotes. It is not the outward powers of man the toolmaker that threaten us. It is a growing danger which has already afflicted vast areas of the world—the danger that we have created an unbearable last idol for our worship. That idol, that uncreated and ruined visage which confronts us daily, is no less than man made natural. Beyond this replica of ourselves, this countenance already grown so distantly inhuman that it terrifies us, still beckons the lonely figure of man's dreams. It is a nature, not of this age, but of the becoming—the light once glimpsed by a creature just over the threshold from a beast, a despairing cry from the dark shadow of a cross on Golgotha long ago.

Man is not totally compounded of the nature we profess to understand. Man is always partly of the future, and the future he possesses a power to shape. "Natural" is a magician's word—and like all such entities, it should be used sparingly lest there arise from it, as now, some unglimpsed, unintended world, some monstrous caricature called into being by the indiscreet articulation of worn syllables. Perhaps, if we are wise, we will prefer to stand like those forgotten humble creatures who poured little gifts of flints into a grave. Perhaps there may come to us then, in some such moment, a ghostly sense that an invisible doorway has been opened—a doorway which, widening out, will take man beyond the nature that he knows.

Bibliography

Blyth, Edward. "An Attempt to Classify the Variations of Animals, etc.," *Magazine of Natural History*, Vol. 8 (1835), pp. 40–53.

Blyth, Edward. "On the Psychological Distinctions Between Man and All Other Animals, etc.," *Magazine of Natural History*, Vol. 1 N.S. (1837), Parts I, II, III.

Brückner, John. *A Philosophical Survey of the Animal Creation*, London, 1768.

Cannon, H. Graham. "What Lamarck Really Said," Proceedings of the Linnean Society of London, 168th Session, 1955–56, Parts I, II.

Coleridge, S. T. *Philosophical Lectures* 1818–1819, London, Pilot Press, 1949.

Darlington, C. D. *Darwin's Place in History*, Oxford, Blackwell, 1959.

Darlington, William. *Memorials of John Bartram and Humphrey Marshall*, Philadelphia, 1849.

Dart, Raymond and Craig, Dennis. *Adventures with the Missing Link*, New York, Harper & Brothers, 1959.

de Beer, Sir Gavin. "Darwin's Notebooks on Transmutation of Species," Bulletin of the British Museum of Natural History, Historical Series, Vol. 2, No. 2 (1960).

Eiseley, Loren. "Charles Darwin, Edward Blyth and the Theory of Natural Selection," *Proceedings of the American Philosophical Society*, Vol. 103 (1959), pp. 94–158.

———. "Charles Lyell," *Scientific American*, Vol. 201 (1959), pp. 98–101.

Falconer, Hugh. *Paleontological Memoirs*, Vols. I and II, London, Robert Hardwicke, 1868.

"James Hutton 1726–1797: Commemoration of the 150th Anniversary of His Death," Proceedings of the Royal Society of Edinburgh, Vol. 63 (1949), pp. 351–400.

Morris, Wright. *The Territory Ahead*, New York, Harcourt, Brace and Company, 1958.

Picard, Max. *The Flight from God*, Chicago, Henry Regnery Co., 1951.

Trow-Smith, Robert. *A History of British Livestock Husbandry 1700–1900*, London, Routledge and Kegan Paul, 1959.

THE UNEXPECTED UNIVERSE

TO WOLF,
who sleeps forever
with an ice age bone
across his heart,
the last gift
of one
who loved him

I wish to thank the sponsors of the William Haas Lectures of Stanford University, where three of these explorations of the unexpected universe were given, my colleagues in the Institute for Research in the Humanities at the University of Wisconsin, where I was a 1967 guest, and my associates at the Menninger Foundation, where I was a similar visitor. To the John Simon Guggenheim Memorial Foundation and its former director Henry Allen Moe for his patience and his faith I am most grateful. I would like also to express appreciation to the editors of Time-Life Books for permitting the reprinting, with modifications, of a passage from an older article of mine, once confined to a specialized purpose, and to *The American Scholar* and *Life*, in which some of this material also previously appeared.

LOREN EISELEY

Wynnewood, Pa.
March 3, 1969

The universe is not only queerer than we suppose, but queerer than we can suppose.

—J. B. S. HALDANE

If you do not expect it, you will not find the unexpected, for it is hard to find and difficult.

—HERACLITUS

ONE

The Ghost Continent

*The winds are mad, they know not whence they come, nor whither
they would go: and those men are maddest of all that go to sea.*
—ROBERT BURTON

EVERY MAN CONTAINS within himself a ghost continent—
a place circled as warily as Antarctica was circled two hundred years ago by Captain James Cook. If, in addition, the
man is a scientist, he will see strange shapes amidst his interior
ice floes and be fearful of exposing to the ridicule of his fellows
what he has seen. To begin such a personal record it may be
well to start with the odyssean voyages of legend and science.
These may defend with something of their own magic the
small story of an observer lost upon the fringes of large events.
Let it be understood that I claim no discoveries. I claim only
the events of a life in science as they were transformed inwardly
into something that was whispered to Odysseus long ago.

Like Odysseus, man seeks his spiritual home and is denied it;
along his path the shape-shifting immortal monsters of his
earlier wanderings assume more sophisticated guises, but they
survive because man himself remains and man has called them
forth. The almost three-thousand-year-old epic of the Odyssey
takes on a particular pertinence today. It possesses a perennial
literary freshness that causes it to be translated anew in every
generation. It involves an extended journey amid magical obstacles and Cyclopean assailants. Moreover, it can be read as
containing the ingredients of both an inward journey of reflection and an outwardly active adventure. Both these journeys
threaten to culminate in our time. Man's urge toward space
has impelled him to circle the planet, and in the week of December 25, 1968, precisely two hundred years after the navigator James Cook's first great voyage into the Pacific, three
American astronauts had returned from the moon. The event
lies more than two million years after the first man-ape picked
up and used a stone.

Nevertheless, throughout this entire pilgrimage, as reflected in his religious and philosophical thinking, man's technological triumphs have frequently been at odds with his hunger for psychological composure and peace. Thus the epic journey of modern science is a story at once of tremendous achievement, loneliness, and terror. Odysseus' passage through the haunted waters of the eastern Mediterranean symbolizes, at the start of the Western intellectual tradition, the sufferings that the universe and his own nature impose upon homeward-yearning man.

In the restless atmosphere of today all the psychological elements of the Odyssey are present to excess: the driving will toward achievement, the technological cleverness crudely manifest in the blinding of Cyclops, the fierce rejection of the sleepy Lotus Isles, the violence between man and man. Yet, significantly, the ancient hero cries out in desperation, "There is nothing worse for men than wandering."

The words could just as well express the revulsion of a modern thinker over the sight of a nation harried by irrational activists whose rejection of history constitutes an equal, if unrecognized, rejection of any humane or recognizable future. We are a society bemused in its purposes and yet secretly homesick for a lost world of inward tranquillity. The thirst for illimitable knowledge now conflicts directly with the search for a serenity obtainable nowhere upon earth. Knowledge, or at least what the twentieth century acclaims as knowledge, has not led to happiness.

Ours is certainly the most time-conscious generation that has ever lived. Our cameras, our television, our archaeological probings, our C^{14} datings, pollen counts, under-water researches, magnetometer readings have resurrected lost cities, placing them accurately in stratigraphic succession. Each Christmas season the art of ice age Lascaux is placed beside that of Rembrandt on our coffee tables. Views of Pompeii share honors with Chichén Itzá upon the television screen in the living room. We unearth obscure ancestral primates and, in the motion picture "2001," watch a struck fragment of bone fly into the air and become a spaceship drifting among the stars, thus telescoping in an instant the whole technological history of man. We expect the average onlooker to comprehend the symbolism; such a civilization, one must assume, should show a deep veneration for the past.

Strangely, the results are quite otherwise. We appear to be living, instead, amidst a meaningless mosaic of fragments. From ape skull to Mayan temple we contemplate the miscellaneous debris of time like sightseers to whom these mighty fragments, fallen gateways, and sunken galleys convey no present instruction.

In our streets and on our campuses there riots an extremist minority dedicated to the now, to the moment, however absurd, degrading, or irrelevant the moment may be. Such an activism deliberately rejects the past and is determined to start life anew—indeed, to reject the very institutions that feed, clothe, and sustain our swarming millions.

A yearning for a life of noble savagery without the accumulated burdens of history seems in danger of engulfing a whole generation, as it did the French *philosophes* and their eighteenth-century followers. Those individuals who persist in pursuing the mind-destroying drug of constant action have not alone confined themselves to an increasingly chaotic present—they are also, by the deliberate abandonment of their past, destroying the conceptual tools and values that are the means of introducing the rational into the oncoming future.

Their world, therefore, becomes increasingly the violent, unpredictable world of the first men simply because, in losing faith in the past, one is inevitably forsaking all that enables man to be a planning animal. For man's story, in brief, is essentially that of a creature who has abandoned instinct and replaced it with cultural tradition and the hard-won increments of contemplative thought. The lessons of the past have been found to be a reasonably secure instruction for proceeding against the unknown future. To hurl oneself recklessly without method upon a future that we ourselves have complicated is a sheer nihilistic rejection of all that history, including the classical world, can teach us.

Odysseus' erratic journey homeward after the sack of Troy to his own kingdom in Ithaca consumed ten years. There is a sense in which this sea-battered wanderer, who, at one point in concealment, calls himself "Nobody," represents the human journey toward eternity. The sea god Poseidon opposes his passage. He is shipwrecked, escapes monsters, evades the be-witchment of goddesses. In the words of Kazantzakis, he appears to have "a wind chart in his breast for heart."

Yet Odysseus, like Cook and Darwin, the scientific voyagers, is shrewd, self-reliant, and persistent. He is farsighted even when on the magical isles, but he could not always save his companions. They frequently caused him trouble because they were concerned solely with the immediate. Scenting treasure, they opened at the wrong moment the bag that loosed all the winds of the sea upon their vessel. Like man in the mass, they were feckless, unstable, and pursuing the will-o'-the-wisp of the moment. In Homer's words, "they wanted to stay with the Lotus-Eaters and forget the way home."

By contrast, Circe, the great enchantress, says coolly to Odysseus, "There is a mind in you no magic will work on." The remark, however, is two-edged. It is circumspect, as one great lord of thought might speak to another. It hints at the dawn in the Greek mind of that intelligence which we of this age choose to call scientific. Nevertheless, beneath the complimentary words can be sensed a veiled warning. For the man whom no magic will charm may, in the end, find himself, by means of a darker sorcery, upon a shore as desolate as that which Odysseus narrowly escaped in passing the Isle of the Sirens. The Sirens had sung sweetly to him of all knowledge, while about them lay dead men's bones. If living for the day and the senses is the folly of the thoughtless, so also is there danger in that insatiable hunger for power which besets the human intellect. Far more than modern men, Homer is wary of that vaulting pride the Greeks called *hubris*, which is an affront to the immortal gods.

It was once said, half in irony, by an ancient geographer that "you will find where Odysseus wandered when you find the cobbler who stitched the bag of the winds." Doubtless this is true, but is not man similarly the product of such an untraceable cobbler? And, even more, is not each individual life a bag full of surging dreams and compulsions imprisoned in human skin by that same cobbler, and equally capable of inadvertent release? Yet for all the vagaries of human voyaging amidst inward and outward tempests, the mariners of three thousand years ago had begun scientifically to watch the stars. Homer himself was acquainted with the guide stars of the seafarers. We know from the *Odyssey* that the constellation of the Great Bear was never wetted by the waves on its night circling above the Mediterranean.

If one now turns to the Odyssean voyages of science in the eighteenth and nineteenth centuries, one will come surprisingly close both to that shadowy Cimmerian land where the sun is hidden and to that rift in time where life leaped backward to confront itself upon its evolutionary road. The voyages that produced these observations were as irretraceable and marvel-filled as any upon the lost sea charts of Odysseus.

II

It has been said of Captain James Cook that no discoverer ever measured his claims with more moderation. Yet among the great mariners none ran greater risks for the purposes of close inshore mapping. None sailed farther, or sailed under weightier secret instructions. Masterful and solitary, but with a superb gift of leadership, he endured the vanities of his scientific associates as he endured plotting natives and the hard diet of the sea.

He was a man supremely indifferent to every circumstance, who brought to the tenth year of his Pacific voyaging the same ingenuity and doggedness that had brought Odysseus home from Troy. Like Odysseus he could practice wise restraint; like Odysseus he could improvise against the future. Unlike the more primitive warriors of the Greek bronze age, however, he was not vengeful. He saw in the looming future the possibilities offered by Australia for civilized settlement, and directed attention to the furtherance of such potentials. So vast was the range of his wanderings and so attractive the Pacific isles he visited and explored that they have veiled from our memory the storms and darkness that hovered over his greatest achievement, the circumnavigation of Antarctica.

Men of today frequently turn to science for such knowledge of the hazardous future as can be gained by mortals. In Homer's time it was believed that this information might be sought among the dead. At Circe's instigation Odysseus reached toward the world's edge the mist-shrouded land of Hades, the dwelling place of the shades:

> hidden in fog and closed, nor does Helios, the radiant
>   sun, ever break through the dark . . .
> always a drear night is spread.

It was there that the dead Theban, Tiresias, foretold the end of Odysseus' voyage, further remarking that in old age "death will come to you from the sea, in some altogether unwarlike way."

Cook, who opened the Pacific frontier to science, is largely associated in the public mind with those Lotus Isles of forgetfulness that lie in Polynesia. In reality, however, he had received, like Odysseus, assignment to a more desperate errand. The adventure was as stark in his day as the adventure into space is in ours. In terms of supporting equipment, it was equally, if not more, dangerous, for Cook's mission was to penetrate the unknown region at the bottom of the globe. In 1768 his public orders were ostensibly to proceed to the Pacific and observe from Tahiti the transit of Venus. On the island he was to open his sealed instructions, which read in part:

> Whereas there is reason to imagine that a Continent or Land of great extent, may be found to . . . the Southward of the Track of any former Navigators . . . you are to proceed to the southward in order to make discovery of the Continent above mentioned. . . .

Whence had emerged this conception of a continent unknown but supposed to be clothed with verdure and inhabited by living people whose wares and activities might be of interest to the Crown? Since the days of Ptolemy a great southern land mass, a continent labeled Terra Incognita, had floated down through the centuries on a succession of maps. Although belief in such a continent wavered and died down, certain sightings of islands in the sixteenth century restimulated the hopes of geographers. Visions were entertained of a rich and habitable continent south and westward of South America. It moved upon the eighteenth-century sea charts as elusively as Melville's great white whale, in all meridians.

By Cook's time, in the late eighteenth century, an ambitious scholar-merchant, Alexander Dalrymple, whose hobby was the memoirs of the early voyagers, had become a believer in the ghost continent. Dalrymple thought this huge mainland must be necessary to balance the earth on its axis and that its human population must be numbered in the millions. Dalrymple wished to lead an expedition there to establish trade relations. Cook, a proven naval commander with long mapping and coastal sailing experience, was chosen instead. Dalrymple was bitterly disappointed and harassed Cook after his first voyage of 1768, saying, "I would not have come back in ignorance." Cook had stated, after long voyaging, "I do not believe any such thing exists, unless [and here he proved prophetic] in a high latitude." Dalrymple succeeded in creating such doubt and confusion that the Admiralty decided on a second expedition, but again Cook was the chosen officer; the ghost continent was once more pursued through the shifting degrees of latitude.

Antarctica is another world. Instead of discovering a living continent, Cook, like Odysseus, came to a land of Cimmerian darkness. Huge icicles hung on the ship's sails and rigging. The pack ice "exhibited such a variety of figures that there was not an animal on earth that was not represented by it." A breeding sow on board farrowed nine pigs, every one of which was killed by cold in spite of attempts to save them. Scurvy appeared. In the after cabin a gentleman died. A sailor dropped from the rigging and vanished beneath the ice floes. Cavernous icebergs, against which the waves resounded, inspired exclamations of admiration and horror. Vast-winged gray albatrosses drifted by in utter silence.

In 1773 Cook first crossed the Antarctic Circle, where, in the words of one of his scientist passengers, "We were . . . wrapped in thick fogs, beaten with showers of rain, sleet, hail and snow . . . and daily ran the risk of being shipwrecked." Cook himself, after four separate and widely removed plunges across the Circle, speaks, as does Homer, of lands "never to yield to the warmth of the sun." His description of an "inexpressibly horrid Antarctica" resounds like an Odyssean line. Terra Incognita Australis had been circumnavigated at last, its population reduced to penguins. If there was land at all beyond the ice barrier, it was the frozen world of another planet. Only the oaths of sailors splintered and re-echoed amidst the pinnacles of ice.

"I can be bold to say," pronounced Cook, turning north

toward the summer isles, "that no man will venture farther than I have done." In eighteenth-century terms he was right, just as now the way beyond the moon lies dark and inexpressibly desolate and costly for men to follow. The mariners in their sea jackets and canvas headgear were not equipped for the long land traverse that, in the twentieth century, took Scott and Amundsen to the pole; even then the fierce ice gods took away one man's homecoming. With grim amusement Cook heard his officers suggest that if Terra Incognita existed it lay north, in sunnier climes. Ignoring them, perhaps with Dalrymple in mind, he tried one more run southward before he swung away. The highborn scientists were frequently to complain uneasily that he never told them where he was going. What was he expected to do, sailing under secret orders in the wastes below the Circle? He might, if he had chosen, have responded like Odysseus, "I am a man. I am not a god."

A little over a half century after Cook's death in Hawaii, an inexperienced voyager into the bottomless sea of time came to anchor in the Galápagos. It was the young Charles Darwin, fresh from his wanderings on the pampas and his Andean ascents. At Tierra del Fuego he had glimpsed from the *Beagle* the stormy, racing waters through which Cook's ships, the *Endeavor* and the *Resolution*, had ridden on their long world journeys. Now the *Beagle* had rung down its anchor at what the Spanish had called, with singular discernment, the "Encantadas," the Enchanted Isles.

Odysseus had come similarly upon Circe's island, only to find his crew transformed into animals—specifically, into pigs. When, at his behest, the changelings were created men once more, they took on a more lively and youthful appearance. By the sixteenth century the Florentine writer Giovanni Battista Gelli had produced his *Circe*, in which a variety of animals refused Odysseus' offer to restore them to their original form. Their arguments for remaining as they were constitute an ingenious commentary on the human condition. From rabbit to lion the animals are united in being done with humanity. Not all the argumentative wiles of Odysseus can talk them back into the shape of *Homo sapiens*. The single exception proves to be a dubious Greek philosopher immured in the body of an elephant. He alone consents to a renewed transformation.

The Encantadas are the means by which the whole Circean labyrinth of organic change was precipitated upon the mind of man. What had appeared to Odysseus as the trick of a goddess was, in actuality, the shape shifting of the incomprehensible universe itself. Among the upthrust volcanic chimneys, which Darwin compared to "the cultivated parts of the Infernal Regions," the young naturalist meditated upon a flora and fauna seemingly distinct, as in the case of the famed Galápagos tortoises, from the living inhabitants of the continents.

Circe kept herself hidden, but it was evident to the wondering Darwin that there was a power hidden in time and isolation that alone could transmute, not just men, but all things living, into wavering shadows. He had entered by the mysterious doorway of the Galápagos into a sea as vast in its own way as the limitless Pacific. Yet even here upon the ocean of time the ghostly sail of Cook's vessel passes by. Cook's surgeon-naturalist, William Anderson, records in his journal of the third and fatal voyage—that voyage which Cook himself had preternaturally referred to as the "last"—that animals and human beings must be attributed to the different stocks "from which they sprung before their arrival in the south sea, or we must believe that at the creation every particular island was furnished with its inhabitants in the same manner as with its peculiar plants." Anderson's words do not precisely express evolutionary ideas, but they hint prophetically at the puzzling thoughts—to a degree, heretical—that had arisen and would not be downed about the role of islands in the biology of the living world.

The momentary phantom of the *Resolution* fades on the infinite waters of the past. William Anderson is dead on the Bering passage, but others will look and ask, and ask again, until the answer arises upon Darwin's return to London—that London which, like Odysseus searching the horizon for Ithaca, he yearns for in a note to his former teacher John Henslow: "Oh, to be home again without one single novel object near me." But the novel objects are there, fixed in the memory, ever to be relived. Each covert of the Enchanted Archipelago will forever resound with the reptilian hisses of antediluvian antiquity.

The memory of the modern scholar will, I suspect, be wrenched by the thought of all that Captain Cook patiently endured as well as learned from Sir Joseph Banks, the well-intentioned, blustering, aristocratic naturalist, and his sometimes complaining colleagues of the first and second voyages. Only once did the restrained Captain allow himself the liberty of a quarter-deck comment, involving the violation of the commander's printing rights by the naturalist Forster and his son. The careless assignment to Cook by the scribe Hawkesworth of some undiplomatic remarks derived from Banks's private journal did nothing to allay the Captain's wrath. "Damn and blast all scientists," he exploded to Lieutenant King on the eve of the third voyage.

Cook had reason for his spleen. Yet it is worth recording that this son of a Yorkshire laborer left a record of his voyages far more accurate than the first version produced by Hawkesworth, who had been assigned by the Admiralty to prettify Cook's work and make it palatable. In modern terms, Cook was, on the whole, a magnificent and tolerant anthropologist who, at every inhabited island, had to improvise his Odyssean role. It was not scientists who died on the beach of Kealakekua in Hawaii. Instead, it was the underrated circumnavigator of Antarctica, the genuine Ahab of the ghost continent. Cook had, times without number, brought his ships off dangerous shoals and, in addition, penetrated the high latitudes of both poles.

We inhabitants of a scientific era may well prize our accomplishments, but it behooves us to acknowledge that without the skills of the relentless Odyssean voyager, Joseph Banks and all his colleagues might well have fed the coral upon some tropic reef. Banks is believed to have anonymously paid Cook a just, belated tribute in the *Morning Chronicle* when the news of his death was announced in London. "Cook's competence changed the face of the world" was to be a mature, twentieth-century judgment.

James Cook himself would have cast a proud, cold eye upon land-bound opinion. He had emerged unheralded and alone from the grimy foreshore of history. The horizon, the pack ice, and the albatross would have been all he asked as memorials of his passage—these and the carefully drawn maps for those who followed. It may be that the space captains, when and if they leave the solar system, will alone come to understand that remote, serene indifference.

There is, however, one thing more to be considered: the Odyssean journey in which the mind turns homeward, seeking surcease from outer triumphs. Perhaps this crossing of the two contradictory impulses in the mind of man is nowhere better expressed than by the Italian poet Giovanni Pascoli in "Ultimo Viaggio" ("The Last Voyage"), published in 1904. Pascoli realized that Odysseus' return to Ithaca, his homeward goal, was in a sense an anticlimax—that the magical spell wrought by Circe would follow the hero into the prosaic world.

Pascoli picks up the Odyssean tale when Odysseus, grown old and restless, drawn on by migratory birds, sets forth to retrace his magical journey, the journey of all men down the pathway of their youth, the road beyond retracing. Circe's isle lies at last before the wanderer in the plain colors of reality. Circe and whatever she represents have vanished. Much as Darwin might have viewed the Galápagos in old age, Odysseus passes the scenes of the marvelous voyage with all the obstacles reduced to trifles. The nostalgia of space, which is what the Greeks meant by nostalgia, that is, the hunger for home, is transmuted by Pascoli into the hunger for lost time, for the forever vanished days. The Sirens no longer sing, but Pascoli's Odysseus, having made his inward journey, understands them. Knowledge without sympathetic perception is barren. Odysseus in his death is carried by the waves to Calypso, who hides him in her hair. "Nobody" has come home to Nothingness.

III

Those archaeologists and students of folklore who frequent the Scandinavian swamps in which bog acids have preserved such bodies as those of Tollund man—men who were alive upon earth in a time almost coeval with Homer—speak of a strange crossroads religion. An earth goddess was driven in a massive coach through gloomy northern forests. She was peripatetic and did not stay, but the ox-drawn vehicle in which she lumbered over abandoned and ancient roads paused, with curtains drawn, at ignorant villages.

Out of the rising fog wisps shambled her priests and obedient worshipers. No one was allowed to thrust his head behind the draped curtains of the ornamented cart or to query the immobile coachman or to touch with worshipful fingers the steaming oxen in the night. After the proper ritual, a selected human sacrifice would be cast down into the waiting bog. The coach would lumber again into the night, across wild heaths and forbidden pathways.

But suppose—the thought strikes me suddenly as I startle awake at midnight, hearing cries and rumblings in my hotel corridors—suppose the wheels of the great car are still revolving before its attendant worshipers. Suppose further—and I sit up at this and shiver uncontrollably as I hear the mead-soaked voices and the running steps—suppose the same awkward coach still lurches through the darker hours of our assembled scientific priesthood.

Suppose that in the ancient car there sits in one age the masked face of Newton, his world machine ticking like a remorseless clock in the dead and confined air; or suppose that Darwin lurks concealed behind the curtains, and all is wild uncertainty and change in the misty features of his company; or that Doctor Freud looks coldly and contemplatively down upon a sea of fleecing goblin faces. Or is it the Abbé Lemaître's followers who hear the alternate expansion and contraction of nature's pounding heart, like a rhythmic drum amidst the receding coals of the night stars? Or imagine that, silhouetted gigantically in the fierce rays of atomic light streaming from the carriage, four sinister horsemen trample impatiently, while within a muffled voice cries out to the assembled masses: "God is dead. All is permitted."

Suppose, I think, lying awake and weary in my blankets, the figure in the coach is a changeling and its true face is no face, as Odysseus was "no one" until he shouted a vengeful name before the Cyclops. Or deduce that behind the concealing drapery, hooded in a faceless cowl, there is caught only the swirling vapor of an untamed void whose vassals we are—we who fancy ourselves as the priesthood of powers safely contained and to be exhibited as evidences of our own usurping godhood.

I hear now, as I have heard before, the far-off, long, premonitory trundling of the dreadful cart at midnight. I think of the face of No Face—protean, dissembling, alternating death's-head and beautiful Circean goddess, forever forbidden to descend and touch common flesh. I think of her acolytes, ourselves, toiling in a hundred laboratories with our secret visions of what is, or may not be, while the wild reality always eludes our grasp.

Long ago a Greek named Plato, who was also a voyager not unfamiliar with Odyssean wanderings, remarked after much thought, "We must take the best and most irrefragable of human doctrines and embark on that, as if it were a raft on which to risk the voyage of life." Plato makes, however, one humble addition to his meditation, one that marks the Greek distaste of overvaulting pride. "Unless," he adds wistfully, "it were possible to find a stronger vessel, some Divine Word on which we might take our journey more surely and with confidence."

Earlier I have dwelt upon the magical incidents that beset Odysseus on his voyage. These are the stories remembered through unnumbered generations, of the island of the Cyclops, of Circe, of Calypso, of the cattle of the sun, of the Lotus-Eaters and the Sirens, of the dreadful bag of the four winds inadvertently loosed in mid-passage. The Odyssey is, as the critic Howard Clarke once observed, the world of the folk tale. It is the arena of uncertain violence that confronts man on the voyage of life—sex, irresponsibility, inordinate hungers that evoke equally monstrous figures from the human subconscious. Such a world is like the Jack-and-the-Beanstalk world of childhood, whose memories linger more clearly than many of the events of later life.

Once it is raised, not only is it difficult to subdue such a world again, but anything that follows it is apt to prove an anticlimax. Odysseus' goal is home, but after the adventures on the mysterious sea where Proteus can be found basking on a rock among his creatures, the suitors and their puerile human rapacity are singularly unattractive. Odysseus himself loses stature when he is reduced to lynching helpless if inobedient women servants. Something vanishes from the tale at this point, no matter how necessary it was for Homer to complete the Odyssean voyage. The poet has been sufficiently perceptive, however, to lay upon Odysseus, through the device of prophecy, the injunction of a further journey.

The poet, or, rather, his hero, has returned in middle age to a household that will obviously not long give him scope to breathe. Odysseus must, in other words, be given a legitimate reason to escape from Ithaca. This has been so strongly felt by generations of readers that from Dante through Tennyson to Kazantzakis poets have been impelled to launch the immortal navigator once more into the realms of dream. Perhaps among these poets Pascoli alone was wise enough to visualize an end in which the trivial and magicless themselves are transmuted by human wisdom into a timeless dimension having its own enchanted reality.

In this connection, there is one event involving Odysseus' homecoming, one episode that, in the mundane world of Ithaca, shines like a far light reflected from Calypso's isle. Oddly enough, that other voyager on more recent seas, Charles Darwin, experienced and recorded a similar episode after his five-year absence from Shrewsbury. Odysseus' great dog Argos, abused and cast out on a dung heap, wagged his tail and was the first to recognize his old master after nineteen years of absence. In a like way Darwin's favorite dog recognized him after his five-year absence on the *Beagle*.

It is not necessary to cavil over the great age of Odysseus' dog. The surprising thing about the story's descent through the millennia is that it comes from a fierce and violent era, yet it bespeaks some recognized bond between man and beast. The tie runs beyond the cities into some remote glade in a far forest where man willingly accepted the help of his animal kin. Though men in the mass forget the origins of their need, they still bring wolfhounds into city apartments, where dog and man both sit brooding in wistful discomfort.

The magic that gleams an instant between Argos and Odysseus is both the recognition of diversity and the need for affection across the illusions of form. It is nature's cry to homeless, far-wandering, insatiable man: "Do not forget your brethren, nor the green wood from which you sprang. To do so is to invite disaster."

Many great writers have written private meanings into their versions of the Odyssean voyage. In the twentieth century particularly, Odysseus has been seen as a symbol of the knowledge-hungry scientist, the Faustian penetrator of space and time. But, as scientists, we have sometimes forgotten the inward journey so poetically expressed by Pascoli in "The Last Voyage," that inward journey whose true meaning was long ago expressed by Circe's cryptic warning. "Magic cannot touch you," she had said to Odysseus, but today we know that the heart untouched by the magic of wonder may come to an impoverished age. Cook, the navigator, died in the Lotus Isles, perhaps fortunate that he never returned to his Penelope. Darwin, who voyaged as deeply into time, came home to Down, where he is said to have slept ill. In fact, it has been reported on good authority that he walked so late he met the foxes trotting home at dawn.

Thus, in the heart of man, and, above all, within this turbulent century, the Odyssean voyage stands as a symbol of both man's homelessness and his power, a power more unregenerate than that which drove Odysseus to string the great bow before the suitors. Long ago, when the time to Homer might still be numbered in centuries, Plotinus wrote of the soul's journey, "It shall come, not to another, but to itself." It is possible to add that for the soul to come to its true self it needs the help and recognition of the dog Argos. It craves that empathy clinging between man and beast, that nagging shadow of remembrance which, try as we may to deny it, asserts our unity with life and does more. Paradoxically, it establishes, in the end, our own humanity. One does not meet oneself until one catches the reflection from an eye other than human.

It has been asserted that we are destined to know the dark beyond the stars before we comprehend the nature of our own journey. This may be true. But we know also that our inward destination lies somewhere a long way past the reef of the Sirens, who sang of knowledge but not of wisdom. Beyond that point, if perchance we reach it, exists the realm of Plato's undiscovered word. It is a divine word, or so Plato gropingly called it, for he hoped against hope that it might suffice to guide our human pilgrimage when, in the ritual language of the *Odyssey*, "the sun has set and all the journeying ways are darkened."

In our time, however, the mind still persists in traveling along those darkened sea paths where all manner of strange creatures swarm. For myself, I have penetrated as far as I could dare among rain-dimmed crags and seascapes. But there is more, assuredly there is still more, as Circe tried to tell Odysseus when she warned that death would come to him from the sea. She meant, I think now, the upwelling of that inner tide which finally engulfs each traveler.

I have listened belatedly to the warning of the great enchantress. I have cast, while there was yet time, my own oracles on the sun-washed deck. My attempt to read the results contains elements of autobiography. I set it down just as the surge begins to lift, towering and relentless, against the reefs of age.

TWO

The Unexpected Universe

Imagine God, as the Poet saith, Ludere in Humanis, *to play but a game at Chesse with this world; to sport Himself with making little things great, and great things nothing; Imagine God to be at play with us, but a gamester. . . .*

—JOHN DONNE

A BRITISH ESSAYIST of distinction, H. J. Massingham, once remarked perceptively that woods nowadays are haunted not by ghosts, but by a silence and man-made desolation that might well take terrifying material forms. There is nothing like a stalled train in a marsh to promote such reflections—particularly if one has been transported just beyond the environs of a great city and set down in some nether world that seems to partake both of nature before man came and of the residue of what will exist after him. It was night when my train halted, but a kind of flame-wreathed landscape attended by shadowy figures could be glimpsed from the window.

After a time, with a companion, I descended and strolled forward to explore this curious region. It turned out to be a perpetually burning city dump, contributing its miasmas and choking vapors to the murky sky above the city. Amidst the tended flames of this inferno I approached one of the grimy attendants who was forking over the rubbish. In the background, other shadows, official and unofficial, were similarly engaged. For a moment I had the insubstantial feeling that must exist on the borders of hell, where everything, wavering among heat waves, is transported to another dimension. One could imagine ragged and distorted souls grubbed over by scavengers for what might usefully survive.

I stood in silence watching this great burning. Sodden papers were being forked into the flames, and after a while it crossed my mind that this was perhaps the place where last year's lace valentines had gone, along with old Christmas trees, and the beds I had slept on in childhood.

"I suppose you get everything here," I ventured to the grimy attendant.

He nodded indifferently and drew a heavy glove across his face. His eyes were red-rimmed from the fire. Perhaps they were red anyhow.

"Know what?" He swept a hand outward toward the flames.

"No," I confessed.

"Babies," he growled in my ear. "Even dead babies sometimes turn up. From there." He gestured contemptuously toward the city and hoisted an indistinguishable mass upon his fork. I stepped back from the flare of light, but it was only part of an old radio cabinet. Out of it had once come voices and music and laughter, perhaps from the twenties. And where were the voices traveling now? I looked at the dangling fragments of wire. They reminded me of something, but the engine bell sounded before I could remember.

I made a parting gesture. Around me in the gloom dark shapes worked ceaselessly at the dampened fires. My eyes were growing accustomed to their light.

"We get it all," the dump philosopher repeated. "Just give it time to travel, we get it all."

"Be seeing you," I said irrelevantly. "Good luck."

Back in my train seat, I remembered unwillingly the flames and the dangling wire. It had something to do with an air crash years ago and the identification of the dead. Anthropologists get strange assignments. I put the matter out of my mind, as I always do, but I dozed and it came back: the box with the dangling wires. I had once fitted a seared and broken skullcap over a dead man's brains, and I had thought, peering into the scorched and mangled skull vault, it is like a beautiful, irreparably broken machine, like something consciously made to be used, and now where are the voices and the music?

"We get it all," a dark figure said in my dreams. I sighed, and the figure in the murk faded into the clicking of the wheels.

One can think just so much, but the archaeologist is awake to memories of the dead cultures sleeping around us, to our destiny, and to the nature of the universe we profess to inhabit. I would speak of these things not as a wise man, with scientific certitude, but from a place outside, in the role, shall we say, of a city-dump philosopher. Nor is this a strained figure of speech.

The archaeologist is the last grubber among things mortal. He puts not men, but civilizations, to bed, and passes upon them final judgments. He finds, if imprinted upon clay, both our grocery bills and the hymns to our gods. Or he uncovers, as I once did in a mountain cavern, the skeleton of a cradled child, supplied, in the pathos of our mortality, with the carefully "killed" tools whose shadowy counterparts were intended to serve a tiny infant through the vicissitudes it would encounter beyond the dark curtain of death. Infinite care had been lavished upon objects that did not equate with the child's ability to use them in this life. Was his spirit expected to grow to manhood, or had this final projection of bereaved parental care thrust into the night, in desperate anxiety, all that an impoverished and simple culture could provide where human affection could not follow?

In a comparable but more abstract way, the modern mind, the scientific mind, concerned as it is with the imponderable mysteries of existence, has sought to equip oncoming generations with certain mental weapons against the terrors of ignorance. Protectively, as in the case of the dead child bundled in a cave, science has proclaimed a universe whose laws are open to discovery and, above all, it has sought, in the words of one of its greatest exponents, Francis Bacon, "not to imagine or suppose, but to *discover* what nature does or may be made to do."

To discover what nature does, however, two primary restrictions are laid upon a finite creature: he must extrapolate his laws from what exists in his or his society's moment of time and, in addition, he is limited by what his senses can tell him of the surrounding world. Later, technology may provide the extension of those senses, as in the case of the microscope and telescope. Nevertheless the same eye or ear with which we are naturally endowed must, in the end, interpret the data derived from such extensions of sight or hearing. Moreover, science since the thirteenth century has clung to the dictum of William of Ockham that hypotheses must not be multiplied excessively; that the world, in essence, is always simple, not complicated, and its secrets accessible to men of astute and sufficiently penetrating intellect. Ironically, in the time of our greatest intellectual and technological triumphs one is forced to say that Ockham's long-honored precepts, however well they have served man, are, from another view, merely a more sophisticated projection of man's desire for order—and for the ability to control, understand, and manipulate his world.

All of these intentions are commendable enough, but perhaps we would approach them more humbly and within a greater frame of reference if we were to recognize what Massingham sensed as lying latent in his wood, or what John Donne implied more than three centuries ago when he wrote:

> I am rebegot
> of absence, darknesse, death:
> Things which are not.

Donne had recognized that behind visible nature lurks an invisible and procreant void from whose incomprehensible magnitude we can only recoil. That void has haunted me ever since I handled the shattered calvarium that a few hours before had contained, in microcosmic dimensions, a similar lurking potency.

Some years previously, I had written a little book of essays in which I had narrated how time had become natural in our thinking, and I had gone on to speak likewise of life and man. In the end, however, I had been forced to ask, How Natural is Natural?—a subject that raised the hackles of some of my scientifically inclined colleagues, who confused the achievements of their disciplines with certitude on a cosmic scale. My very question thus implied an ill-concealed heresy. That heresy it is my intent to pursue further. It will involve us, not in the denigration of science, but, rather, in a farther stretch of the imagination as we approach those distant and wooded boundaries of thought where, in the words of the old fairy tale, the fox and the hare say good night to each other. It is here that predictability ceases and the unimaginable begins—or, as a final heretical suspicion, we might ask ourselves whether our own little planetary fragment of the cosmos has all along concealed a mocking refusal to comply totally with human conceptions of order and secure prediction.

The world contains, for all its seeming regularity, a series of surprises resembling those that in childhood terrorized us by erupting on springs from closed boxes. The world of primitive man is not dissimilar. Lightning leaps from clouds, something invisible rumbles in the air, the living body, spilling its mysterious red fluid, lies down in a sleep from which it cannot waken. There are night cries in the forest, talking waters, guiding omens, or portents in the fall of a leaf. No longer, as with the animal, can the world be accepted as given. It has to be perceived and consciously thought about, abstracted, and considered. The moment one does so, one is outside of the natural; objects are each one surrounded with an aura radiating meaning to man alone. To a universe already suspected of being woven together by unseen forces, man brings the organizing power of primitive magic. The manikin that is believed to control the macrocosm by some sympathetic connection is already obscurely present in the poppet thrust full of needles by the witch. Crude and imperfect, magic is still man's first conscious abstraction from nature, his first attempt to link disparate objects by some unseen attraction between them.

II

If we now descend into the early years of modern science, we find the world of the late eighteenth and early nineteenth centuries basking comfortably in the conception of the balanced world machine. Newton had established what appeared to be the reign of universal order in the heavens. The planets—indeed, the whole cosmic engine—were self-regulatory. This passion for order controlled by a divinity too vast to be concerned with petty miracle was slowly extended to earth. James Hutton glimpsed, in the long erosion and renewal of the continents by subterranean uplift, a similar "beautiful machine" so arranged that recourse to the "preternatural," or "destructive accident," such as the Mosaic Deluge, was unnecessary to account for the physical features of the planet.

Time had lengthened, and through those eons, law, not chaos, reigned. The imprint of fossil raindrops similar to those of today had been discovered upon ancient shores. The marks of fossil ripples were also observable in uncovered strata, and buried trees had absorbed the sunlight of far millennia. The remote past was one with the present and, over all, a lawful similarity decreed by a Christian Deity prevailed.

In the animal world, a similar web of organization was

believed to exist, save by a few hesitant thinkers. The balanced Newtonian clockwork of the heavens had been transferred to earth and, for a few decades, was destined to prevail in the world of life. Plants and animals would be frozen into their existing shapes; they would compete but not change, for change in this system was basically a denial of law. Hutton's world renewed itself in cycles, just as the oscillations observable in the heavens were similarly self-regulatory.

Time was thus law-abiding. It contained no novelty and was self-correcting. It was, as we have indicated, a manifestation of divine law. That law was a comfort to man. The restive world of life fell under the same dominion as the equally restive particles of earth. Organisms oscillated within severely fixed limits. The smallest animalcule in a hay infusion carried a message for man; the joints of an insect assured him of divine attention. "In every nature and every portion of nature which we can descry," wrote William Paley in a book characteristic of the period, "we find attention bestowed upon even the minutest parts. The hinges in the wing of an earwig . . . are as highly wrought as if the creator had nothing else to finish. We see no signs of diminution of care by multiplicity of objects, or distraction of thought by variety. We have no reason to fear, therefore, our being forgotten, or overlooked, or neglected." Written into these lines in scientific guise is the same humanly protective gesture that long ago had heaped skin blankets, bone needles, and a carved stick for killing rabbits into the burial chamber of a child.

This undeviating balance in which life was locked was called "natural government" by the great anatomist John Hunter. It was, in a sense, like the cyclic but undeviating life of the planet earth itself. That vast elemental creature felt the fall of raindrops on its ragged flanks, was troubled by the drift of autumn leaves or the erosive work of wind throughout eternity. Nevertheless, the accounts of nature were strictly kept. If a continent was depressed at one point, its equivalent arose elsewhere. Whether the item in the scale was the weight of a raindrop or a dislodged boulder on a mountainside, a dynamic balance kept the great beast young and flourishing upon its course.

And as it was with earth, so also with its inhabitants. "There is an equilibrium kept up among the animals by themselves,"

Hunter went on to contend. They kept their own numbers pruned and in proportion. Expansion was always kept within bounds. The struggle for existence was recognized before Darwin, but only as the indefinite sway of a returning pendulum. Life was selected, but it was selected for but one purpose: vigor and consistency in appearance. The mutative variant was struck down. What had been was; what would be already existed. As in the case of that great animal the earth, of the living flora and fauna it could be said that there was to be found "no vestige of a beginning,—no prospect of an end." An elemental order lay across granite, sea, and shore. Each individual animal peered from age to age out of the same unyielding sockets of bone. Out of no other casements could he stare; the dweller within would see leaf and bird eternally the same. This was the scientific doctrine known as uniformitarianism. It had abolished magic as it had abolished the many changes and shape shiftings of witch doctors and medieval necromancers. At last the world was genuinely sane under a beneficent Deity. Then came Darwin.

III

At first, he was hailed as another Newton who had discovered the laws of life. It was true that what had once been deemed independent creations—the shells in the collector's cabinet, the flowers pressed into memory books—were now, as in the abandoned magic of the ancient past, once more joined by invisible threads of sympathy and netted together by a common ancestry. The world seemed even more understandable, more natural than natural. The fortuitous had become fashionable, and the other face of "natural government" turned out to be creation. Life's pendulum of balance was an illusion.

Behind the staid face of that nature we had worshiped for so long we were unseen shapeshifters. Viewed in the long light of limitless time, we were optical illusions whose very identity was difficult to fix. Still, there was much talk of progress and perfection. Only later did we begin to realize that what Charles Darwin had introduced into nature was not Newtonian predictability but absolute random novelty. Life was bent, in the phrase of Alfred Russel Wallace, upon "indefinite departure."

No living thing, not even man, understood upon what journey he had embarked. Time was no longer cyclic or monotonously repetitious.* It was historic, novel, and unreturning. Since that momentous discovery, man has, whether or not he realizes or accepts his fate, been moving in a world of contingent forms.

Even in the supposedly stable universe of matter, as it was viewed by nineteenth-century scientists, new problems constantly appear. The discovery by physicists of antimatter particles having electric charges opposite to those that compose our world and unable to exist in concert with known matter raises the question of whether, after all, our corner of the universe is representative of the entire potentialities that may exist elsewhere. The existence of antimatter is unaccounted for in present theories of the universe, and such peculiarities as the primordial atom and the recently reported flash of the explosion at the birth of the universe, as recorded in the radio spectrum, lead on into unknown paths.

If it were not for the fact that familiarity leads to assumed knowledge, we would have to admit that the earth's atmosphere of oxygen appears to be the product of a biological invention, photosynthesis, another random event that took place in Archeozoic times. That single "invention," for such it was, determined the entire nature of life on this planet, and there is no possibility at present of calling it preordained. Similarly, the stepped-up manipulation of chance, in the shape of both mutation and recombination of genetic factors, which is one result of the sexual mechanism, would have been unprophesiable.

The brain of man, that strange gray iceberg of conscious and unconscious life, was similarly unpredictable until its appearance. A comparatively short lapse of geological time has evolved a humanity that, beginning in considerable physical diversity, has increasingly converged toward a universal biological similarity, marked only by a lingering and insignificant racial differentiation. With the rise of *Homo sapiens* and the final perfection of the human brain as a manipulator of symbolic

---

*For purposes of space I have chosen to ignore the short-lived doctrine of the early century known as catastrophism, since I have treated it at length elsewhere.

thought, the spectrum of man's possible social behavior has widened enormously. What is essentially the same brain biologically can continue to exist in the simple ecological balance of the Stone Age or, on the other hand, may produce those enormous inflorescences known as civilizations. These growths seemingly operate under their own laws and take distinct and irreversible pathways. In an analogous way, organisms mutate and diverge through adaptive radiation from one or a few original forms.

In the domain of culture, man's augmented ability to manipulate abstract ideas and to draw in this fashion enormous latent stores of energy from his brain has led to an intriguing situation: the range of his *possible* behavior is greater and more contradictory than that which can be contained within the compass of a single society, whether tribal or advanced. Thus, as man's penetration into the metaphysical and abstract has succeeded, so has his capacity to follow, in the same physical body, a series of tangential roads into the future. Likeness in body has, paradoxically, led to diversity in thought. Thought, in turn, involves such vast institutional involutions as the rise of modern science, with its intensified hold upon modern society.

All past civilizations of men have been localized and have had, therefore, the divergent mutative quality to which we have referred. They have offered choices to men. Ideas have been exchanged, along with technological innovations, but never on so vast, overwhelming, and single-directed a scale as in the present. Increasingly, there is but one way into the future: the technological way. The frightening aspect of this situation lies in the constriction of human choice. Western technology has released irrevocable forces, and the "one world" that has been talked about so glibly is frequently a distraught conformity produced by the centripetal forces of Western society. So great is its power over men that any other solution, any other philosophy, is silenced. Men, unknowingly, and whether for good or ill, appear to be making their last decisions about human destiny. To pursue the biological analogy, it is as though, instead of many adaptive organisms, a single gigantic animal embodied the only organic future of the world.

IV

Archaeology is the science of man's evening, not of his midday triumphs. I have spoken of my visit to a flame-wreathed marsh at nightfall. All in it had been substance, matter, trailing wires and old sandwich wrappings, broken toys and iron bedsteads. Yet there was nothing present that science could not reduce into its elements, nothing that was not the product of the urban world whose far-off towers had risen gleaming in the dusk beyond the marsh. There on the city dump had lain the shabby debris of life: the waxen fragment of an old record that had stolen a human heart, wilted flowers among smashed beer cans, the castaway knife of a murderer, along with a broken tablespoon. It was all a maze of invisible, floating connections, and would be until the last man perished. These forlorn materials had all been subjected to the dissolving power of the human mind. They had been wrenched from deep veins of rock, boiled in great crucibles, and carried miles from their origins. They had assumed shapes that, though material enough, had existed first as blueprints in the profound darkness of a living brain. They had been defined before their existence, named and given shape in the puff of air that we call a word. That word had been evoked in a skull box which, with all its contained powers and lurking paradoxes, has arisen in ways we can only dimly retrace.

Einstein is reputed to have once remarked that he refused to believe that God plays at dice with the universe. But as we survey the long backward course of history, it would appear that in the phenomenal world the open-endedness of time is unexpectedly an essential element of His creation. Whenever an infant is born, the dice, in the shape of genes and enzymes and the intangibles of chance environment, are being rolled again, as when that smoky figure from the fire hissed in my ear the tragedy of the cast-off infants of the city. Each one of us is a statistical impossibility around which hover a million other lives that were never destined to be born—but who, nevertheless, are being unmanifest, a lurking potential in the dark storehouse of the void.

Today, in spite of that web of law, that network of forces which the past century sought to string to the ends of the universe, a strange unexpectedness lingers about our world. This change in viewpoint, which has frequently escaped our attention, can be illustrated in the remark of Heinrich Hertz, the nineteenth-century experimenter in the electromagnetic field. "The most important problem which our conscious knowledge of nature should enable us to solve," Hertz stated, "is the anticipation of future events, so that we may arrange our present affairs in accordance with such anticipation."

There is an attraction about this philosophy that causes it to linger in the lay mind and, as a short-term prospect, in the minds of many scientists and technologists. It implies a tidiness that is infinitely attractive to man, increasingly a homeless orphan lost in the vast abysses of space and time. Hertz's remark seems to offer surcease from uncertainty, power contained, the universe understood, the future apprehended before its emergence. The previous Elizabethan age, by contrast, had often attached to its legal documents a humble obeisance to life's uncertainties expressed in the phrase "by the mutability of fortune and favor." The men of Shakespeare's century may have known less of science, but they knew only too well what unexpected overthrow was implied in the frown of a monarch or a breath of the plague.

The twentieth century, on the other hand, surveys a totally new universe. That our cosmological conceptions bear a relationship to the past is obvious, that some of the power of which Hertz dreamed lies in our hands is all too evident, but never before in human history has the mind soared higher and seen less to cheer its complacency. We have heard much of science as the endless frontier, but we whose immediate ancestors were seekers of gold among great mountains and gloomy forests are easily susceptible to a simplistic conception of the word *frontier* as something conquerable in its totality. We assume that, with enough time and expenditure of energy, the ore will be extracted and the forests computed in board feet of lumber. A tamed wilderness will subject itself to man.

Not so the wilderness beyond the stars or concealed in the infinitesimal world beneath the atom. Wise reflection will lead us to recognize that we have come upon a different and less conquerable region. Forays across its border already suggest that man's dream of mastering all aspects of nature takes no account of his limitations in time and space or of his own senses, augmented though they may be by his technological devices. Even the thought that he can bring to bear upon that frontier is limited in quantity by the number of trained minds that can sustain such an adventure. Ever more expensive grow the tools with which research can be sustained, ever more diverse the social problems which that research, in its technological phase, promotes. To take one single example: who would have dreamed that a tube connecting two lenses of glass would pierce into the swarming depths of our being, force upon us incredible feats of sanitary engineering, master the plague, and create that giant upsurge out of unloosened nature that we call the population explosion?

The Roman Empire is a past event in history, yet by analogy it presents us with a small scale model comparable to the endless frontier of science. A great political and military machine had expanded outward to the limits of the known world. Its lines of communication grew ever more tenuous, taxes rose fantastically, the disaffected and alienated within its borders steadily increased. By the time of the barbarian invasions the vast structure was already dying of inanition. Yet that empire lasted far longer than the world of science has yet endured.

But what of the empire of science? Does not its word leap fast as light, is it not a creator of incalculable wealth, is not space its plaything? Its weapons are monstrous; its eye is capable of peering beyond millions of light-years. There is one dubious answer to this buoyant optimism: science is human; it is of human devising and manufacture. It has not prevented war; it has perfected it. It has not abolished cruelty or corruption; it has enabled these abominations to be practiced on a scale unknown before in human history.

Science is a solver of problems, but it is dealing with the limitless, just as, in a cruder way, were the Romans. Solutions to problems create problems; their solutions, in turn, multiply into additional problems that escape out of scientific hands like noxious insects into the interstices of the social fabric. The rate of growth is geometric, and the vibrations set up can even now be detected in our institutions. This is what the Scottish biologist D'Arcy Thompson called the evolution of contingency. It is no longer represented by the long, slow turn of world time as the geologist has known it. Contingency has escaped into human hands and flickers unseen behind every whirl of our machines, every pronouncement of political policy.

Each one of us before his death looks back upon a childhood whose ways now seem as remote as those of Rome. "Daddy," the small daughter of a friend of mine recently asked, "tell me how it was in olden days." As my kindly friend groped amidst his classical history, he suddenly realized with a slight shock that his daughter wanted nothing more than an account of his own childhood. It was forty years away and it was already "olden days." "There was a time," he said slowly to the enchanted child, "called the years of the Great Depression. In that time there was a very great deal to eat, but men could not buy it. Little girls were scarcer than now. You see," he said painfully, "their fathers could not afford them, and they were not born." He made a half-apologetic gesture to the empty room, as if to a gathering of small reproachful ghosts. "There was a monster we never understood called Overproduction. There were," and his voice trailed hopelessly into silence, "so many dragons in that time you could not believe it. And there was a very civilized nation where little girls were taken from their parents. . . ." He could not go on. The eyes from Auschwitz, he told me later, would not permit him.

V

Recently, I passed a cemetery in a particularly bleak countryside. Adjoining the multitude of stark upthrust gray stones was an incongruous row of six transparent telephone booths erected in that spot for reasons best known to the communications industry. Were they placed there for the midnight convenience of the dead, or for the midday visitors who might attempt speech with the silent people beyond the fence? It was difficult to determine, but I thought the episode suggestive of our dilemma.

An instrument for communication, erected by a powerful unseen intelligence, was at my command, but I suspect—although I was oddly averse to trying to find out—that the wires did not run in the proper direction, and that there was something disconnected or disjointed about the whole endeavor. It was, I fear, symbolic of an unexpected aspect of our universe, a universe that, however strung with connecting threads, is endowed with an open-ended and perverse quality we shall never completely master. Nature contains that which does not concern us, and has no intention of taking us into its confidence. It may provide us with receiving boxes of white bone as cunning in their way as the wired booths in the cemetery, but, like these, they appear to lack some essential ingredient of genuine connection. As we consider what appears to be the chance emergence of photosynthesis, which turns the light of a far star into green leaves, or the creation of the phenomenon of sex that causes the cards at the gaming table of life to be shuffled with increasing frequency and into ever more diverse combinations, it should be plain that nature contains the roiling unrest of a tornado. It is not the self-contained stately palace of the eighteenth-century philosophers, a palace whose doorstep was always in precisely the same position.

From the oscillating universe, beating like a gigantic heart, to the puzzling existence of antimatter, order, in a human sense, is at least partially an illusion. Ours, in reality, is the order of a time, and of an insignificant fraction of the cosmos, seen by the limited senses of a finite creature. Behind the appearance, as even one group of primitive philosophers, the Hopi, have grasped, lurks being unmanifest, whose range and number exceed the real. This is why the unexpected will always confront us; this is why the endless frontier is really endless. This is why the half-formed chaos of the marsh moved me as profoundly as though a new prophetic shape induced by us had risen monstrously from dangling wire and crumpled cardboard.

We are more dangerous than we seem and more potent in our ability to materialize the unexpected that is drawn from our own minds. "Force maketh Nature more violent in the Returne," Francis Bacon had once written. In the end, this is her primary quality. Her creature man partakes of that essence, and it is well that he consider it in contemplation and not always in action. To the unexpected nature of the universe man owes his being. More than any other living creature he contains, unknowingly, the shapes and forms of an uncreated future to be drawn from his own substance. The history of this unhappy century should prove a drastic warning of his powers of dissolution, even when directed upon himself. Waste, uncertain marshes, lie close to reality in our heads. Shapes as yet unevoked had best be left lying amidst those spectral bog lights, lest the drifting smoke of dreams merge imperceptibly, as once it did, with the choking real fumes from the ovens of Belsen and Buchenwald.

"It is very unhappy, but too late to be helped," Emerson had noted in his journal, "the discovery we have made that we exist. That discovery is called the Fall of Man. Ever afterwards we suspect our instruments. We have learned that we do not see directly." Wisdom interfused with compassion should be the consequence of that discovery, for at the same moment one aspect of the unexpected universe will have been genuinely revealed. It lies deep-hidden in the human heart, and not at the peripheries of space. Both the light we seek and the shadows that we fear are projected from within. It is through ourselves that the organic procession pauses, hesitates, or renews its journey. "We have learned to ask terrible questions," exclaimed that same thinker in the dawn of Victorian science. Perhaps it is just for this that the Unseen Player in the void has rolled his equally terrible dice. Out of the self-knowledge gained by putting dreadful questions man achieves his final dignity.

THREE

The Hidden Teacher

Sometimes the best teacher teaches only once to a single child or to a grownup past hope.

—ANONYMOUS

THE PUTTING OF formidable riddles did not arise with to-day's philosophers. In fact, there is a sense in which the experimental method of science might be said merely to have widened the area of man's homelessness. Over two thousand years ago, a man named Job, crouching in the Judean desert, was moved to challenge what he felt to be the injustice of his God. The voice in the whirlwind, in turn, volleyed pitiless questions upon the supplicant—questions that have, in truth, precisely the ring of modern science. For the Lord asked of Job by whose wisdom the hawk soars, and who had fathered the rain, or entered the storehouses of the snow.

A youth standing by, one Elihu, also played a role in this drama, for he ventured diffidently to his protesting elder that it was not true that God failed to manifest Himself. He may speak in one way or another, though men do not perceive it. In consequence of this remark perhaps it would be well, whatever our individual beliefs, to consider what may be called the hidden teacher, lest we become too much concerned with the formalities of only one aspect of the education by which we learn.

We think we learn from teachers, and we sometimes do. But the teachers are not always to be found in school or in great laboratories. Sometimes what we learn depends upon our own powers of insight. Moreover, our teachers may be hidden, even the greatest teacher. And it was the young man Elihu who observed that if the old are not always wise, neither can the teacher's way be ordered by the young whom he would teach.

For example, I once received an unexpected lesson from a spider.

It happened far away on a rainy morning in the West. I had come up a long gulch looking for fossils, and there, just at eye level, lurked a huge yellow-and-black orb spider, whose web was moored to the tall spears of buffalo grass at the edge of the arroyo. It was her universe, and her senses did not extend beyond the lines and spokes of the great wheel she inhabited. Her extended claws could feel every vibration throughout that delicate structure. She knew the tug of wind, the fall of a raindrop, the flutter of a trapped moth's wing. Down one spoke of the web ran a stout ribbon of gossamer on which she could hurry out to investigate her prey.

Curious, I took a pencil from my pocket and touched a strand of the web. Immediately there was a response. The web, plucked by its menacing occupant, began to vibrate until it was a blur. Anything that had brushed claw or wing against that amazing snare would be thoroughly entrapped. As the vibrations slowed, I could see the owner fingering her guidelines for signs of struggle. A pencil point was an intrusion into this universe for which no precedent existed. Spider was circumscribed by spider ideas; its universe was spider universe. All outside was irrational, extraneous, at best, raw material for spider. As I proceeded on my way along the gully, like a vast impossible shadow, I realized that in the world of spider I did not exist.

Moreover, I considered, as I tramped along, that to the phagocytes, the white blood cells, clambering even now with some kind of elementary intelligence amid the thin pipes and tubing of my body—creatures without whose ministrations I could not exist—the conscious "I" of which I was aware had no significance to these amoeboid beings. I was, instead, a kind of chemical web that brought meaningful messages to them, a natural environment seemingly immortal if they could have thought about it, since generations of them had lived and perished, and would continue to so live and die, in that odd fabric which contained my intelligence—a misty light that was beginning to seem floating and tenuous even to me.

I began to see that among the many universes in which the world of living creatures existed, some were large, some small, but that all, including man's, were in some way limited or finite. We were creatures of many different dimensions passing through each other's lives like ghosts through doors.

In the years since, my mind has many times returned to that far moment of my encounter with the orb spider. A message has arisen only now from the misty shreds of that webbed universe. What was it that had so troubled me about the incident? Was it that spidery indifference to the human triumph?

If so, that triumph was very real and could not be denied. I saw, had many times seen, both mentally and in the seams of exposed strata, the long backward stretch of time whose recovery is one of the great feats of modern science. I saw the drifting cells of the early seas from which all life, including our own, has arisen. The salt of those ancient seas is in our blood, its lime is in our bones. Every time we walk along a beach some ancient urge disturbs us so that we find ourselves shedding shoes and garments, or scavenging among seaweed and whitened timbers like the homesick refugees of a long war.

And war it has been indeed—the long war of life against its inhospitable environment, a war that has lasted for perhaps three billion years. It began with strange chemicals seething under a sky lacking in oxygen; it was waged through long ages until the first green plants learned to harness the light of the nearest star, our sun. The human brain, so frail, so perishable, so full of inexhaustible dreams and hungers, burns by the power of the leaf.

The hurrying blood cells charged with oxygen carry more of that element to the human brain than to any other part of the body. A few moments' loss of vital air and the phenomenon we know as consciousness goes down into the black night of inorganic things. The human body is a magical vessel, but its life is linked with an element it cannot produce. Only the green plant knows the secret of transforming the light that comes to us across the far reaches of space. There is no better illustration of the intricacy of man's relationship with other living things.

The student of fossil life would be forced to tell us that if we take the past into consideration the vast majority of earth's creatures—perhaps over ninety per cent—have vanished. Forms that flourished for a far longer time than man has existed upon earth have become either extinct or so transformed that their descendants are scarcely recognizable. The specialized perish with the environment that created them, the tooth of the tiger fails at last, the lances of men strike down the last mammoth.

In three billion years of slow change and groping effort only one living creature has succeeded in escaping the trap of specialization that has led in time to so much death and wasted endeavor. It is man, but the word should be uttered softly, for his story is not yet done.

With the rise of the human brain, with the appearance of a creature whose upright body enabled two limbs to be freed for the exploration and manipulation of his environment, there had at last emerged a creature with a specialization—the brain—that, paradoxically, offered escape from specialization. Many animals driven into the nooks and crannies of nature have achieved momentary survival only at the cost of later extinction.

Was it this that troubled me and brought my mind back to a tiny universe among the grass-blades, a spider's universe concerned with spider thought?

Perhaps.

The mind that once visualized animals on a cave wall is now engaged in a vast ramification of itself through time and space. Man has broken through the boundaries that control all other life. I saw, at last, the reason for my recollection of that great spider on the arroyo's rim, fingering its universe against the sky.

The spider was a symbol of man in miniature. The wheel of the web brought the analogy home clearly. Man, too, lies at the heart of a web, a web extending through the starry reaches of sidereal space, as well as backward into the dark realm of prehistory. His great eye upon Mount Palomar looks into a distance of millions of light-years, his radio ear hears the whisper of even more remote galaxies, he peers through the electron microscope upon the minute particles of his own being. It is a web no creature of earth has ever spun before. Like the orb spider, man lies at the heart of it, listening. Knowledge has given him the memory of earth's history beyond the time of his emergence. Like the spider's claw, a part of him touches a world he will never enter in the flesh. Even now, one can see him reaching forward into time with new machines, computing, analyzing, until elements of the shadowy future will also compose part of the invisible web he fingers.

Yet still my spider lingers in memory against the sunset sky.

Spider thoughts in a spider universe—sensitive to raindrop and moth flutter, nothing beyond, nothing allowed for the unexpected, the inserted pencil from the world outside.

Is man at heart any different from the spider, I wonder: man thoughts, as limited as spider thoughts, contemplating now the nearest star with the threat of bringing with him the fungus rot from earth, wars, violence, the burden of a population he refuses to control, cherishing again his dream of the Adamic Eden he had pursued and lost in the green forests of America. Now it beckons again like a mirage from beyond the moon. Let man spin his web, I thought further; it is his nature. But I considered also the work of the phagocytes swarming in the rivers of my body, the unresting cells in their mortal universe. What is it we are a part of that we do not see, as the spider was not gifted to discern my face, or my little probe into her world?

We are too content with our sensory extensions, with the fulfillment of that ice age mind that began its journey amidst the cold of vast tundras and that pauses only briefly before its leap into space. It is no longer enough to see as a man sees— even to the ends of the universe. It is not enough to hold nuclear energy in one's hand like a spear, as a man would hold it, or to see the lightning, or times past, or time to come, as a man would see it. If we continue to do this, the great brain— the human brain—will be only a new version of the old trap, and nature is full of traps for the beast that cannot learn.

It is not sufficient any longer to listen at the end of a wire to the rustlings of galaxies; it is not enough even to examine the great coil of DNA in which is coded the very alphabet of life. These are our extended perceptions. But beyond lies the great darkness of the ultimate Dreamer, who dreamed the light and the galaxies. Before act was, or substance existed, imagination grew in the dark. Man partakes of that ultimate wonder and creativeness. As we turn from the galaxies to the swarming cells of our own being, which toil for something, some entity beyond their grasp, let us remember man, the self-fabricator who came across an ice age to look into the mirrors and the magic of science. Surely he did not come to see himself or his wild visage only. He came because he is at heart a listener and a searcher for some transcendent realm beyond himself. This he has worshiped by many names, even in the dismal caves of his beginning. Man, the self-fabricator, is so by reason of gifts he had no part in devising—and so he searches as the single living cell in the beginning must have sought the ghostly creature it was to serve.

II

The young man Elihu, Job's counselor and critic, spoke simply of the "Teacher," and it is of this teacher I speak when I refer to gifts man had no part in devising. Perhaps—though it is purely a matter of emotional reactions to words—it is easier for us today to speak of this teacher as "nature," that omnipresent all which contained both the spider and my invisible intrusion into her carefully planned universe. But nature does not simply represent reality. In the shapes of life, it prepares the future; it offers alternatives. Nature teaches, though what it teaches is often hidden and obscure, just as the voice from the spinning dust cloud belittled Job's thought but gave back no answers to its own formidable interrogation.

A few months ago I encountered an amazing little creature on a windy corner of my local shopping center. It seemed, at first glance, some long-limbed, feathery spider teetering rapidly down the edge of a store front. Then it swung into the air and, as hesitantly as a spider on a thread, blew away into the parking lot. It returned in a moment on a gust of wind and ran toward me once more on its spindly legs with amazing rapidity.

With great difficulty I discovered the creature was actually a filamentous seed, seeking a hiding place and scurrying about with the uncanny surety of a conscious animal. In fact, it *did* escape me before I could secure it. Its flexible limbs were stiffer than milkweed down, and, propelled by the wind, it ran rapidly and evasively over the pavement. It was like a gnome scampering somewhere with a hidden packet—for all that I could tell, a totally new one: one of the jumbled alphabets of life.

A new one? So stable seem the years and all green leaves, a botanist might smile at my imaginings. Yet bear with me a moment. I would like to tell a tale, a genuine tale of childhood. Moreover, I was just old enough to know the average of my kind and to marvel at what I saw. And what I saw was straight from the hidden Teacher, whatever be his name.

It is told in the Orient of the Hindu god Krishna that his mother, wiping his mouth when he was a child, inadvertently peered in and beheld the universe, though the sight was mercifully and immediately veiled from her. In a sense, this is what happened to me. One day there arrived at our school a newcomer, who entered the grade above me. After some days this lad, whose look of sleepy-eyed arrogance is still before me as I write, was led into my mathematics classroom by the principal. Our class was informed severely that we should learn to work harder.

With this preliminary exhortation, great rows of figures were chalked upon the blackboard, such difficult mathematical problems as could be devised by adults. The class watched in helpless wonder. When the preparations had been completed, the young pupil sauntered forward and, with a glance of infinite boredom that swept from us to his fawning teachers, wrote the answers, as instantaneously as a modern computer, in their proper place upon the board. Then he strolled out with a carelessly exaggerated yawn.

Like some heavy-browed child at the wood's edge, clutching the last stone hand ax, I was witnessing the birth of a new type of humanity—one so beyond its teachers that it was being used for mean purposes while the intangible web of the universe in all its shimmering mathematical perfection glistened untaught in the mind of a chance little boy. The boy, by then grown self-centered and contemptuous, was being dragged from room to room to encourage us, the paleanthropes, to duplicate what, in reality, our teachers could not duplicate. He was too precious an object to be released upon the playground among us, and with reason. In a few months his parents took him away.

Long after, looking back from maturity, I realized that I had been exposed on that occasion, not to human teaching, but to the Teacher, toying with some sixteen billion nerve cells interlocked in ways past understanding. Or, if we do not like the anthropomorphism implied in the word *teacher*, then nature, the old voice from the whirlwind fumbling for the light. At all events, I had been the fortunate witness to life's unbounded creativity—a creativity seemingly still as unbalanced and chance-filled as in that far era when a black scaled creature had broken from an egg and the age of the giant reptiles, the creatures of the prime, had tentatively begun.

Because form cannot be long sustained in the living, we collapse inward with age. We die. Our bodies, which were the product of a kind of hidden teaching by an alphabet we are only beginning dimly to discern, are dismissed into their elements. What is carried onward, assuming we have descendants, is the little capsule of instructions such as I encountered hastening by me in the shape of a running seed. We have learned the first biological lesson: that in each generation life passes through the eye of a needle. It exists for a time molecularly and in no recognizable semblance to its adult condition. It *instructs* its way again into man or reptile. As the ages pass, so do variants of the code. Occasionally, a species vanishes on a wind as unreturning as that which took the pterodactyls.

Or the code changes by subtle degrees through the statistical altering of individuals; until I, as the fading Neanderthals must once have done, have looked with still-living eyes upon the creature whose genotype was quite possibly to replace me. The genetic alphabets, like genuine languages, ramify and evolve along unreturning pathways.

If nature's instructions are carried through the eye of a needle, through the molecular darkness of a minute world below the field of human vision and of time's decay, the same, it might be said, is true of those monumental structures known as civilizations. They are transmitted from one generation to another on invisible puffs of air known as words—words that can also be symbolically incised on clay. As the delicate printing on the mud at the water's edge retraces a visit of autumn birds long since departed, so the little scrabbled tablets in perished cities carry the seeds of human thought across the deserts of millennia. In this instance the teacher is the social brain but it, too, must be compressed into minute hieroglyphs, and the minds that wrought the miracle efface themselves amidst the jostling torrent of messages, which, like the genetic code, are shuffled and reshuffled as they hurry through eternity. Like a mutation, an idea may be recorded in the wrong time, to lie latent like a recessive gene and spring once more to life in an auspicious era.

Occasionally, in the moments when an archaeologist lifts the slab over a tomb that houses a great secret, a few men gain a unique glimpse through that dark portal out of which all men living have emerged, and through which messages again must pass. Here the Mexican archaeologist Ruz Lhuillier speaks of his first penetration of the great tomb hidden beneath dripping stalactites at the pyramid of Palenque: "Out of the dark shadows, rose a fairy-tale vision, a weird ethereal spectacle from another world. It was like a magician's cave carved out of ice, with walls glittering and sparkling like snow crystals." After shining his torch over hieroglyphs and sculptured figures, the explorer remarked wonderingly: "We were the first people for more than a thousand years to look at it."

Or again, one may read the tale of an unknown Pharaoh who had secretly arranged that a beloved woman of his household should be buried in the tomb of the god-king—an act of compassion carrying a personal message across the millennia in defiance of all precedent.

Up to this point we have been talking of the single hidden teacher, the taunting voice out of that old Biblical whirlwind which symbolizes nature. We have seen incredible organic remembrance passed through the needle's eye of a microcosmic world hidden completely beneath the observational powers of creatures preoccupied and ensorcelled by dissolution and decay. We have seen the human mind unconsciously seize upon the principles of that very code to pass its own societal memory forward into time. The individual, the momentary living cell of the society, vanishes, but the institutional structures stand, or if they change, do so in an invisible flux not too dissimilar from that persisting in the stream of genetic continuity.

Upon this world, life is still young, not truly old as stars are measured. Therefore it comes about that we minimize the role of the synapsid reptiles, our remote forerunners, and correspondingly exalt our own intellectual achievements. We refuse to consider that in the old eye of the hurricane we may be, and doubtless are, in aggregate, a slightly more diffuse and dangerous dragon of the primal morning that still enfolds us.

Note that I say "in aggregate." For it is just here, among men, that the role of messages, and, therefore, the role of the individual teacher—or, I should say now, the hidden teachers —begin to be more plainly apparent and their instructions become more diverse. The dead Pharaoh, though unintentionally, by a revealing act, had succeeded in conveying an impression of human tenderness that has outlasted the trappings of a vanished religion.

Like most modern educators I have listened to student demands to grade their teachers. I have heard the words repeated until they have become a slogan, that no man over thirty can teach the young of this generation. How would one grade a dead Pharaoh, millennia gone, I wonder, one who did not intend to teach, but who, to a few perceptive minds, succeeded by the simple nobility of an act.

Many years ago, a student who was destined to become an internationally known anthropologist sat in a course in linguistics and heard his instructor, a man of no inconsiderable wisdom, describe some linguistic peculiarities of Hebrew words. At the time, the young student, at the urging of his family, was contemplating a career in theology. As the teacher warmed to his subject, the student, in the back row, ventured excitedly:

"I believe I can understand that, sir. It is very similar to what exists in Mohegan."

The linguist paused and adjusted his glasses. "Young man," he said, "Mohegan is a dead language. Nothing has been recorded of it since the eighteenth century. Don't bluff."

"But sir," the young student countered hopefully, "it can't be dead so long as an old woman I know still speaks it. She is Pequot-Mohegan. I learned a bit of vocabulary from her and could speak with her myself. She took care of me when I was a child."

"Young man," said the austere, old-fashioned scholar, "be at my house for dinner at six this evening. You and I are going to look into this matter."

A few months later, under careful guidance, the young student published a paper upon Mohegan linguistics, the first of a long series of studies upon the forgotten languages and ethnology of the Indians of the Northeastern forests. He had changed his vocation and turned to anthropology because of the attraction of a hidden teacher. But just who was the teacher? The young man himself, his instructor, or that solitary speaker of a dying tongue who had so yearned to hear her people's voice that she had softly babbled it to a child?

Later, this man was to become one of my professors. I absorbed much from him, though I hasten to make the reluctant confession that he was considerably beyond thirty. Most of what I learned was gathered over cups of coffee in a dingy campus restaurant. What we talked about were things some centuries older than either of us. Our common interest lay in snakes, scapulimancy, and other forgotten rites of benighted forest hunters.

I have always regarded this man as an extraordinary individual, in fact, a hidden teacher. But alas, it is all now so old-fashioned. We never protested the impracticality of his quaint subjects. We were all too ready to participate in them. He was an excellent canoe-man, but he took me to places where I fully expected to drown before securing my degree. To this day, fragments of his unused wisdom remain stuffed in some back attic of my mind. Much of it I have never found the opportunity to employ, yet it has somehow colored my whole adult existence. I belong to that elderly professor in somewhat the same way that he, in turn, had become the wood child of a hidden forest mother.

There are, however, other teachers. For example, among the hunting peoples there were the animal counselors who appeared in prophetic dreams. Or, among the Greeks, the daemonic supernaturals who stood at the headboard while a man lay stark and listened—sometimes to dreadful things. "You are asleep," the messengers proclaimed over and over again, as though the man lay in a spell to hear his doom pronounced. "You, Achilles, you, son of Atreus. You are asleep, asleep," the hidden ones pronounced and vanished.

We of this modern time know other things of dreams, but we know also that they can be interior teachers and healers as well as the anticipators of disaster. It has been said that great art is the night thought of man. It may emerge without warning from the soundless depths of the unconscious, just as supernovae may blaze up suddenly in the farther reaches of void space. The critics, like astronomers, can afterward triangulate such worlds but not account for them.

A writer friend of mine with bitter memories of his youth and estranged from his family, who, in the interim, had died, gave me this account of the matter in his middle years. He had been working, with an unusual degree of reluctance, upon a novel that contained certain autobiographical episodes. One night he dreamed; it was a very vivid and stunning dream in its detailed reality.

He found himself hurrying over creaking snow through the blackness of a winter night. He was ascending a familiar path through a long-vanished orchard. The path led to his childhood home. The house, as he drew near, appeared dark and uninhabited, but, impelled by the power of the dream, he stepped upon the porch and tried to peer through a dark window into his own old room.

"Suddenly," he told me, "I was drawn by a strange mixture of repulsion and desire to press my face against the glass. I knew intuitively they were all there waiting for me within, if I could but see them. My mother and my father. Those I had loved and those I hated. But the window was black to my gaze. I hesitated a moment and struck a match. For an instant in that freezing silence I saw my father's face glimmer wan and remote behind the glass. My mother's face was there, with the hard, distorted lines that marked her later years.

"A surge of fury overcame my cowardice. I cupped the match before me and stepped closer, closer toward that dreadful confrontation. As the match guttered down, my face was pressed almost to the glass. In some quick transformation, such as only a dream can effect, I saw that it was my own face into which I stared, just as it was reflected in the black glass. My father's haunted face was but my own. The hard lines upon my mother's aging countenance were slowly reshaping themselves upon my living face. The light burned out. I awoke sweating from the terrible psychological tension of that nightmare. I was in a far port in a distant land. It was dawn. I could hear the waves breaking on the reef."

"And how do you interpret the dream?" I asked, concealing a sympathetic shudder and sinking deeper into my chair.

"It taught me something," he said slowly, and with equal slowness a kind of beautiful transfiguration passed over his features. All the tired lines I had known so well seemed faintly to be subsiding.

"Did you ever dream it again?" I asked out of a comparable experience of my own.

"No, never," he said, and hesitated. "You see, I had learned it was just I, but more, much more, I had learned that I was they. It makes a difference. And at the last, much too late, it was all right. I understood. My line was dying, but I understood. I hope they understood, too." His voice trailed into silence.

"It is a thing to learn," I said. "You were seeking something and it came." He nodded, wordless. "Out of a tomb," he added after a silent moment, "my kind of tomb—the mind."

On the dark street, walking homeward, I considered my friend's experience. Man, I concluded, may have come to the end of that wild being who had mastered the fire and the lightning. He can create the web but not hold it together, not save himself except by transcending his own image. For at the last, before the ultimate mystery, it is himself he shapes. Perhaps it is for this that the listening web lies open: that by knowledge we may grow beyond our past, our follies, and ever closer to what the Dreamer in the dark intended before the dust arose and walked. In the pages of an old book it has been written that we are in the hands of a Teacher, nor does it yet appear what man shall be.

FOUR

The Star Thrower

Who is the man walking in the Way?
An eye glaring in the skull.

—SECCHO

IT HAS EVER BEEN my lot, though formally myself a teacher, to be taught surely by none. There are times when I have thought to read lessons in the sky, or in books, or from the behavior of my fellows, but in the end my perceptions have frequently been inadequate or betrayed. Nevertheless, I venture to say that of what man may be I have caught a fugitive glimpse, not among multitudes of men, but along an endless wave-beaten coast at dawn. As always, there is this apparent break, this rift in nature, before the insight comes. The terrible question has to translate itself into an even more terrifying freedom.

If there is any meaning to this book, it began on the beaches of Costabel with just such a leap across an unknown abyss. It began, if I may borrow the expression from a Buddhist sage, with the skull and the eye. I was the skull. I was the inhumanly stripped skeleton without voice, without hope, wandering alone upon the shores of the world. I was devoid of pity, because pity implies hope. There was, in this desiccated skull, only an eye like a pharos light, a beacon, a search beam revolving endlessly in sunless noonday or black night. Ideas like swarms of insects rose to the beam, but the light consumed them. Upon that shore meaning had ceased. There were only the dead skull and the revolving eye. With such an eye, some have said, science looks upon the world. I do not know. I know only that I was the skull of emptiness and the endlessly revolving light without pity.

Once, in a dingy restaurant in the town, I had heard a woman say: "My father reads a goose bone for the weather." A modern primitive, I had thought, a diviner, using a method older than Stonehenge, as old as the arctic forests.

"And where does he do that?" the woman's companion had asked amusedly.

"In Costabel," she answered complacently, "in Costabel." The voice came back and buzzed faintly for a moment in the dark under the revolving eye. It did not make sense, but nothing in Costabel made sense. Perhaps that was why I had finally found myself in Costabel. Perhaps all men are destined at some time to arrive there as I did.

I had come by quite ordinary means, but I was still the skull with the eye. I concealed myself beneath a fisherman's cap and sunglasses, so that I looked like everyone else on the beach. This is the way things are managed in Costabel. It is on the shore that the revolving eye begins its beam and the whispers rise in the empty darkness of the skull.

The beaches of Costabel are littered with the debris of life. Shells are cast up in windrows; a hermit crab, fumbling for a new home in the depths, is tossed naked ashore, where the waiting gulls cut him to pieces. Along the strip of wet sand that marks the ebbing and flowing of the tide death walks hugely and in many forms. Even the torn fragments of green sponge yield bits of scrambling life striving to return to the great mother that has nourished and protected them.

In the end the sea rejects its offspring. They cannot fight their way home through the surf which casts them repeatedly back upon the shore. The tiny breathing pores of starfish are stuffed with sand. The rising sun shrivels the mucilaginous bodies of the unprotected. The seabeach and its endless war are soundless. Nothing screams but the gulls.

In the night, particularly in the tourist season, or during great storms, one can observe another vulturine activity. One can see, in the hour before dawn on the ebb tide, electric torches bobbing like fireflies along the beach. It is the sign of the professional shellers seeking to outrun and anticipate their less aggressive neighbors. A kind of greedy madness sweeps over the competing collectors. After a storm one can see them hurrying along with bundles of gathered starfish, or, toppling and overburdened, clutching bags of living shells whose hidden occupants will be slowly cooked and dissolved in the outdoor kettles provided by the resort hotels for the cleaning of specimens. Following one such episode I met the star thrower.

As soon as the ebb was flowing, as soon as I could make out in my sleeplessness the flashlights on the beach, I arose and dressed in the dark. As I came down the steps to the shore I could hear the deeper rumble of the surf. A gaping hole filled with churning sand had cut sharply into the breakwater. Flying sand as light as powder coated every exposed object like snow. I made my way around the altered edges of the cove and proceeded on my morning walk up the shore. Now and then a stooping figure moved in the gloom or a rain squall swept past me with light pattering steps. There was a faint sense of coming light somewhere behind me in the east.

Soon I began to make out objects, upended timbers, conch shells, sea wrack wrenched from the far-out kelp forests. A pink-clawed crab encased in a green cup of sponge lay sprawling where the waves had tossed him. Long-limbed starfish were strewn everywhere, as though the night sky had showered down. I paused once briefly. A small octopus, its beautiful dark-lensed eyes bleared with sand, gazed up at me from a ragged bundle of tentacles. I hesitated, and touched it briefly with my foot. It was dead. I paced on once more before the spreading whitecaps of the surf.

The shore grew steeper, the sound of the sea heavier and more menacing, as I rounded a bluff into the full blast of the offshore wind. I was away from the shellers now and strode more rapidly over the wet sand that effaced my footprints. Around the next point there might be a refuge from the wind. The sun behind me was pressing upward at the horizon's rim—an ominous red glare amidst the tumbling blackness of the clouds. Ahead of me, over the projecting point, a gigantic rainbow of incredible perfection had sprung shimmering into existence. Somewhere toward its foot I discerned a human figure standing, as it seemed to me, within the rainbow, though unconscious of his position. He was gazing fixedly at something in the sand.

Eventually he stooped and flung the object beyond the breaking surf. I labored toward him over a half mile of uncertain footing. By the time I reached him the rainbow had receded ahead of us, but something of its color still ran hastily in many changing lights across his features. He was starting to kneel again.

In a pool of sand and silt a starfish had thrust its arms up stiffly and was holding its body away from the stifling mud.

"It's still alive," I ventured.

"Yes," he said, and with a quick yet gentle movement he picked up the star and spun it over my head and far out into the sea. It sank in a burst of spume, and the waters roared once more.

"It may live," he said, "if the offshore pull is strong enough." He spoke gently, and across his bronzed worn face the light still came and went in subtly altering colors.

"There are not many come this far," I said, groping in a sudden embarrassment for words. "Do you collect?"

"Only like this," he said softly, gesturing amidst the wreckage of the shore. "And only for the living." He stooped again, oblivious of my curiosity, and skipped another star neatly across the water.

"The stars," he said, "throw well. One can help them."

He looked full at me with a faint question kindling in his eyes, which seemed to take on the far depths of the sea.

"I do not collect," I said uncomfortably, the wind beating at my garments. "Neither the living nor the dead. I gave it up a long time ago. Death is the only successful collector." I could feel the full night blackness in my skull and the terrible eye resuming its indifferent journey. I nodded and walked away, leaving him there upon the dune with that great rainbow ranging up the sky behind him.

I turned as I neared a bend in the coast and saw him toss another star, skimming it skillfully far out over the ravening and tumultuous water. For a moment, in the changing light, the sower appeared magnified, as though casting larger stars upon some greater sea. He had, at any rate, the posture of a god.

But again the eye, the cold world-shriveling eye, began its inevitable circling in my skull. He is a man, I considered sharply, bringing my thought to rest. The star thrower is a man, and death is running more fleet than he along every seabeach in the world.

I adjusted the dark lens of my glasses and, thus disguised, I paced slowly back by the starfish gatherers, past the shell collectors, with their vulgar little spades and the stick-length shelling pincers that eased their elderly backs while they snatched at treasures in the sand. I chose to look full at the steaming kettles in which beautiful voiceless things were being boiled alive. Behind my sunglasses a kind of litany began and refused to die down. *"As I came through the desert thus it was, as I came through the desert."*

In the darkness of my room I lay quiet with the sunglasses removed, but the eye turned and turned. In the desert, an old monk had once advised a traveler, the voices of God and the Devil are scarcely distinguishable. Costabel was a desert. I lay quiet, but my restless hand at the bedside fingered the edge of an invisible abyss. "Certain coasts," the remark of a perceptive writer came back to me, "are set apart for shipwreck." With unerring persistence I had made my way thither.

II

There is a difference in our human outlook, depending on whether we have been born upon level plains, where one step reasonably leads to another, or whether, by contrast, we have spent our lives amidst glacial crevasses and precipitous descents. In the case of the mountaineer, one step does not always lead rationally to another save by a desperate leap over a chasm or by an even more hesitant tiptoeing across precarious snow bridges.

Something about these opposed landscapes has its analogue in the mind of man. Our prehistoric life, one might say, began amidst enforested gloom with the abandonment of the protected instinctive life of nature. We sought, instead, an adventurous existence amidst the crater lands and ice fields of self-generated ideas. Clambering onward, we have slowly made our way out of a maze of isolated peaks into the level plains of science. Here, one step seems definitely to succeed another, the universe appears to take on an imposed order, and the illusions through which mankind has painfully made its way for many centuries have given place to the enormous vistas of past and future time. The encrusted eye in the stone speaks to us of undeviating sunlight; the calculated elliptic of Halley's comet no longer forecasts world disaster. The planet plunges on through a chill void of star years, and there is little or nothing that remains unmeasured.

Nothing, that is, but the mind of man. Since boyhood I had been traveling across the endless co-ordinated realms of science, just as, in the body, I was a plains dweller, accustomed to plodding through distances unbroken by precipices. Now that I come to look back, there was one contingent aspect of that landscape I inhabited whose significance, at the time, escaped me. "Twisters," we called them locally. They were a species of cyclonic, bouncing air funnel that could suddenly loom out of nowhere, crumpling windmills or slashing with devastating fury through country towns. Sometimes, by modest contrast, more harmless varieties known as dust devils might pursue one in a gentle spinning dance for miles. One could see them hesitantly stalking across the alkali flats on a hot day, debating, perhaps, in their tall, rotating columns, whether to ascend and assume more formidable shapes. They were the trickster part of an otherwise pedestrian landscape.

Infrequent though the visitations of these malign creations of the air might be, all prudent homesteaders in those parts had provided themselves with cyclone cellars. In the careless neighborhood in which I grew up, however, we contented ourselves with the queer yarns of cyclonic folklore and the vagaries of weather prophecy. As a boy, aroused by these tales and cherishing a subterranean fondness for caves, I once attempted to dig a storm cellar. Like most such projects this one was never completed. The trickster element in nature, I realize now, had so buffeted my parents that they stoically rejected planning. Unconsciously, they had arrived at the philosophy that foresight merely invited the attention of some baleful intelligence that despised and persecuted the calculating planner. It was not until many years later that I came to realize that a kind of maleficent primordial power persists in the mind as well as in the wandering dust storms of the exterior world.

A hidden dualism that has haunted man since antiquity runs across his religious conceptions as the conflict between good and evil. It persists in the modern world of science under other guises. It becomes chaos versus form or antichaos. Form, since the rise of the evolutionary philosophy, has itself taken on an illusory quality. Our apparent shapes no longer have the stability of a single divine fiat. Instead, they waver and dissolve into the unexpected. We gaze backward into a contracting cone of life until words leave us and all we know is dissolved into the simple circuits of a reptilian brain. Finally, sentience subsides into an animalcule.

Or we revolt and refuse to look deeper, but the void remains. We are rag dolls made out of many ages and skins, changelings who have slept in wood nests or hissed in the uncouth guise of waddling amphibians. We have played such roles for infinitely longer ages than we have been men. Our identity is a dream. We are process, not reality, for reality is an illusion of the daylight—the light of our particular day. In a fortnight, as aeons are measured, we may lie silent in a bed of stone, or, as has happened in the past, be figured in another guise. Two forces struggle perpetually in our bodies: Yam, the old sea dragon of the original Biblical darkness, and, arrayed against him, some wisp of dancing light that would have us linger, wistful, in our human form. "Tarry thou, till I come again"—an old legend survives among us of the admonition given by Jesus to the Wandering Jew. The words are applicable to all of us. Deep-hidden in the human psyche there is a similar injunction no longer having to do with the longevity of the body but, rather, a plea to wait upon some transcendent lesson preparing in the mind itself.

Yet the facts we face seem terrifyingly arrayed against us. It is as if at our backs, masked and demonic, moved the trickster as I have seen his role performed among the remnant of a savage people long ago. It was that of the jokester present at the most devout of ceremonies. This creature never laughed; he never made a sound. Painted in black, he followed silently behind the officiating priest, mimicking, with the added flourish of a little whip, the gestures of the devout one. His timed and stylized posturings conveyed a derision infinitely more formidable than actual laughter.

In modern terms, the dance of contingency, of the indeterminable, outwits us all. The approaching, fateful whirlwind on the plain had mercifully passed me by in youth. In the moment when I had witnessed that fireside performance I knew with surety that primitive man had lived with a dark message. He had acquiesced in the admission into his village of a cosmic messenger. Perhaps the primitives were wiser in the ways of the trickster universe than ourselves; perhaps they knew, as we do not, how to ground or make endurable the lightning.

At all events, I had learned, as I watched that half-understood drama by the leaping fire, why man, even modern man, reads goose bones for the weather of his soul. Afterward I had gone out, a troubled unbeliever, into the night. There was a shadow I could not henceforth shake off, which I knew was posturing and would always posture behind me. That mocking shadow looms over me as I write. It scrawls with a derisive pen and an exaggerated flourish. I know instinctively it will be present to caricature the solemnities of my deathbed. In a quarter of a century it has never spoken.

Black magic, the magic of the primeval chaos, blots out or transmogrifies the true form of things. At the stroke of twelve the princess must flee the banquet or risk discovery in the rags of a kitchen wench; coach reverts to pumpkin. Instability lies at the heart of the world. With uncanny foresight folklore has long toyed symbolically with what the nineteenth century was to proclaim a reality, namely, that form is an illusion of the time dimension, that the magic flight of the pursued hero or heroine through frogskin and wolf coat has been, and will continue to be, the flight of all men.

Goethe's genius sensed, well before the publication of the *Origin of Species*, the thesis and antithesis that epitomize the eternal struggle of the immediate species against its dissolution into something other: in modern terms, fish into reptile, ape into man. The power to change is both creative and destructive— a sinister gift, which, unrestricted, leads onward toward the formless and inchoate void of the possible. This force can only be counterbalanced by an equal impulse toward specificity. Form, once arisen, clings to its identity. Each species and each individual holds tenaciously to its present nature. Each strives to contain the creative and abolishing maelstrom that pours unseen through the generations. The past vanishes; the present momentarily persists; the future is potential only. In this specious present of the real, life struggles to maintain every manifestation, every individuality, that exists. In the end, life always fails, but the amorphous hurrying stream is held and diverted into new organic vessels in which form persists, though the form may not be that of yesterday.

The evolutionists, piercing beneath the show of momentary stability, discovered, hidden in rudimentary organs, the discarded rubbish of the past. They detected the reptile under the lifted feathers of the bird, the lost terrestrial limbs dwindling beneath the blubber of the giant cetaceans. They saw life rushing outward from an unknown center, just as today the astronomer senses the galaxies fleeing into the infinity of darkness. As the spinning galactic clouds hurl stars and worlds across the night, so life, equally impelled by the centrifugal powers lurking in the germ cell, scatters the splintered radiance of consciousness and sends it prowling and contending through the thickets of the world.

All this devious, tattered way was exposed to the ceaselessly turning eye within the skull that lay hidden upon the bed in Costabel. Slowly that eye grew conscious of another eye that searched it with equal penetration from the shadows of the room. It may have been a projection from the mind within the skull, but the eye was, nevertheless, exteriorized and haunting. It began as something glaucous and blind beneath a web of clinging algae. It altered suddenly and became the sand-smeared eye of the dead cephalopod I had encountered upon the beach. The transformations became more rapid with the concentration of my attention, and they became more formidable. There was the beaten, bloodshot eye of an animal from somewhere within my childhood experience. Finally, there was an eye that seemed torn from a photograph, but that looked through me as though it had already raced in vision up to the steep edge of nothingness and absorbed whatever terror lay in that abyss. I sank back again upon my cot and buried my head in the pillow. I knew the eye and the circumstance and the question. It was my mother. She was long dead, and the way backward was lost.

III

Now it may be asked, upon the coasts that invite shipwreck, why the ships should come, just as we may ask the man who pursues knowledge why he should be left with a revolving search beam in the head whose light falls only upon disaster or the flotsam of the shore. There is an answer, but its way is not across the level plains of science, for the science of remote abysses no longer shelters man. Instead, it reveals him in vaporous metamorphic succession as the homeless and unspecified one, the creature of the magic flight.

Long ago, when the future was just a simple tomorrow, men had cast intricately carved game counters to determine its course, or they had traced with a grimy finger the cracks on the burnt shoulder blade of a hare. It was a prophecy of tomorrow's hunt, just as was the old farmer's anachronistic reading of the weather from the signs on the breastbone of a goose. Such quaint almanacs of nature's intent had sufficed mankind since antiquity. They would do so no longer, nor would formal apologies to the souls of the game men hunted. The hunters had come, at last, beyond the satisfying supernatural world that had always surrounded the little village, into a place of homeless frontiers and precipitous edges, the indescribable world of the natural. Here tools increasingly revenged themselves upon their creators and tomorrow became unmanageable. Man had come in his journeying to a region of terrible freedoms.

It was a place of no traditional shelter, save those erected with the aid of tools, which had also begun to achieve a revolutionary independence from their masters. Their ways had grown secretive and incalculable. Science, more powerful than the magical questions that might be addressed by a shaman to a burnt shoulder blade, could create these tools but had not succeeded in controlling their ambivalent nature. Moreover, they responded all too readily to that urge for tampering and dissolution which is part of our primate heritage.

We had been safe in the enchanted forest only because of our weakness. When the powers of that gloomy region were given to us, immediately, as in a witch's house, things began to fly about unbidden. The tools, if not science itself, were linked intangibly to the subconscious poltergeist aspect of man's nature. The closer man and the natural world drew together, the more erratic became the behavior of each. Huge shadows leaped triumphantly after every blinding illumination. It was a magnified but clearly recognizable version of the black trickster's antics behind the solemn backs of the priesthood. Here, there was one difference. The shadows had passed out of all human semblance; no societal ritual safely contained their posturings, as in the warning dance of the trickster. Instead, unseen by many because it was so gigantically real, the multiplied darkness threatened to submerge the carriers of the light.

Darwin, Einstein, and Freud might be said to have released the shadows. Yet man had already entered the perilous domain that henceforth would contain his destiny. Four hundred years ago Francis Bacon had already anticipated its dual nature. The individuals do not matter. If they had not made their discoveries, others would have surely done so. They were good men, and they came as enlighteners of mankind. The tragedy was only that at their backs the ritual figure with the whip was invisible. There was no longer anything to subdue the pride of man. The world had been laid under the heavy spell of the natural; henceforth, it would be ordered by man.

Humanity was suddenly entranced by light and fancied it reflected light. Progress was its watchword, and for a time the shadows seemed to recede. Only a few guessed that the retreat of darkness presaged the emergence of an entirely new and less tangible terror. Things, in the words of G. K. Chesterton, were to grow incalculable by being calculated. Man's powers were finite; the forces he had released in nature recognized no such limitations. They were the irrevocable monsters conjured up by a completely amateur sorcerer.

But what, we may ask, was the nature of the first discoveries that now threaten to induce disaster? Preeminent among them was, of course, the perception to which we have already referred: the discovery of the interlinked and evolving web of life. The great Victorian biologists saw, and yet refused to see, the war between form and formlessness, chaos and antichaos, which the poet Goethe had sensed contesting beneath the smiling surface of nature. "The dangerous gift from above," he had termed it, with uneasy foresight.

By contrast, Darwin, the prime student of the struggle for existence, sought to visualize in a tangled bank of leaves the silent and insatiable war of nature. Still, he could imply with a veiled complacency that man might "with some confidence" look forward to a secure future "of inappreciable length." This he could do upon the same page in the *Origin of Species* where he observes that "of the species now living very few will transmit progeny to a far distant futurity." The contradiction escaped him; he did not wish to see it. Darwin, in addition, saw life as a purely selfish struggle, in which nothing is modified for the good of another species without being directly advantageous to its associated form.

If, he contended, one part of any single species had been formed for the exclusive good of another, "it would annihilate my theory." Powerfully documented and enhanced though the statement has become, famine, war, and death are not the sole instruments biologists today would accept as the means toward that perfection of which Darwin spoke. The subject is subtle and intricate, however, and one facet of it must be reserved for another chapter. Let it suffice to say here that the sign of the dark cave and the club became so firmly fixed in human thinking that in our time it has been invoked as signifying man's true image in books selling in the hundreds of thousands.

From the thesis and antithesis contained in Darwinism we come to Freud. The public knows that, like Darwin, the master of the inner world took the secure, stable, and sunlit province of the mind and revealed it as a place of contending furies. Ghostly transformations, flitting night shadows, misshapen changelings existed there, as real as anything that haunted the natural universe of Darwin. For this reason, appropriately, I had come as the skull and the eye to Costabel—the coast demanding shipwreck. Why else had I remembered the phrase, except for a dark impulse toward destruction lurking somewhere in the subconscious? I lay on the bed while the agonized eye in the remembered photograph persisted at the back of my closed lids.

It had begun when, after years of separation, I had gone dutifully home to a house from which the final occupant had departed. In a musty attic—among old trunks, a broken aquarium, and a dusty heap of fossil shells collected in childhood—I found a satchel. The satchel was already a shabby antique, in whose depths I turned up a jackknife and a "rat" of hair such as women wore at the beginning of the century. Beneath these lay a pile of old photographs and a note—two notes, rather, evidently dropped into the bag at different times. Each, in a thin, ornate hand, reiterated a single message that the writer had believed important. "This satchel belongs to my son, Loren Eiseley." It was the last message. I recognized the trivia. The jackknife I had carried in childhood. The rat of hair had belonged to my mother, and there were also two incredibly pointed slippers that looked as though they had been intended for a formal ball, to which I knew well my mother would never in her life have been invited. I undid the rotted string around the studio portraits.

Mostly they consisted of stiff, upright bearded men and heavily clothed women equally bound to the formalities and ritual that attended upon the photography of an earlier generation. No names identified the pictures, although here and there a reminiscent family trait seemed faintly evident. Finally I came upon a less formal photograph, taken in the eighties of the last century. Again no names identified the people, but a commercial stamp upon the back identified the place: Dyersville, Iowa. I had never been in that country town, but I knew at once it was my mother's birthplace.

Dyersville, the thought flashed through my mind, making the connection now for the first time: the dire place. I recognized at once the two sisters at the edge of the photograph, the younger clinging reluctantly to the older. Six years old, I thought, turning momentarily away from the younger child's face. Here it began, her pain and mine. The eyes in the photograph were already remote and shadowed by some inner turmoil. The poise of the body was already that of one miserably departing the peripheries of the human estate. The gaze was mutely clairvoyant and lonely. It was the gaze of a child who knew unbearable difference and impending isolation.

I dropped the notes and pictures once more into the bag. The last message had come from Dyersville: "my son." The child in the photograph had survived to be an ill-taught prairie artist. She had been deaf. All her life she had walked the precipice of mental breakdown. Here on this faded porch it had begun—the long crucifixion of life. I slipped downstairs and out of the house. I walked for miles through the streets.

Now at Costabel I put on the sunglasses once more, but the face from the torn photograph persisted behind them. It was as though I, as man, was being asked to confront, in all its overbearing weight, the universe itself. "Love not the world," the Biblical injunction runs, "neither the things that are in the world." The revolving beam in my mind had stopped, and the insect whisperings of the intellect. There was, at last, an utter stillness, a waiting as though for a cosmic judgment. The eye, the torn eye, considered me.

"But I *do* love the world," I whispered to a waiting presence in the empty room. "I love its small ones, the things beaten in the strangling surf, the bird, singing, which flies and falls and is not seen again." I choked and said, with the torn eye still upon me, "I love the lost ones, the failures of the world." It was like the renunciation of my scientific heritage. The torn eye surveyed me sadly and was gone. I had come full upon one of the last great rifts in nature, and the merciless beam no longer was in traverse around my skull.

But no, it was not a rift but a joining: the expression of love projected beyond the species boundary by a creature born of Darwinian struggle, in the silent war under the tangled bank. "There is no boon in nature," one of the new philosophers had written harshly in the first years of the industrial cities. Nevertheless, through war and famine and death, a sparse mercy had persisted, like a mutation whose time had not yet come. I had seen the star thrower cross that rift and, in so doing, he had reasserted the human right to define his own frontier. He had moved to the utmost edge of natural being, if not across its boundaries. It was as though at some point the supernatural had touched hesitantly, for an instant, upon the natural.

Out of the depths of a seemingly empty universe had grown an eye, like the eye in my room, but an eye on a vastly larger scale. It looked out upon what I can only call itself. It searched the skies and it searched the depths of being. In the shape of man it had ascended like a vaporous emanation from the depths of night. The nothing had miraculously gazed upon the nothing and was not content. It was an intrusion into, or a projection out of, nature for which no precedent existed. The act was, in short, an assertion of value arisen from the domain of absolute zero. A little whirlwind of commingling molecules had succeeded in confronting its own universe.

Here, at last, was the rift that lay beyond Darwin's tangled bank. For a creature, arisen from that bank and born of its contentions, had stretched out its hand in pity. Some ancient, inexhaustible, and patient intelligence, lying dispersed in the planetary fields of force or amidst the inconceivable cold of interstellar space, had chosen to endow its desolation with an apparition as mysterious as itself. The fate of man is to be the ever recurrent, reproachful Eye floating upon night and solitude. The world cannot be said to exist save by the interposition of that inward eye—an eye various and not under the restraints to be apprehended from what is vulgarly called the natural.

I had been unbelieving. I had walked away from the star thrower in the hardened indifference of maturity. But thought mediated by the eye is one of nature's infinite disguises. Belatedly, I arose with a solitary mission. I set forth in an effort to find the star thrower.

IV

Man is himself, like the universe he inhabits, like the demoniacal stirrings of the ooze from which he sprang, a tale of desolations. He walks in his mind from birth to death the long resounding shores of endless disillusionment. Finally, the commitment to life departs or turns to bitterness. But out of such desolation emerges the awesome freedom to choose—to choose beyond the narrowly circumscribed circle that delimits the animal being. In that widening ring of human choice, chaos and order renew their symbolic struggle in the role of titans. They contend for the destiny of a world.

Somewhere far up the coast wandered the star thrower beneath his rainbow. Our exchange had been brief because upon that coast I had learned that men who ventured out at dawn resented others in the greediness of their compulsive collecting. I had also been abrupt because I had, in the terms of my profession and experience, nothing to say. The star thrower was mad, and his particular acts were a folly with which I had not chosen to associate myself. I was an observer and a scientist. Nevertheless, I had seen the rainbow attempting to attach itself to earth.

On a point of land, as though projecting into a domain beyond us, I found the star thrower. In the sweet rain-swept morning, that great many-hued rainbow still lurked and wavered tentatively beyond him. Silently I sought and picked up a still-living star, spinning it far out into the waves. I spoke once briefly. "I understand," I said. "Call me another thrower."

Only then I allowed myself to think, He is not alone any longer. After us there will be others.

We were part of the rainbow—an unexplained projection into the natural. As I went down the beach I could feel the drawing of a circle in men's minds, like that lowering, shifting realm of color in which the thrower labored. It was a visible model of something toward which man's mind had striven, the circle of perfection.

I picked and flung another star. Perhaps far outward on the rim of space a genuine star was similarly seized and flung. I could feel the movement in my body. It was like a sowing—the sowing of life on an infinitely gigantic scale. I looked back across my shoulder. Small and dark against the receding rainbow, the star thrower stooped and flung once more. I never looked again. The task we had assumed was too immense for gazing. I flung and flung again while all about us roared the insatiable waters of death.

But we, pale and alone and small in that immensity, hurled back the living stars. Somewhere far off, across bottomless abysses, I felt as though another world was flung more joyfully. I could have thrown in a frenzy of joy, but I set my shoulders and cast, as the thrower in the rainbow cast, slowly, deliberately, and well. The task was not to be assumed lightly, for it was men as well as starfish that we sought to save. For a moment, we cast on an infinite beach together beside an unknown hurler of suns. It was, unsought, the destiny of my kind since the rituals of the ice age hunters, when life in the Northern Hemisphere had come close to vanishing. We had lost our way, I thought, but we had kept, some of us, the memory of the perfect circle of compassion from life to death and back again to life—the completion of the rainbow of existence. Even the hunters in the snow, making obeisance to the souls of the hunted, had known the cycle. The legend had come down and lingered that he who gained the gratitude of animals gained help in need from the dark wood.

I cast again with an increasingly remembered sowing motion and went my lone way up the beaches. Somewhere, I felt, in a great atavistic surge of feeling, somewhere the Thrower knew. Perhaps he smiled and cast once more into the boundless pit of darkness. Perhaps he, too, was lonely, and the end toward which he labored remained hidden—even as with ourselves.

I picked up a star whose tube feet ventured timidly among my fingers while, like a true star, it cried soundlessly for life. I saw it with an unaccustomed clarity and cast far out. With it, I flung myself as forfeit, for the first time, into some unknown dimension of existence. From Darwin's tangled bank of unceasing struggle, selfishness, and death, had arisen, incomprehensibly, the thrower who loved not man, but life. It was the subtle cleft in nature before which biological thinking had faltered. We had reached the last shore of an invisible island—yet, strangely, also a shore that the primitives had always known. They had sensed intuitively that man cannot exist spiritually without life, his brother, even if he slays. Somewhere, my thought persisted, there is a hurler of stars, and he walks, because he chooses, always in desolation, but not in defeat.

In the night the gas flames under the shelling kettles would continue to glow. I set my clock accordingly. Tomorrow I would walk in the storm. I would walk against the shell collectors and the flames. I would walk remembering Bacon's forgotten words "for the uses of life." I would walk with the knowledge of the discontinuities of the unexpected universe. I would walk knowing of the rift revealed by the thrower, a hint that there looms, inexplicably, in nature something above the role men give her. I knew it from the man at the foot of the rainbow, the starfish thrower on the beaches of Costabel.

FIVE

The Angry Winter

As to what happened next, it is possible to maintain that the hand of heaven was involved, and also possible to say that when men are desperate no one can stand up to them.

—XENOPHON

A TIME COMES when creatures whose destinies have crossed somewhere in the remote past are forced to appraise each other as though they were total strangers. I had been huddled beside the fire one winter night, with the wind prowling outside and shaking the windows. The big shepherd dog on the hearth before me occasionally glanced up affectionately, sighed, and slept. I was working, actually, amidst the debris of a far greater winter. On my desk lay the lance points of ice age hunters and the heavy leg bone of a fossil bison. No remnants of flesh attached to these relics. The deed lay more than ten thousand years remote. It was represented here by naked flint and by bone so mineralized it rang when struck. As I worked on in my little circle of light, I absently laid the bone beside me on the floor. The hour had crept toward midnight. A grating noise, a heavy rasping of big teeth diverted me. I looked down.

The dog had risen. That rock-hard fragment of a vanished beast was in his jaws and he was mouthing it with a fierce intensity I had never seen exhibited by him before.

"Wolf," I exclaimed, and stretched out my hand. The dog backed up but did not yield. A low and steady rumbling began to rise in his chest, something out of a long-gone midnight. There was nothing in that bone to taste, but ancient shapes were moving in his mind and determining his utterance. Only fools gave up bones. He was warning me.

"Wolf," I chided again.

As I advanced, his teeth showed and his mouth wrinkled to strike. The rumbling rose to a direct snarl. His flat head swayed low and wickedly as a reptile's above the floor. I was the most loved object in his universe, but the past was fully alive in him now. Its shadows were whispering in his mind. I knew he was not bluffing. If I made another step he would strike.

Yet his eyes were strained and desperate. "Do not," something pleaded in the back of them, some affectionate thing that had followed at my heel all the days of his mortal life, "do not force me. I am what I am and cannot be otherwise because of the shadows. Do not reach out. You are a man, and my very god. I love you, but do not put out your hand. It is midnight. We are in another time, in the snow."

"The *other* time," the steady rumbling continued while I paused, "the other time in the snow, the big, the final, the terrible snow, when the shape of this thing I hold spelled life. I will not give it up. I cannot. The shadows will not permit me. Do not put out your hand."

I stood silent, looking into his eyes, and heard his whisper through. Slowly I drew back in understanding. The snarl diminished, ceased. As I retreated, the bone slumped to the floor. He placed a paw upon it, warningly.

And were there no shadows in my own mind, I wondered. Had I not for a moment, in the grip of that savage utterance, been about to respond, to hurl myself upon him over an invisible haunch ten thousand years removed? Even to me the shadows had whispered—to me, the scholar in his study.

"Wolf," I said, but this time, holding a familiar leash, I spoke from the door indifferently. "A walk in the snow." Instantly from his eyes that other visitant receded. The bone was left lying. He came eagerly to my side, accepting the leash and taking it in his mouth as always.

A blizzard was raging when we went out, but he paid no heed. On his thick fur the driving snow was soon clinging heavily. He frolicked a little—though usually he was a grave dog—making up to me for something still receding in his mind. I felt the snowflakes fall upon my face, and stood thinking of another time, and another time still, until I was moving from midnight to midnight under ever more remote and vaster snows. Wolf came to my side with a little whimper. It was he who was civilized now. "Come back to the fire," he nudged gently, "or you will be lost." Automatically I took the leash he offered. He led me safely home and into the house.

"We have been very far away," I told him solemnly. "I think

there is something in us that we had both better try to forget."
Sprawled on the rug, Wolf made no response except to thump
his tail feebly out of courtesy. Already he was mostly asleep and
dreaming. By the movement of his feet I could see he was
running far upon some errand in which I played no part.

Softly I picked up his bone—our bone, rather—and replaced
it high on a shelf in my cabinet. As I snapped off the light the
white glow from the window seemed to augment itself and
shine with a deep, glacial blue. As far as I could see, nothing
moved in the long aisles of my neighbor's woods. There was
no visible track, and certainly no sound from the living. The
snow continued to fall steadily, but the wind, and the shadows
it had brought, had vanished.

II

Vast desolation and a kind of absence in nature invite the
emergence of equally strange beings or spectacular natural
events. An influx of power accompanies nature's every hesitation; each pause is succeeded by an uncanny resurrection. The
evolution of a lifeless planet eventually culminates in green
leaves. The altered and oxygenated air hanging above the
continents presently invites the rise of animal apparitions compounded of formerly inert clay.

Only after long observation does the sophisticated eye succeed in labeling these events as natural rather than miraculous.
There frequently lingers about them a penumbral air of mystery
not easily dispersed. We seem to know much, yet we frequently
find ourselves baffled. Humanity itself constitutes such a mystery, for our species arose and spread in a time of great extinctions. We are the final product of the Pleistocene period's
millennial winters, whose origin is still debated. Our knowledge of this ice age is only a little over a century old, and the
time of its complete acceptance even less. Illiterate man has lost
the memory of that vast snowfall from whose depths he has
emerged blinking.

"Nature is a wizard," Thoreau once said. The self-styled inspector of snowstorms stood in awe of the six-pointed perfection of a
snowflake. The air, even thin air, was full of genius. The poetic
naturalist to whom, in our new-found scientism, we grudgingly

accord a literary name but whom we dismiss as an indifferent investigator, made a profound observation about man during a moment of shivering thought on a frozen river. "The human brain," meditated the snowbound philosopher, "is the kernel which the winter itself matures." The winter, he thought, tended to concentrate and extend the power of the human mind.

"The winter," Thoreau continued, "is thrown to us like a bone to a famishing dog, and we are expected to get the marrow out of it." In foreshortened perspective Thoreau thus symbolically prefigured man's passage through the four long glacial seasons, from which we have indeed painfully learned to extract the marrow. Although Thoreau had seen the scratches left by the moving ice across Mount Monadnock, even to recording their direction, he was innocent of their significance. What he felt was a sign of his intuitive powers alone. He sensed uncannily the opening of a damp door in a remote forest, and he protested that nature was too big for him, that it was, in reality, a playground of giants.

Nor was Thoreau wrong. Man is the product of a very unusual epoch in earth's history, a time when the claws of a vast dragon, the glacial ice, groped fumbling toward him across a third of the world's land surface and blew upon him the breath of an enormous winter. It was a world of elemental extravagance, assigned by authorities to scarcely one per cent of earth's history and labeled "geo-catastrophic." For over a million years man, originally a tropical orphan, has wandered through age-long snowdrifts or been deluged by equally giant rains.

He has been present at the birth of mountains. He has witnessed the disappearance of whole orders of life and survived the cyclonic dust clouds that blew in the glacial winds off the receding ice fronts. In the end it is no wonder that he himself has retained a modicum of that violence and unpredictability which lie sleeping in the heart of nature. Modern man, for all his developed powers and his imagined insulation in his cities, still lives at the mercy of those giant forces that created him and can equally decree his departure. These forces are revealed in man's simplest stories—the stories in which the orphaned and abused prince evades all obstacles and, through the assistance of some benign sorcerer, slays the dragon and enters into his patrimony.

As the ice age presents a kind of caricature or sudden concentration of those natural forces that normally govern the world, so man, in the development of that awful instrument, his brain, himself partakes of the same qualities. Both his early magic and his latest science have magnified and frequently distorted the powers of the natural world, stirring its capricious and evil qualities. The explosive force of suns, once safely locked in nature, now lies in the hand that long ago dropped from a tree limb into the upland grass.

We have become planet changers and the decimators of life, including our own. The sorcerer's gift of fire in a dark cave has brought us more than a simple kingdom. Like so many magical gifts it has conjured up that which cannot be subdued but henceforth demands unceasing attention lest it destroy us. We are the genuine offspring of the sleeping ice, and we have inherited its power to magnify the merely usual into the colossal. The nature we have known throughout our venture upon earth has not been the stable, drowsy summer of the slow reptilian ages. Instead, we are the final product of a seemingly returning cycle, which comes once in two hundred and fifty million years—about the time, it has been estimated, that it takes our sun to make one full circle of the galactic wheel.

That circle and its recurrent ice have been repeated back into dim pre-Cambrian eras, whose life is lost to us. When our first tentative knowledge of the cold begins, the time is Permian. This glaciation, so far as we can determine, was, in contrast to the ice age just past, confined primarily to the southern hemisphere. Like the later Pleistocene episode, which saw the rise of man, it was an epoch of continental uplift and of a steeper temperature gradient from the poles to the equator. It produced a crisis in the evolution of life that culminated in the final invasion of the land by the reptilian vertebrates. More significantly, so far as our own future was concerned, it involved the rise of those transitional twilight creatures, the mammal-like reptiles whose remote descendants we are.

They were moving in the direction of fur, warm blood, and controlled body temperature, which, in time, would give the true mammals the mastery of the planet. Their forerunners were the first vertebrate responses to the recurrent menace of the angry dragon. Yet so far away in the past, so dim and distant was the breath of that frosty era that the scientists of the nineteenth century, who believed in a constant heat loss from a once fiery earth, were amazed when A. C. Ramsay, in 1854, produced evidence that the last great winter, from which man is only now emerging, had been long preceded in the Permian by a period of equally formidable cold.

Since we live on the borders of the Pleistocene, an ice age that has regressed but not surely departed, it is perhaps well to observe that the older Permian glaciation is the only one of whose real duration we can form a reasonable estimate. Uncertain traces of other such eras are lost in ancient strata or buried deep in the pre-Cambrian shadows. For the Permian glaciation, however, we can derive a rough estimate of some twenty-five to thirty million years, during which the southern continents periodically lay in the grip of glacial ice. Philip King, of the U.S. Geological Survey, has observed that in Australia the period of Permian glaciation was prolonged, and that in eastern Australia boulder beds of glacial origin are interspersed through more than ten thousand feet of geological section. The temperature gradient of that era would never again be experienced until the onset of the cold that accompanied the birth of man.

If the cause of these glacial conditions, with the enormous intervals between them, is directed by recurrent terrestrial or cosmic conditions, then man, unknowingly, is huddling memoryless in the pale sunshine of an interstadial spring. Ice still lies upon the poles; the arctic owl, driven south on winter nights, drifts white and invisible over the muffled countryside. He is a survival from a vanished world, a denizen of the long cold of which he may yet be the returning harbinger.

I have said that the earlier Permian glaciation appears to have fluctuated over perhaps thirty million years. Some two hundred and fifty million years later, the Pleistocene—the ice age we call our own—along with four interglacial summers (if we include the present), has persisted a scant million years.* So

---

*Late discoveries have extended the Tertiary time range of the protohuman line. *Homo sapiens* may have existed for a time contemporaneously with the last of the heavy-browed forms of man, well back in the Pleistocene. If so, however, there is suggestive evidence that fertile genetic mixture between the two types existed. The human interminglings of hundreds of thousands of years of prehistory are not to be clarified by a single generation of archaeologists.

recent is it that its two earlier phases yield little evidence of animal adjustment to the cold. "The origin of arctic life," remarks one authority, "is shrouded in darkness." Only the last two ice advances have given time, apparently, for the emergence of a fauna of arctic aspect—the woolly mammoths, white bears, and tundra-grazing reindeer, who shared with man the experience of the uttermost cold.

The arctic, in general, has been the grave of life rather than the place of its primary development. Man is the survivor among many cold-adapted creatures who streamed away at last with the melting glaciers. So far, the Pleistocene, in which geologically we still exist, has been a time of great extinctions. Its single new emergent, man, has himself contributed to making it what Alfred Russel Wallace has called a "zoologically impoverished world." Judging from the Permian record, if we were to experience thirty million more years of alternate ice and sun across a third of the earth's surface, man's temperate-zone cities would be ground to powder, his numbers decimated, and he himself might die in bestial savagery and want. Or, in his new-found scientific cleverness, he might survive his own unpredictable violence and live on as an archaic relic, a dropped pebble from a longer geological drama.

Already our own kind, *Homo sapiens*, with the assistance of the last two ice advances, appears to have eliminated, directly or indirectly, a sizable proportion of the human family. One solitary, if fertile, species, lost in internecine conflict, confronts the future even now. Man's survival record, for all his brains, is not impressive against the cunning patience of the unexpired Great Winter. In fact, we would do well to consider the story of man's past and his kinship with the planetary dragon—for of this there is no doubt at all. "The association of unusual physical conditions with a crisis in evolution is not likely to be pure coincidence," George Gaylord Simpson, a leading paleontologist, has declared. "Life and its environment are interdependent and evolve together."

The steps to man begin before the ice age. Just as in the case of the ancestral mammals, however, they are heralded by the oncoming cold of the late Age of Mammals, the spread of grass, the skyward swing of the continents, and the violence of mountain upheaval. The Pleistocene episode, so long unguessed and as insignificant as a pinprick on the earth's great time scale, signifies also, as did the ice of the late Paleozoic, the rise of a new organic world. In this case it marked in polycentric waves, distorted and originally hurled back by the frost, the rise of a creature not only new, but also one whose head contained its own interior lights and shadows and who was destined to reflect the turbulence and beauty of the age in which it was born.

It was an age in which the earth, over a third of its surface, over millions of square miles of the Northern Hemisphere, wore a mantle of blue ice stolen from the shrinking seas. And as that mantle encased and covered the final strata of earth, so, in the brain of man, a similar superimposed layer of crystalline thought substance superseded the dark, forgetful pathways of the animal brain. Sounds had their origin there, strange sounds that took on meaning in the air, named stones and gods. For the first time in the history of the planet, living men received names. For the first time, also, men wept bitterly over the bodies of their dead.

There was no longer a single generation, which bred blindly and without question. Time and its agonizing nostalgia would touch the heart each season and be seen in the fall of a leaf. Or, most terrible of all, a loved face would grow old. Kronos and the fates had entered into man's thinking. Try to escape as he might, he would endure an interior ice age. He would devise and unmake fables and at last, and unwillingly, comprehend an intangible abstraction called space-time and shiver inwardly before the endless abysses of space as he had once shivered unclothed and unlighted before the earthly frost.

As Thoreau anticipated, man has been matured by winter; he has survived its coming, and has eaten of its marrow. But its cold is in his bones. The child will partake always of the parent, and that parent is the sleeping dragon whose kingdom we hold merely upon sufferance, and whose vagaries we have yet to endure.

III

A few days ago I chanced to look into a rain pool on the walk outside my window. For a long time, because I was dreaming of other things, I saw only the occasional spreading ripple from a raindrop or the simultaneous overlapping circles as the rain fell faster. Then, as the beauty and the strange rhythm of the extending and concentric wavelets entered my mind, I saw that I was looking symbolically upon the whole history of life upon our globe. There, in a wide, sweeping circle, ran the early primates from whom we are descended; here, as a later drop within the rim of the greater circle, emerged the first men. I saw the mammoths pass in a long, slow, world-wide surge, but the little drop of man changed into a great hasty wave that swept them under.

There were sudden little ringlets, like the fauna of isolated islands, that appeared and disappeared with rapidity. Sometimes so slow were the drops that the pool was almost quiet, like the intense, straining silence of a quiescent geological period. Sometimes the rain, like the mutations in animal form, came so fast that the ripples broke, mixed, or kept their shapes with difficulty and did not spread far. Jungles, I read in my mystical water glass, microfaunas changing rapidly but with little spread.

Watch instead, I thought, for the great tides—it is they that contain the planet's story. As the rain hastened or dripped slowly, the pictures in the little pool were taken into my mind as though from the globe of a crystal-gazer. How often, if we learn to look, is a spider's wheel a universe, or a swarm of summer midges a galaxy, or a canyon a backward glance into time. Beneath our feet is the scratched pebble that denotes an ice age, or above us the summer cloud that changes form in one afternoon as an animal might do in ten million windy years.

All of these perceptive insights that we obtain from the natural world around us depend upon painfully accumulated knowledge. Otherwise, much as to our ancestors, the pebble remains a pebble, the pool but splashing water, the canyon a deep hole in the ground. Increasingly, the truly perceptive man must know that where the human eye stops, and hearing terminates, there still vibrates an inconceivable and spectral world of which we learn only through devised instruments. Through such instruments measuring atomic decay we have learned to probe the depths of time before our coming and to gauge temperatures long vanished.

Little by little, the orders of life that had characterized the earlier Age of Mammals ebbed away before the oncoming cold of the Pleistocene, interspersed though this cold was by interglacial recessions and the particularly long summer of the second interglacial. There were times when ice accumulated over Britain; in the New World, there were times when it stretched across the whole of Canada and reached southward to the fortieth parallel of latitude, in what today would be Kansas.

Manhattan Island and New Jersey felt its weight. The giant, long-horned bison of the middle Pleistocene vanished before man had entered America. Other now extinct but less colossal forms followed them. By the closing Pleistocene, it has been estimated, some seventy per cent of the animal life of the Western world had perished. Even in Africa, remote from the ice centers, change was evident. Perhaps the devastation was a partial response to the Pluvials, the great rains that in the tropics seem to have accompanied or succeeded the ice advances in the north.

The human groups that existed on the Old World land mass were alternately squeezed southward by advancing ice, contracted into pockets, or released once more to find their way northward. Between the alternate tick and tock of ice and sun, man's very bones were changing. Old species passed slowly away in obscure refuges or fell before the weapons devised by sharper minds under more desperate circumstances. Perhaps, since the rise of mammals, life had been subjected to no more drastic harassment, no more cutting selective edge, no greater isolation and then renewed genetic commingling. Yet we know that something approximating man was on the ground before the ice commenced and that naked man is tropical in origin. What, then, has ice to do with his story?

It has, in fact, everything. The oncoming chill caught him early in his career; its forces converged upon him even in the tropics; its influence can be seen in the successive human waves that edge farther and ever farther north until at last they spill across the high latitudes at Bering Strait and descend the length of the two Americas. Only then did the last southwestern mammoths perish in the shallow mud of declining lakes, the last mastodons drop their tired bones in the New Jersey bogs on the receding drift.

The story can best be seen from the map, as time, ice, and the sorcerer's gift of fire run like the concentric ripples of the falling rain across the zones of temperature. The tale is not confined to ice alone. As one glaciologist, J. K. Charlesworth, has written: "The Pleistocene . . . witnessed earth-movements on a considerable, even catastrophic scale. There is evidence that it created mountains and ocean deeps of a size previously unequalled. . . . The Pleistocene represents one of the crescendi in the earth's tectonic history."

I have spoken of the fact that, save for violent glacial episodes, the world's climate has been genial. The planet has been warmer than today—"acryogenic," as the specialists would say. Both earlier and later, warm faunas reached within the Arctic Circle, and a much higher percentage of that fauna represented forest forms. Then in the ice phases, world temperature dipped, even in the tropics; the mountain snowlines crept downward. In the north, summers were "short and false," periods of "dry cold"—again to quote the specialists. Snow blanketed the high ground in winter, and that winter covered half the year and was extremely harsh.

With our short memory, we accept the present climate as normal. It is as though a man with a huge volume of a thousand pages before him—in reality, the pages of earth time—should read the final sentence on the last page and pronounce it history. The ice has receded, it is true, but world climate has not completely rebounded. We are still on the steep edge of winter or early spring. Temperature has reached mid-point. Like refugees, we have been dozing memoryless for a few scant millennia before the windbreak of a sun-warmed rock. In the European Lapland winter that once obtained as far south as Britain, the temperature lay eighteen degrees Fahrenheit lower than today.

On a world-wide scale this cold did not arrive unheralded. Somewhere in the highlands of Africa and Asia the long Tertiary descent of temperature began. It was, in retrospect, the prelude to the ice. One can trace its presence in the spread of grasslands and the disappearance over many areas of the old forest browsers. The continents were rising. We know that by Pliocene time, in which the trail of man ebbs away into the grass, man's history is more complicated than the simple late descent, as our Victorian forerunners sometimes assumed, of a chimpanzee from a tree. The story is one whose complications we have yet to unravel.

Avoiding complexities and adhering as we can to our rain-pool analogy, man, subman, protoman, the euhominid, as we variously denote him, was already walking upright on the African grasslands more than two million years ago. He appears smaller than modern man, pygmoid and light of limb. Giantism comes late in the history of a type and sometimes foretells extinction. Man is now a giant primate and, where food is plentiful, growing larger, but he is a unique creature whose end is not yet foreseeable.

Three facts can be discerned as we examine the earliest bipedal man-apes known to us. First, they suggest, in their varied dentition and skull structure, a physical diversity implying, as Alfred Russel Wallace theorized long ago, an approach to that vanished era in which protoman was still being molded by natural selective processes unmediated by the softening effect of cultural defenses. He was, in other words, scant in numbers and still responding genetically to more than one ecological niche. Heat and cold were direct realities; hunger drove him, and, on the open savanna into which he had descended, vigilance was the price of life. The teeth of the great carnivores lay in wait for the old, the young, and the unwary.

Second, at the time we encounter man, the long descent of the world's climate toward the oncoming Pleistocene cold had already begun. It is not without interest that all man's most primitive surviving relatives—living fossils, we would call them—are tree dwellers hidden in the tropical rain forests of Africa, Madagascar, and the islands of southeastern Asia. They are the survivors of an older and a warmer world—the incubation time that was finally to produce, in some unknown fashion, the world-encircling coils of the ice dragon.

In the last of Tertiary time, grasslands and high country were spreading even in the tropics. Savanna parkland interspersed with trees clothed the uplands of East Africa; North China grew colder and more arid. Steppe- and plains-loving animals became predominant. Even the seas grew colder, and the tropical zone narrowed. Africa was to remain the least glaciated of the continents, but even here the lowering of temperature drew on, and the mountain glaciations finally began to descend into their valleys. As for Asia, the slow, giant upthrust of the Himalayas had brought with it the disappearance of jungles harboring the old-fashioned tree climbers.

Of the known regions of late-Tertiary primate development, whether African or Asian, both present the spectacle of increasing grasslands and diminishing forest. The latter, as in southeastern Asia, offered a refuge for the arboreal conservatives, such as the gibbon and orang, but the Miocene-Pliocene parklands and savannas must have proved an increasing temptation to an intelligent anthropoid sufficiently unspecialized and agile to venture out upon the grass. Our evidence from Africa is more complete at present, but fragmentary remains that may prove to be those of equally bipedal creatures are known from pre-Pleistocene and less explored regions below the Himalayas.

Third, and last of the points to be touched on here, the man-apes, in venturing out erect upon the grass, were leaving forever the safety of little fruit-filled niches in the forest. They were entering the open sunlight of a one-dimensional world, but they were bringing to that adventure a freed forelimb at the conscious command of the brain, and an eye skillfully adjusted for depth perception. Increasingly they would feed on the rich proteins provided by the game of the grasslands; by voice and primitive projectile weapon, man would eventually become a space leaper more deadly than the giant cats.

In the long, chill breath that presaged the stirring of the world dragon, the submen drifted naked through an autumnal haze. They were, in body, partly the slumbering product of the earth's long summer. The tropical heat had warmed their bones. Thin-furred and hungry, old-fashioned descendants from the forest attic, they clung to the tropical savannas. Unlike the light gazelle, they could neither bound from enemies nor graze on the harsh siliceous grasses. With a minimum of fragmentary chips and stones, and through an intensified group co-operation, they survived.

The first human wave, however, was a little wave, threatening to vanish. A patch in Africa, a hint in the Siwalik beds below the Himalayas—little more. Tremendous bodily adjustments were in process, and, in the low skull vault, a dream animal was in the process of development, a user of invisible symbols. In its beginnings, and ever more desperately, such a being walks the knife-edge of extinction. For a creature who dreams outside of nature, but is at the same time imprisoned within reality, has acquired, in the words of the psychiatrist Leonard Sillman, "one of the cruelest and most generous endowments ever given to a species of life by a mysterious providence."

On that one most recent page of life from which we can still read, it is plain that the second wave of man ran onward into the coming of the ice. In China a pithecanthropine creature with a cranial capacity of some 780 cubic centimeters has been recently retrieved from deposits suggesting a warm grassland fauna of the lower Pleistocene, perhaps over 700,000 years remote from the present. The site lies in Shensi province in about thirty degrees north latitude. Man is moving northward. His brain has grown, but he appears still to lack fire.

IV

In the legendary cycles of the Blackfoot Indians there is an account of the early people, who were poor and naked and did not know how to live. Old Man, their maker, said: "Go to sleep and get power. Whatever animals appear in your dream, pray and listen." "And," the story concludes, "that was how the first people got through the world, by the power of their dreams."

Man was not alone young and ignorant in the morning of his world; he also died young. Much of what he grasped of the world around him he learned like a child from what he imagined, or was gleaned from his own childlike parents. The remarks of Old Man, though clothed in myth, have an elemental ring. They tell the story of an orphan—man—bereft of instinctive instruction and dependent upon dream, upon, in the end, his own interpretation of the world. He had to seek animal helpers because they alone remembered what was to be done.

And so the cold gathered and man huddled, dreaming, in the lightless dark. Lightning struck, the living fire ran from volcanoes in the fury of earth's changes, and still man slumbered. Twice the ice ground southward and once withdrew, but no fire glimmered at a cave mouth. Humanly flaked flints were heavier and better-made. Behind that simple observation lies the unknown history of drifting generations, the children of the dreamtime.

At about the forty-fifth parallel of latitude, in the cave vaults at Choukoutien, near Peking, a heavy-browed, paleoanthropic form of man with a cranial capacity as low, in some instances, as 860 cubic centimeters, gnawed marrow bones and chipped stone implements. The time lies 500,000 years remote; the hour is late within the second cold, the place northward and more bleak than Shensi.

One thing strikes us immediately. This creature, with scarcely two-thirds of modern man's cranial capacity, was a fire user. Of what it meant to him beyond warmth and shelter we know nothing; with what rites, ghastly or benighted, it was struck or maintained, no word remains. We know only that fire opened to man the final conquest of the earth.

I do not include language here, in spite of its tremendous significance to humanity, because the potentiality of language is dependent upon the germ plasm. Its nature, not its cultural expression, is written into the motor centers of the brain, into high auditory discrimination and equally rapid neuromuscular response in tongue, lips, and palate. We are biologically adapted for the symbols of speech. We have determined its forms, but its potential is not of our conscious creation. Its mechanisms are written in our brain, a simple gift from the dark powers behind nature. Speech has made us, but it is a human endowment not entirely of our conscious devising.

By contrast, the first fires flickering at a cave mouth are our own discovery, our own triumph, our grasp upon invisible chemical power. Fire contained, in that place of brutal darkness and leaping shadows, the crucible and the chemical retort, steam and industry. It contained the entire human future.

Across the width of the Old World land mass near what is now Swanscombe, England, a better-brained creature of almost similar dating is also suspected of using fire, though the evidence, being from the open, is not so clear. But at last the sorcerer-priest, the stealer from the gods, the unknown benefactor remembered in a myriad legends around the earth, had done his work. He had supplied man with an overmastering magic. It would stand against the darkness and the cold.

In the frontal and temporal lobes, anatomy informs us, lie areas involved with abstract thought. In modern man the temporal lobes in particular are "hazardously supplied with blood through tenuous arteries. They are protected by a thin skull and crowded against a shelf of bone. They are more commonly injured than any other higher centers." The neurologist Frederick Gibbs goes on to observe that these lobes are attached to the brain like dormer windows, jammed on as an afterthought of nature. In the massive armored cranium of Peking man those lobes had already lit the fires that would knit family ties closer, promote the more rapid assimilation of wild food, and increase the foresight that goes into the tending of fires always. Fire is the only natural force on the planet that can both feed and travel. It is strangely like an animal; that is, it has to be tended and fed. Moreover, it can also rage out of control.

Man, long before he trained the first dog, had learned to domesticate fire. Its dancing midnight shadows and the comfort it gave undoubtedly enhanced the opportunities for brain growth. The fourth ice would see man better clothed and warmed. In our own guise, as the third and last great human wave, man would pursue the trail of mammoths across the Arctic Circle into America. The animal counselors that once filled his dreams would go down before him. Thus, inexorably, he would be forced into a new and profound relationship with plants. If one judges by the measures of civilization, it was all for the best. There are, however, lingering legends that carry a pathetic symbolism: that it was fire that separated man from the animals. It is perhaps a last wistful echo from a time when the chasm between ourselves and the rest of life did not yawn so impassably.

V

They tell an old tale in camping places, where men still live in the open among stones and trees. Always, in one way or another, the tale has to do with messages, messages that the gods have sent to men. The burden of the stories is always the same. Someone, man or animal, is laggard or gets the message wrong, and the error is impossible to correct; thus have illness and death intruded in the world.

Mostly the animals understand their roles, but man, by comparison, seems troubled by a message that, it is often said, he cannot quite remember, or has gotten wrong. Implied in this is our feeling that life demands an answer from us, that an essential part of man is his struggle to remember the meaning of the message with which he has been entrusted, that we are, in fact, message carriers. We are not what we seem. We have had a further instruction.

There is another story that is sometimes told of the creator in the morning of the world. After he had created the first two beings, which he pronounced to be "people," the woman, standing by the river, asked: "Shall we always live?" Now the god had not considered this, but he was not unwilling to grant his new creations immortality. The woman picked up a stone and, gesturing toward the stream, said: "If it floats we shall always live, but if it sinks, people must die so that they shall feel pity and have compassion." She tossed the stone. It sank. "You have chosen," said the creator.

Many years ago, as a solitary youth much given to wandering, I set forth on a sullen November day for a long walk that would end among the fallen stones of a forgotten pioneer cemetery on the High Plains. The weather was threatening, and only an unusual restlessness drove me into the endeavor. Snow was on the ground and deepening by the hour. There was a rising wind of blizzard proportions sweeping across the land.

Late in a snow-filled twilight, I reached the cemetery. The community that placed it there had long vanished. Frost and snow, season by season, had cracked and shattered the flat, illegible stones till none remained upright. It was as though I, the last living man, stood freezing among the dead. I leaned across a post and wiped the snow from my eyes.

It was then I saw him—the only other living thing in that bleak countryside. We looked at each other. We had both come across a way so immense that neither my immediate journey nor his seemed of the slightest importance. We had each passed over some immeasurably greater distance, but whatever the word we had carried, it had been forgotten between us.

He was nothing more than a western jack rabbit, and his ribs were gaunt with hunger beneath his skin. Only the storm contained us equally. That shrinking, long-eared animal, cowering beside a slab in an abandoned graveyard, helplessly expected the flash of momentary death, but it did not run. And I, with the rifle so frequently carried in that day and time, I also stood while the storm—a real blizzard now—raged over and between us, but I did not fire.

We both had a fatal power to multiply, the thought flashed on me, and the planet was not large. Why was it so, and what was the message that somehow seemed spoken from a long way off beyond an ice field, out of all possible human hearing?

The snow lifted and swirled between us once more. He was going to need that broken bit of shelter. The temperature was falling. For his frightened, trembling body in all the million years between us, there had been no sorcerer's aid. He had survived alone in the blue nights and the howling dark. He was thin and crumpled and small.

Step by step I drew back among the dead and their fallen stones. Somewhere, if I could follow the fence lines, there would be a fire for me. For a moment I could see his ears nervously recording my movements, but I was a wraith now, fading in the storm.

"There are so few tracks in all this snow," someone had once protested. It was true. I stood in the falling flakes and pondered it. Even my own tracks were filling. But out of such desolation had arisen man, the desolate. In essence, he is a belated phantom of the angry winter. He carried, and perhaps will always carry, its cruelty and its springtime in his heart.

SIX

The Golden Alphabet

A creature without memory cannot discover the past; one without expectation cannot conceive a future.

—GEORGE SANTAYANA

"WISDOM," THE ESKIMO SAY, "can be found only far from man, out in the great loneliness." These people speak from silences we will not know again until we set foot upon the moon. Perhaps our track is somehow rounding evocatively backward into another version of the giant winter out of which we emerged ten thousand years ago. Perhaps it is our destiny to have plunged across it only to re-enter it once more.

Of all the men of the nineteenth century who might be said to have been intimates of that loneliness and yet, at the same time, to have possessed unusual prophetic powers, Henry David Thoreau and Charles Darwin form both a spectacular comparison and a contrast. Both Thoreau and Darwin were voyagers. One confined himself to the ever widening ripples on a pond until they embraced infinity. The other went around the world and remained for the rest of his life a meditative recluse in an old Victorian house in the English countryside.

The two men shared a passion for odd facts. In much else they differed. Darwin, after long travel, had immured himself at home. Thoreau could only briefly tolerate a dwelling, and his journals suggest that he suffered from claustrophobic feelings that a house was a disguised tomb, from which he had constantly to escape into the open. "There is no circulation there," he once protested.

Both men were insatiable readers and composers of works not completely published in their individual lifetimes. Both achieved a passionate satisfaction out of their association with the wilderness. Each in his individual way has profoundly influenced the lives of the generations that followed him. Darwin achieved fame through a great biological synthesis—what Thoreau would have called the demoniacal quality of the man who can discern a law, or couple two facts. Thoreau, by contrast, is known as much for what he implied as for what he spoke. His life, like Darwin's, is known but in many ways hidden. As he himself intimated cryptically, he had long ago lost a bay horse, a hound dog, and a turtledove, for which he was searching. It is not known that he ever came upon them or precisely what they represented.

All his life Thoreau dwelt along the edge of that visible nature of which Darwin assumed the practical mastery. Like the owls Thoreau described in *Walden*, he himself represented the stark twilight of a nature "behind the ordinary," which has passed unrecognized. As he phrased it, "We live on the outskirts of that region. . . . We are still being born, and have as yet but a dim vision."

Both men forfeited the orthodox hopes that had sustained, through many centuries, the Christian world. Yet, at the last, the one transcends the other's vision, or amplifies it. Darwin remains, though sometimes hesitantly, the pragmatic scientist, content with what his eyes have seen. The other turns toward an unseen spring beyond the wintry industrialism of the nineteenth century, with its illusions of secular progress. The two views, even the two lives, can be best epitomized in youthful expressions that have come down to us. The one, Darwin's, is sure, practical, and exuberant. The other reveals an exploring, but wary, nature.

Darwin, the empiricist, wrote from Valparaiso in 1834: "I have just got scent of some fossil bones of a MAMMOTH; what they may be I do not know, but if gold or galloping will get them they shall be mine." Thoreau, by nature more skeptical of what can be captured in this world, mused, in his turn, "I cannot lean so hard on any arm as on a sunbeam." It was one of the first of many similar enigmatic expressions that were finally to lead his well-meaning friend, Ellery Channing, to venture sadly, "I have never been able to understand what he meant by his life. . . . Why was he so disappointed with everybody else? Why was he so interested in the river and the woods . . . ? Something peculiar here I judge."

Channing was not wrong. There *was* something peculiar about Thoreau, just as there was something equally peculiar about Darwin. The difference between them lies essentially in the nature of man himself, the creature who persists in drawing sharp, definitive lines across the indeterminate face of nature. Essentially, the problem may be easily put. It is its varied permutations and combinations that each generation finds so defeating, and that our own time is busy, one might say horribly busy, in re-creating.

One may begin with what we all remember from childhood —the emerald light in the wonderful city of Oz. Those who lived in the city wore spectacles that were locked on by night and day. Oz had so ordered it when the city was built. Now Oz, it was explained to the simple ones who came there, was a great wizard who could take on any form he wished. "But," as one denizen of the city explained, "who the real Oz is when he is in his own form, no living person can tell."

Among the visitors to that city came several creatures, only one of whom was human, but all of whom dealt with great questions couched in very simple form. One was the Tin Woodman, in search of a heart. One was the Cowardly Lion, who had not the courage to keep tramping forever without getting anywhere at all. Then there was also the little girl, Dorothy, from Kansas, who was sure that if they walked far enough they would sometime come to some place. Particularly pertinent here is that appealing character, the Scarecrow, who, with his straw-filled head and patient good nature has always represented the better, more humble side of man. Scarecrow had been made out of straw instead of the clay so frequently utilized in the creation of man, and perhaps he proved the better for it. At any rate, his only recorded comment upon his existence in the fields was, "It was a lonely life to lead, for I had nothing to think of, having been made such a little while before."

The whole story of humanity is basically that of a journey toward the Emerald City, and of an effort to learn the nature of Oz, who, perhaps wisely, keeps himself concealed. In each human heart exists the Cowardly Lion and the little girl who was sure that the solution to life lay in just walking far enough. Finally, among our great discoverers are those with precious straw-filled heads who have to make up their own thoughts because each knows he has been made such a little while before, and has stood alone in the fields. Darwin and Thoreau are two such oddly opposed, yet similar, scarecrows. As it turned out, they came to two different cities, or at least vistas. They discovered something of the nature of Oz, and, rightly understood, their views are complementary to each other.

I shall treat first of Darwin and then of Thoreau, because, though contemporaries, they were distinct in temperament. Thoreau, who died young, perhaps trudged farther toward the place which the little girl Dorothy was so sure existed, and thus, in a sense, he may be a messenger from the future. Since futures do not really exist until they are present, it might be more cautious to say that Thoreau was the messenger of a *possible* future in some way dependent upon ourselves.

Neither of the two men ever discovered the nature of Oz himself. The one, Darwin, learned much about his ways—so much, indeed, that I suspect he came to doubt the existence of Oz. The other, Thoreau, leaned perhaps too heavily upon his sunbeam, and in time it faded, but not surely, because to the last he clung to the fields and heard increasingly distant echoes. Both men wore spectacles of sorts, for this is a rule that Oz has decreed for all men. Moreover, there are diverse kinds of spectacles.

There are, for example, the two different pairs through which philosophers may look at the world. Through one we see ourselves in the light of the past; through the other, in the light of the future. If we fail to use both pairs of spectacles equally, our view of ourselves and of the world is apt to be distorted, since we can never see completely without the use of both. The historical sciences have made us very conscious of our past, and of the world as a machine generating successive events out of foregoing ones. For this reason some scholars tend to look totally backward in their interpretation of the human future. It is, unconsciously, an exercise much favored in our time.

Like much else, this attitude has a history.

When science, early in the nineteenth century, began to ask what we have previously termed "the terrible questions" because they involved the nature of evil, the age of the world, the origins of man, of sex, or even of language itself, a kind of Pandora's box had been opened. People could classify giraffes and porcupines but not explain them—much less a man. Everything stood in isolation, and therefore the universe of life was bound to appear a little ridiculous to the honestly enquiring mind. What was needed was the kind of man of whom Thoreau had spoken, who could couple two seemingly unrelated facts and reduce the intractable chaos of the world. Such a man was about to appear. In fact, he had already had his forerunners.

II

Robert Fitzroy was a captain with a conscience. Another of that great breed of English navigators of whom Cook stands as the epitome, Fitzroy had been appointed at twenty-three to the command of H.M.S. *Beagle* on a mapping and exploring voyage around Cape Horn. This event preceded the famous expedition in which Charles Darwin was to participate. Mapping the Strait of Magellan, the ambitious young officer discovered, was rather like mapping the stars in the heavens. Perhaps Fitzroy's wry comment to this effect was an unconscious omen of what was to prove the task of Charles Darwin, venturing upon the greater waters of time and change.

The second voyage might never have taken place had it not been that in his adventures about the Strait Fitzroy had acquired four savage Indians, whom he brought home to London in 1830 with the quixotic idea of familiarizing them with Christian civilization and then returning them to their native land. One man died of smallpox—the other three, one woman and two men, survived. The troubled captain, maintaining and attempting to educate these people at his own expense, decided upon their return, even if he personally had to charter a vessel to see them safely home.

Ironically, it was because of this touch of zealous missionary spirit on the part of Fitzroy that mankind was to find itself eventually displaced, biologically, from the center of the universe. The aristocratic Fitzroy exerted influence upon the Admiralty, and that body, in turn, assented to a second voyage, under renewed instructions for further mapping and exploration.

Fitzroy, at heart a lonely young captain, sought a companion. In the Cook tradition he decided upon a naturalist. Charles Darwin secured the post through the good offices of his botanical instructor at Cambridge, John Henslow. Fitzroy, a passionately religious early Victorian, had taken aboard his vessel a man who by training and inclination carried with him the liberal enquiring attitudes of the Enlightenment—the spirit that had perished in the excesses of the French Revolution.

The shadow of those excesses had fallen darkly across Britain and had accentuated the conservatism of the upper classes. There was a strong tendency to excoriate or ridicule French thinkers, particularly if their ideas appeared religiously unorthodox. Even their rare defenders preferred to remain anonymous. The result, for the historian of science, is unfortunate. Published innovations in some instances remain unidentifiable with their advocates. In other cases, anticipations of what were later to emerge as significant ideas have been hidden, deliberately or by accident, under innocuous titles, or interjected into seemingly guileless and innocent works upon stultifying subjects.

We know perfectly well, for example, that the name of the French evolutionist Jean Baptiste Lamarck was known in Edinburgh University circles when young Charles Darwin was a medical student there. We know, even, that one of Darwin's instructors, Robert Grant, was an avowed follower of the French naturalist. Yet so fixed was this isolated British set of mind that as late as 1863, after the publication of the *Origin of Species*, we find Darwin writing to Sir Charles Lyell on the subject of the *Philosophie Zoologique*, "to me it was an absolutely useless book," owing, he says, to his search for "facts."

In the same letter he dismisses his grandfather's ideas with equal abruptness. Not only are these remarks scarcely borne out by a careful examination of Darwin's work, but the harsh emphasis upon "facts" comes a little oddly from a scholar who could also protest, "Forfend me from a man who weighs every expression with Scotch prudence." Elsewhere he intimates he can scarcely abide facts without attempting to tie them together.

The youth who went aboard the *Beagle* in December 1831 was a great deal more clever than his academic record at both Edinburgh and Cambridge might suggest. The ingenuity with which he went about securing his father's permission for the voyage in itself indicates the dedicated persistence with which he could overcome obstacles. There remained in the motherless young man a certain wary reserve, which would finally draw him into total seclusion. In the first edition of the *Origin of Species* he was to write: "When on board H.M.S. *Beagle*, as naturalist, I was much struck with certain facts. . . . These facts seemed to . . . throw light on the origin of species. . . ." The remark is true, but it is also ingenuous. Young Charles's first knowledge of evolution did not emerge spontaneously aboard the *Beagle*, however much that conception was to be strengthened in the wild lands below the equator.

Instead, its genesis in Darwin's mind lies mysteriously back amidst unrecorded nights in student Edinburgh and lost in the tracery of spider tracks over thousands of dusty volumes after his return. For this is the secret of Charles Darwin the naturalist-voyager, the modern Odysseus who came to Circe's island of change in the Galápagos: he was the product of two odysseys, not one. He lives in the public mind partly by the undoubted drama of a great voyage whose purpose, as defined by the chief hydrographer, was the placing of a chain of meridians around the world. While those meridians were being established through Fitzroy's efforts, another set was being posted by Darwin in the haunted corridors of the past.

But the second odyssey, the one most solitary, secretive, and hidden, is the Merlin-like journey which had no ending save at death. It is the groping through webby corridors of books in smoky London—the kind of journey in which men are accountable only to themselves and by which the public is not at all enlivened. No waves burst, no seaman falls from the masthead, no icy continent confronts the voyager. Within the mind, however, all is different. There are ghost fires burning over swampy morasses of books, confusing trails, interceptions of the lost, the endless weaving and unweaving of floating threads of thought drawn from a thousand sources.

"Lord," briefly writes the man with the increasingly worn face and heavy brow, "in what a medley the origin of cultivated plants is. I have been reading on strawberries, and I can find hardly two botanists agree what are the wild forms; but I pick out horticultural books here and there, with queer cases of variation." Or, being the man he was, again surfacing for a moment from the least expected place: "I sat one evening in a gin palace among pigeon fanciers."

A doubt, a shrinking terrible doubt such as can cause a man's hands to shake and rustle amidst the leaves of folios, came on him in 1854. "How awfully flat I shall feel, if when I get my notes together . . . the whole thing explodes like an empty puff-ball."

References to old transactions, old travels, old gardener's magazines bestrew his letters. Ideas are pressed away like botanical specimens between boards. "The Bishop," he says, "makes a very telling case against me." "Hurrah! A seed has just germinated after twenty-one and a half hours in an owl's stomach." "I am like a gambler and love a wild experiment."

"I am horribly afraid." "I trust to a sort of instinct and God knows can seldom give any reason for my remarks." And then, pathetically, "All nature is perverse and will not do as I wish it. I wish I had my old barnacles to work at, and nothing new."

The confidences go on like the running stream of the years. Volumes of them exist. There are other volumes that are lost— the one, for example, that might explain what strange compulsion drove Darwin, in at least one recorded instance, from bed at midnight to come downstairs to chatting guests and correct a trivial shift in opinion that could affect no one.

The floating threads of all the ideas, all the thinking, all the nightmares of hours spent in the endless galleries of books, meet and are gathered up at last in the great book of 1859— that book termed the *Origin of Species*, which Darwin to the last maintained was only a hasty abstract.

Do not judge harshly, he importuned no one in particular. Wait for the Big Book, the real book, where everything will be explained and more, so much more, will be adequately interpreted. And if not that book, then the endless regression through ever larger volumes seems to occupy his mind. His odyssey is endless. There is the earthworm, the orchid, the pigeons from the sporting gentry in the gin palace, the shapes of leaves, the carnivorous sundew feeding like an animal. There is the weird groping of vines. There is man himself, the subject of subjects. Year after year he brings his treasures forward like a child and proffers them.

In one or another guise he is hiding from the public. He has become, save for his family, almost a recluse. To his astonishment, he is a legend in his lifetime. The "abstract" he scorned has become a classic of the world's literature. His name is coupled with that of Newton.

Variation—that subtle, unnoted shifting of the shapes of men and leaves, bird beaks and turtles, that he had pondered over far off in the Circean Galápagos—was now seen to link the seemingly ridiculous and chaotic world of life into a single whole. Selection was the living screen through which all life must pass. No fact could be left a fact; somewhere in the world it was tied to something else. What made a tuft of feathers suddenly appear on a cock's head or induce the meaningless gyrations of a tumbler pigeon? What, in this final world of the fortuitous, the sad eyes questioned, had convinced a tailless ape that he was the object of divine attention?

The year 1859 was gone. The British Association had met at Oxford in 1860 and Thomas Huxley, Darwin's devastatingly caustic defender, had clashed with Bishop Samuel Wilberforce. A lady had fainted. Captain Robert Fitzroy, himself a Fellow of the Royal Society, now a meteorologist and advocate of storm warnings, had arisen angrily to protest the violation of the first chapter of the Book of Genesis.

In a few more years, worn out with his attempts to convince the indifferent Admiralty and public that weather might be foretold and disasters minimized, Fitzroy would go alone upstairs in the dawn with a razor cold against his throat while his family slumbered. Was he, too, feeling the sickness of that grinding human displacement? Did he, too, feel his solid Victorian world slipping beneath his feet, and, irony of ironies, all because he, Robert Fitzroy, had invited an earnest young man from Cambridge to go on a sea voyage years and years ago? In the dim light the razor glinted. The most important storm warning of all had failed Fitzroy. He was a scientific pioneer who had the misfortune to die unrecognized. He had pleaded for the use of the telegraph in following the weather. Radio and aircraft would prove his wisdom in the following century. Fitzroy had the misfortune to be both behind and ahead of his time. Such men are always subject to injustice.

III

We today know the result of Darwin's endeavors—the knitting together of the vast web of life until it is seen like the legendary tree of Igdrasil, reaching endlessly up through the dead geological strata with living and related branches still glowing in the sun. Bird is no longer bird but can be made to leap magically backward into reptile; man is hidden in the lemur, lemur in tree shrew, tree shrew in reptile; reptile is finally precipitated into fish.

But then there intrudes another problem: Mouse is trying to convert all organic substance into mouse. Black snake is trying to convert mouse into snake. Man maintains factories to convert cattle into human substance. It is an ingenious but hardly edifying spectacle in which nothing really wins, and through which whole orders of life have perished. If our tempo of seeing could be speeded, life would appear and disappear as a chaos of evanescent and writhing forms, possessing the impermanence of the fairy mushroom circles that spring up on our lawns at midnight.

But this is not all; there is something more terrible at the heart of this seething web, something that caused even Darwin's nerves to shudder. Let me illustrate from an experience of my own. I had been far outward on an open prairie swept by wind and sun. The day was fair. It was good to be alive. On all that wild upland I finally saw one human figure far away. He, like myself, seemed wandering and peering. Eventually he approached. We exchanged amenities, and I learned he was a brother scientist, though from a different discipline. He carried dissecting instruments and a bottle.

There was something living in the bottle amidst the juices drained from a rabbit's belly. The man was happy; his day had been successful. He showed me his treasure there in the bright sun on the turf of that sterile upland where there had seemed to lurk no evil. The sight lies a quarter of a century away and still I remember it—the dreadful pulsing object in the bottle and the grinning delight of its possessor.

It was a parasite—a new and unnamed parasite—explained the happy government entomologist, or so the man had represented himself. But there in the bottle, alternately expanding and contracting, searching blindly for the living flesh from which it had been torn, groped an enormous, glistening worm. I was younger then, and my mind clearer. Within it arose the memory of perhaps the most formidable words Darwin was ever to write—words never intended for publication. His personal discovery, in what must have been all its dire immediacy, lay in that moment before me, just as his words re-echoed in my brain. "What a book," he had written with unaccustomed savagery, "a devil's chaplain might write on the clumsy, wasteful, blundering and horribly cruel works of Nature."

Evil has always been one of the difficult questions associated theologically with the fall of man, but now it had been found in the heart of nature itself. It was as though the pulse of the universe had been transferred to that obscene, monstrous body that was swelling as though to engulf the world. The more I looked, the more it appeared to grow. The high clean air of that lonely upland only made the event more unnatural, the collector's professional joy more maniacal.

Finally he went away, bearing his precious bottle with its pulsing, unspeakable life. I watched him until he dwindled to a trudging speck on the distant horizon. It is plain, however, in the light of decades, that that remote, gesticulating figure within my mind is still obscurely laughing in his long descent. If it were not so, I would not see him, for the plain itself was gigantic. This, too, I remember. It would be hard, it seems in retrospect, for anything at all to be found there—just anything, including man. I do not comprehend to this day what it was that ordained our meeting.

Except one thing, perhaps: the fall of man. In that fall, gazing upon the creature in the bottle, I, too, had participated. For man *did* fall; even to an unbeliever and an evolutionist like Darwin. Man fell from the grace of instinct into a confused and troubled cultural realm beyond nature, much as in the old theology man fell from a state of innocence into carnal knowledge. The idea is implicit in Darwin's work and has been commented upon by the critic Stanley Hyman, who has noted its recognition by a late Victorian scholar. This reviewer had commented that "Mr. Darwin [in *The Descent of Man*] finds himself compelled to reintroduce a new doctrine of the fall of man. He shows that the instincts of the higher animals are far nobler than the habits of the savage races of men, and he finds himself, therefore, compelled to reintroduce—in a form whose substantial orthodoxy appears to be quite unconscious—the doctrine that man's gain of knowledge was the cause of the temporary but long enduring moral deterioration . . . of savage tribes." The slow climb back to respectability seems, as one studies Darwin's work, to have culminated in Victorian civilization. The savages of Cape Horn were hardly accorded the graces of domesticated animals.

Darwin had been gazing backward upon the ways of pigeons, apes, and earthworms in extravagant profusion. Upon these creatures and their origins he had expended a sizable proportion of his life. Nature he loved, but by his own words he had become a hermit. Man he achingly endured, as he endured the visitors at Down. He was looking back upon an increasingly remote and violent past, through spectacles few men had raised to their eyes before, and none before him so effectively. Though he occasionally rendered lip service to the idea of human progress toward perfection, cultural man was really a disturbing element in his system, an obstruction difficult to account for, and introducing strange vagaries into Darwin's own version of the Newtonian world machine. In spite of a vast world journey, enormous reading, and a wonderful glimpse, as through the mosaic of a stained-glass window, at the imperfect changing quality of life, Darwin remained an observer held in the bonds of the European social system of his day, and overimpressed by Malthusian struggle. The oncoming world of the indeterminate and the possible that he had helped to initiate he never fully grasped. He looked, and his spectacles brought him light, but it was sometimes the half-light with which Oz has so frequently chosen to shade the eyes of men.

IV

Thoreau had loved nature as intensely as Darwin and perhaps more personally. He had seen with another set of glasses. He was, in an opposite sense to Darwin, a dweller along the edge of the known, a place where the new begins. Thoreau carries a hint of that newness. He dwelt, without being quite consciously aware of it, in the age after tomorrow. His friends felt universally baffled by Thoreau and labeled him "almost another species." One contemporary wrote: "His eyes slipped into every tuft of meadow or beach grass and went winding in and out of the thickest undergrowth, like some slim, silent, cunning animal." It has been said that he was not a true naturalist. What was he, then? The account just quoted implies a man similar to Darwin, and, in his own way, as powerfully motivated.

Of all strange and unaccountable things, Thoreau admits his efforts at his *Journal* to be the strangest. Even in youth he is beset by a prescient sadness. The companions who beguile his way will leave him, he already knows, at the first turn of the road. He was basically doomed all his life to be the Scarecrow of Oz, and if he seems harsher than that genial figure, it is because the city he sought was more elusive and he did not have even the Cowardly Lion for company. He knew only that by approaching nature he would be consulting, in every autumn-leaf fall, not alone those who had gone before him, but also those who would come after. He was writing before the *Origin of Species*, but someone had sewn amazing eyes upon the Concord Scarecrow.

There is a delicacy in him that is all his own. His search for support in nature is as diligent as that of a climbing vine he had once watched with fascinated attention groping eerily toward an invisible branch. Yet, like Darwin, he had witnessed the worst that nature could do. On his deathbed he had asked, still insatiable, to be lifted up in order that he could catch through the window a glimpse of one more spring.

In one passage in the *Journal* he had observed that the fishers' nets strung across the transparent river were no more intrusive than a cobweb in the sun. "It is," he notes, "a very slight and refined outrage at most." In their symmetry, he realizes, they are a beautiful memento of man's presence in nature, as wary a discovery as the footprint upon Crusoe's isle. Moreover, this little symbol of the fishers' seine defines precisely that delicately woven fabric of human relationships in which man, as a social animal, is so thoroughly enmeshed. There are times when, intellectually, Darwin threshed about in that same net as though trapped by a bird spider in his own forested Brazil.

For the most part, the untraveled man in Concord managed to slip in and out of similar meshes with comparative ease. Like some lean-bodied fish he is there, he is curiously observant, but he floats, oddly detached and unfrightened, in the great stream. "If we see nature as pausing," Thoreau remarks more than once, "immediately all mortifies and decays." In that nature is man, merely another creature in perspective, if one does not come too close, his civilizations like toadstools springing up by the road. Everything is in the flowing, not the past.

Museums, by contrast, are catacombs, the dead nature of dead men. Thoreau does not struggle so hard as Darwin in his phylogenies to knit the living world together. Unlike the moderns, Thoreau was not constantly seeking nostalgically for men on other planets. He respected the proud solitude of diversity, as when he watched a sparrow hawk amusing itself with aerial acrobatics. "It appeared to have no companion in the universe and to need none but the morning," he remarked, unconsciously characterizing himself. "It was not lonely but it made all the earth lonely beneath it."

Or again, he says plainly, "fox belongs to a different order of things from that which reigns in the village." Fox is alone. That is part of the ultimate secret shared between fox and scarecrow. They are creatures of the woods' edge. One of Thoreau's peculiar insights lies in his recognition of the creative loneliness of the individual, the struggle of man the evolved animal to live "a supernatural life." In a sense, it is a symbolic expression of the equally creative but microcosmic loneliness of the mutative gene. "Some," he remarks, "record only what has happened to them; but others how *they* have happened to the universe."

To this latter record Thoreau devoted the *Journal* that mystified his friends. Though, like Darwin, he was a seeker who never totally found what he sought, he had found a road, though no one appeared to be walking in it. Nevertheless, he seems to have been interiorly informed that it was a way traversed at long intervals by great minds. Thoreau, the physical stay-at-home, was an avid searcher of travel literature, but he was not a traveler in the body. Indeed, there are times when he seems to have regarded that labyrinth—for so he called it—with some of the same feeling he held toward a house—as a place to escape from. The nature we profess to know never completely contained him.

"I am sensible of a certain doubleness," he wrote, "by which I can stand as remote from myself as from another. . . . When the play—it may be the tragedy of life—is over, the spectator goes his way. It was a kind of fiction." This man does suggest another species, perhaps those cool, removed men of a far, oncoming century who can both live their lives and order them like great art. The gift is rare now, and not wholly enticing to earthbound creatures like ourselves.

Once, while surveying, Thoreau had encountered an unusual echo. After days with humdrum companions, he recorded with surprise and pleasure this generosity in nature. He wanted to linger and call all day to the air, to some voice akin to his own. There needs must be some actual doubleness like this in nature, he reiterates, "for if the voices which we commonly hear were all that we ever heard, what then? Echoes . . . are the only kindred voices that I hear."

v

Here, in Thoreau's question, is the crux, the sum total of the human predicament. This is why I spoke of our figuratively winding our way backward into a spiritual winter, why I quoted an Eskimo upon wisdom. On the eve of the publication of the *Origin of Species*, Thoreau, not by any means inimical to the evolutionary philosophy, had commented: "It is ebb tide with the scientific reports."

In some quarters this has aroused amusement. But what did Thoreau mean? Did he sense amidst English utilitarian philosophy, of which some aspects of Darwinism are an offshoot, an oncoming cold, a muffling of snow, an inability to hear echoes? Paradoxically, Thoreau, who delighted in simplicity of living, was averse to the parsimonious nature of Victorian science. It offended his transcendental vision of man. Lest I seem to exaggerate this conflict, read what Darwin himself admitted of his work in later years:

"I did not formerly consider sufficiently the existence of structures," he confesses, "which as far as we can . . . judge, are neither beneficial nor injurious, and this I believe to be one of the greatest oversights as yet detected in my work. This led to my tacit assumption that every detail of structure was of some special though unrecognized service."

We know that Thoreau already feared that man was becoming the tool of his tools, which can, alas, include ideas. Even now, forgetting Darwin's belated caution, those with the backward-reaching spectacles tell us eagerly, if not arrogantly, in the name of evolution, how we are born to behave and the limitations placed upon us—we who have come the far way from a wood nest in a Paleocene forest. Figuratively, these pronouncements have about them the enlarging, man-destroying evil of the pulsing worm. They stop man at an imagined border of himself. Man suffers, in truth, from a magical worm genuinely enlarged by a certain color of spectacles. It is a part, but not the whole, of the magic of Oz.

Is it not significant, in contrast to certain of these modern prophets, that Thoreau spoke of the freedom he felt to go and come in nature; that what is peculiar to the life of man "consists not in his obedience, but his opposition to his instincts"? The very behavior of the other animals toward mankind, Thoreau knew, revealed that man was not yet the civilized creature he pretended to be.

One must summarize the two philosophies of evolution and then let the Eskimo speak once more. In the Viking Eddas it is written:

> Hard it is on earth . . .
> Ax-time, sword-time . . .
> Wind-time, wolf-time, ere the world falls
> Nor ever shall men each other spare.

Through these lines comes the howl of the world-devouring Fenris-wolf, waiting his moment under the deep-buried rocket silos of today. In the last pages of *Walden* one of Thoreau's wisest remarks is upon the demand scientific intellectuals sometimes make, that one must speak so as to be always understood. "Neither men," he says, "nor toadstools grow so." There is a constant emergent novelty in nature that does not lie totally behind us, or we would not be what we are.

Here is where Thoreau's sensitivity to echoes emerges powerfully: It is onflowing man, not past evolutionary man, who concerns him. He wants desperately to know to what degree the human mind is capable of inward expansion. "If the condition of things which we were made for is not yet at hand," he questions anxiously, "what can we substitute?" The echoes he senses are reverberating from the future.

Finally, he compresses into a single passage the answer to the wolf-time philosophy, whether expressed by the Viking freebooters, or by certain of their modern descendants. "After," he says, "the germs of virtue have thus been prevented many times from developing themselves, the beneficent breath of evening does not suffice to preserve them. . . . Then the nature of man does not differ much from that of the brute."

Does this last constriction contain the true and natural condition of man? No, Thoreau would contend, for nature lives always in anticipation. Thoreau was part of the future. He walked toward it, knowing also that in the case of man it must emerge from within by means of his own creation. That was why Thoreau saw the double nature of the tool and eyed it with doubt.

The soul of the universe, the Upholder, reported Rasmussen of the Alaskan Eskimo, is never seen. Its voice, however, may be heard on occasion, through innocent children. Or in storms. Or in sunshine. Both Darwin and Thoreau had disavowed the traditional paradise, and it has been said of Thoreau that he awaited a Visitor who never came. Nevertheless, he had felt the weight of an unseen power. What it whispers, said the men of the high cold, is, "Be not afraid of the universe."

Man, since the beginning of his symbol-making mind, has sought to read the map of that same universe. Do not believe those serious-minded men who tell us that writing began with economics and the ordering of jars of oil. Man is, in reality, an oracular animal. Bereft of instinct, he must search constantly for meanings. We forget that, like a child, man was a reader before he became a writer, a reader of what Coleridge once called the mighty alphabet of the universe. Long ago, our forerunners knew, as the Eskimo still know, that there is an instruction hidden in the storm or dancing in auroral fires. The future can be invoked by the pictures impressed on a cave wall or in the cracks interpreted by a shaman on the incinerated shoulder blade of a hare. The very flight of birds is a writing to be read. Thoreau strove for its interpretation on his pond, as Darwin, in his way, sought equally to read the message written in the beaks of Galápagos finches.

But the messages, like all the messages in the universe, are elusive.

Some months ago, walking along the shore of a desolate island off the Gulf Coast, I caught a glimpse of a beautiful shell, imprinted with what appeared to be strange writing, rolling in the breakers. Impelled by curiosity, I leaped into the surf and salvaged it. Golden characters like Chinese hieroglyphs ran in symmetrical lines around the cone of the shell. I lifted it up with the utmost excitement, as though a message had come to me from the green depths of the sea.

Later I unwrapped the shell before a dealer in antiquities in the back streets of a seaport town.

"*Conus spurius atlanticus*," he diagnosed for me with brisk efficiency, "otherwise known as the alphabet shell."

But why spurious? I questioned inwardly as I left the grubby little shop, warily refusing an offer for my treasure. The shell, I was sure, contained a message. We *live* by messages—all true scientists, all lovers of the arts, indeed, all true men of any stamp. Some of the messages cannot be read, but man will always try. He hungers for messages, and when he ceases to seek and interpret them he will be no longer man.

The little cone lies now upon my desk, and I handle it as reverently as I would the tablets of a lost civilization. It transmits tidings as real as the increasingly far echoes heard by Thoreau in his last years.

Perhaps I would never have stumbled into so complete a revelation save that the shell was *Conus spurius*, carrying the appellation given it by one who had misread, most painfully misread, a true message from the universe. Each man deciphers from the ancient alphabets of nature only those secrets that his own deeps possess the power to endow with meaning. It had been so with Darwin and Thoreau. The golden alphabet, in whatever shape it chooses to reveal itself, is never spurious. From its inscrutable lettering is created man and all the towering cloudland of his dreams.

SEVEN

The Invisible Island

I say by Sorcery he got this Isle;
From me he got it.

—CALIBAN

IN LATITUDE SIXTY degrees south, under the looming shadow of the antarctic ice, from somewhere deep in the freezing planetary swirl of the Humboldt Current, the great whale surfaced and was struck for the last time. "We found embedded in her hide," the record of the old logbook runs, "divers harpoons of antique shapes, ours being her death."

What shafts have equally been hurled in epochs gone against the quarry earth? What gaping prehistoric jaws, subsiding in oblivion, have struck and splintered in their turn upon her stone? What giant plinths and cromlechs buried in her soil still mark the deeds of conquerors long since undone? What mile-deep drills and missiles wrought by hands more violent now probe at earth's defenses? Yet she flies on regardless. Her hour has not yet come. The blade that will split her heart has not been forged.

Earth is the mightiest of the creatures. She contains beneath her furry hide the dark heart of nothingness, from which springs all that lives. She is the wariest and most complete of animals, for she lends herself to no particular form and in the end she soundlessly forsakes them all. She is the one complete island of being. The rest, including man, are in some degree fragmented and illusory. Yet it is these phantoms, even of whales come up from the deep gulfs, that mark our thought.

Of true places Herman Melville once remarked that "they are never down on any map." He also observed of those great amphibians, those islands of living flesh of which I have just spoken, that it is not known whether the small eyes placed yards apart on opposite sides of their huge heads can co-ordinate the meaning of two sights at once. If not, that great solitary beast, with all its hoary antiquity written in the weapons ranged along its back, saw, as it floundered in its death throes, two separate worlds. It saw with one dimming affectionate eye the ancient mother, the heaving expanse of the universal sea. With the other it glimpsed with indescribable foreboding the approaching shape of man, the messenger of death and change. So it was that the dying whale in its dissociated vision had arrived, if momently, at the one true place where the nature of the past mingles with the onrush of the future and is borne down forever into darkness. It is an instant to remember, for it will come, in turn, to man.

Indeed, this fusing of the past and future has already come and made of man the thing he is, an invisible island, as surely as the great whale was an island, as surely as volcanic clinkered isles produce monstrosities, dwarfs, and giants in secret shifting latitudes where no navigator is able to take readings. It is a somber reflection upon human nature that so much has been written about the triumph of the fittest and so little about the survival of the failures who have changed, if not deranged, the world.

Man is one such creature, but the story begins long before him, a story of only incidental triumphs. The universe, as we have seen, is a place of unexpected events. A major portion of the world's story appears to be that of fumbling little creatures of seemingly no great potential, falling, like the helpless little girl Alice, down a rabbit hole or an unexpected crevice into some new and topsy-turvy realm. Such beings sometimes find themselves cast unhappily in the role of unanticipated giants. It is not struggle and survival alone that have so marked life's exuberant pathway; rather, the mastery has often come after the event and almost as a prelude to extinction. It is as though everything alive had in it a tug of antigravity, a revulsion from the central fire or the mother sea. If stars and galaxies hurtle outward in headlong flight, the urge for dispersal seems equally and unexplainably written into the living substance itself. The myth of Eden, the myth of Babel, and now our evolving science itself all testify to this rent in nature, as though, if we could but see it, everything might be scurrying and hopping toward a myriad exits. Solitude may strike self-conscious man as an affliction, but his march is away from his origins, and even his art is increasingly abstract and self-centered. From the solitude of the wood he has passed to the more dreadful solitude of the heart.

In a university lecture hall some months ago, it was brought home to me just how far this alienation had progressed. I had been speaking, by way of illustrating a point, about a tiny deer mouse, a wonderfully new and radiant little creature of white feet and investigative fervor whom I had seen come into a basement seminar upon the Byzantine Empire. After a time the mouse in its innocent pleasure had actually ascended into an empty chair and perched upright with trembling inquisitive gravity while an internationally renowned historian continued to address the group. By no least sign did he reveal that an eager anticipatory face had appeared among his students.

After my own lecture I was approached and chided by a young lady who informed me with severity that I was betraying evidence of a foolish anthropomorphism, which would certainly place me under disfavor and suspicion in the psychological circles she frequented.

I sighed and reluctantly confessed that perhaps the mouse, since he was obviously a very young mouse recently come from the country, could not have understood every word of the entire lecture. Nevertheless, it was gratifyingly evident to my weary colleague, the great historian, that the mouse had at least tried.

"You see," I explained carefully, "we may have witnessed something like Alice in reverse. The mouse came through a crevice in the wall, a chink in nature. Man in his time has come more than once through similar chinks. I admit that the creatures do not always work out and that the chances seemed rather against this one, but who is to say what may happen when a mouse gets a taste for Byzantium rather beyond that of the average graduate student? It takes time, generations even, for this kind of event to mature.

"If I may be pardoned for being so bold," I remonstrated with the young lady, "what do you think your chances might have been of charging *me* with anthropomorphism when we were both floundering about in a mud puddle, or, for that matter, testing whether an incipient backbone might enable us to wriggle upstream? You must remember," I continued, "that these are all figurative entrances and exits with sometimes kingdoms at the bottom of them. Or disaster, or even both together."

"But not," the young lady protested venomously, "the Byzantine Empire."

"My friend was quite hopeful," I persisted. "The creature was so evidently concerned and alert beyond the average. After all, how must the first two invented words have stirred your ancestors," I appealed, "and there was nobody, absolutely nobody, to give them a lecture on Byzantium, because, well, because Byzantium was"—I was getting out of my depth now—"up there in the future." I gestured uncertainly with one finger.

"It was not down a hole then," said the literal young lady. Triumphantly she drove home her point. "It wasn't anywhere." With this she walked out.

"This woman is evidently part of a conspiracy to keep things just as they are," I later wrote to my friend. "This is what biologically we may call the living screen, the net that keeps things firmly in place, a place called now.

"It doesn't always work," I added in encouragement. "Things get through. We ourselves are an example. Perhaps a bad one. About the mouse . . ."

The answer came back in a few days, lugubriously.

"The Exterminators have come. Your chink is closed. Definitely."

The Exterminators. I turned the harsh word over in my mind. Great God, what were we doing? The net was drawing tighter. Man had his hands upon it. The effects were terrifying. I thought of a sparkling stream where I had played as a boy. There had been sunfish in it, turtles. Now it ran sludge and oil. It was true. The net was tightening. All over the earth it was tightening. Even the ice chinks at the poles were under man's surveillance.

One might say that the surveillance began in 1835, when young Charles Darwin, on the round-the-world cruise of H.M.S. *Beagle*, dropped anchor in the Galápagos Archipelago, six hundred miles off the west coast of South America. Ashore, young Darwin observed with a speculative eye a bird strange to his experience of the continents. He jotted into his notebook an observation that was not to bear full fruit until after the passage of over a quarter of a century. Such island differences among the creatures of the Archipelago, the young man meditated, "*will be well worth examining; for such facts would undermine the stability of species.*"

II

I have often looked with speculative interest upon those delicate insects that row with feathery feet upon the waters of a brook. They make but a slight dimple upon the film of sliding water. They breathe the air; they rove in an immense freedom over ponds and watercourses. The insubstantial film upon which they float resembles the surface tension of the living screen of life, in which every organism, like the forces in the atom, exerts an enormous hold, directly or indirectly, upon every other living thing. The water striders have evolved a way of outwitting the water film that entraps the heavy-footed. In their small way they have risen superior to a dangerous medium and have diverted its tensions to their own advantage.

Man has similarly defeated and diverted the entire web of life and dances, dimpling, over it. Like the water strider he possesses the freedom of a dangerous element. Even the water strider's freedom is relative, however, and contained within nature. One cannot help but dwell upon the hidden powers that produce so delicate a balance between freedom and catastrophe. For freedom of this nature is rare, and in man it is more than rare—it is unique—for he dances upon shadows, the shadows in his brain.

Before turning to that realm of shadows, it is well to define what we mean by the web of life. Some time ago I had occasion one summer morning to visit a friend's grave in a country cemetery. The event made a profound impression upon me. By some trick of midnight circumstance a multitude of graves in the untended grass were covered and interwoven together in a shimmering sheet of gossamer, whose threads ran indiscriminately over sunken grave mounds and headstones.

It was as if the dead were still linked as in life, as if that frail network, touched by the morning sun, had momentarily succeeded in bringing the inhabitants of the grave into some kind of persisting relationship with the living. The night-working spiders had produced a visual facsimile of that intricate web in which past life is intertwined with all that lives and in which the living constitute a subtle, though not totally inescapable, barrier to any newly emergent creature that might attempt to break out of the enveloping strands of the existing world.

As I watched, a gold-winged glittering fly, which had been resting below the net, essayed to rise into the sun. Its wings were promptly entrapped. The analogy was complete. Life *did* bear a relationship to the past and was held in the grasp of the present. The dead in the grass were, figuratively, the sustaining base that controlled the direction of the forces exerted throughout the living web.

The golden fly would perish. Nevertheless, there would be days when the wind would sweep the net away, or when snow would hold the sleeping cemetery in a different, though even more profound, embrace. Everything, in the words of John Muir, tended to be hitched to everything else. Yet this was not all, or life would not have originally engaged itself upon that exploratory adventure we call evolution. The strands of the living web were real; they did check, in degree, the riotous extremes of variation. Fortunately for the advancement of life, the tight-drawn strands sometimes snapped; the archaic reptile gave place to the warm-blooded mammal.

Sometimes it was less the snapping of a thread, even of gossamer, than the secretive discovery of a hidden doorway, found by some blundering creature pushed to the wall by savage competitors in the seemingly impermeable web. An old-fashioned, bog-trapped fish had managed to totter ashore on its fins and find itself safely alone to develop in a new element. A clumsy, leaping reptile had later managed to flounder into the safety of the air.

This is the fascinating rabbit-hole aspect of the living world. Most of its experiments are small, at best localized adjustments that, if anything, draw the mesh of the living screen by degrees ever tighter. Man, once having successfully escaped the net, has busied himself in drawing the meshes together with a strangling intensity exerted upon the rest of life. He now appears to possess the power to tear the net at his pleasure. Yet as I stood in wonder before the entangled tombstone of my friend, it struck me that not even man can escape completely the Laocoön embrace of the living web. Sir Francis Bacon in the early dawn of experimental science had called attention to the fact that nature was capable on occasion of extreme "exuberances," such as would be well exemplified by human expansion today. It did not escape the Elizabethan philosopher that such irregularities are by the nature of things eventually, in their turn, trapped and confined.

Life as we know it is not limitless in its capacities, above all, in the higher organisms. Actually, its manifestations are confined to a small range on the thermometer. Moreover, life demands water for the maintenance of its interior environment, in quantities hard to come by, if this solar system can be taken as a typical example. Long food chains, from microscopic infusoria to whales, sustain and nourish all manner of strange animals and plants. Man himself is not free from these food chains, though increasingly, and sometimes to his peril, he tampers experimentally with his environment—thus inviting that "violence in the Returne" of which Bacon warned.

Life has survived by distributing itself over innumerable tiny environments. It has mastered such extremes of temperature and pressure as are represented by the abyssal depths of the sea, by deserts where water must be hoarded behind impermeable plant walls, or where breath must be drawn painfully upon Himalayan heights. In these circumstances one fact is self-evident: under such trials life tends to thin out to the point of disappearance.

In that disappearance the forms of the highest nervous complexity are the first to vanish. All mutative changes tend to be directed toward holding the unfavorable outside world at bay and preserving by formidable barriers the life within. The almost complete rejection of that outside world, however—no matter how cleverly the organism has adapted its resources for the purpose of sustaining itself under such inhospitable conditions—spells the end of intense conscious interaction with the environment. One is left with drought-resisting plants, infusoria that can survive desiccation, crabs teetering gingerly over nightmare depths of mud, or a few desert burrowers capable of synthesizing minute quantities of water in their tissues. Still, the tenuous lines of the web of life run up into mountainous heights and descend into the depths of the sea.

The living screen known as the biosphere may best be pictured by dipping some rough-skinned fruit, such as an orange, into water. Upon its withdrawal the wet film adhering to the depressions of the orange gives some idea of the relative thickness of the living membrane that covers most of the planet. It will attenuate toward the poles and upon ice-covered peaks; it will slink underground or disappear in the most extreme and shifting deserts.

It is a tenuous film, nothing more—yet a film of incredible complexity. Life survives through the fact that we possess an oxygen-charged atmosphere, a modest temperature, and the seas so beautifully photographed by the homing astronauts. We rotate in the beneficent light of a minor star. On freezing Pluto, our outermost planet, that star would appear about the size of a nailhead. Even in the auspicious environment of earth, the major portion of the ancestral past is represented only in eroding strata. The net, the living film that at any given moment of earth time seems to hold all the verdant world in an everlasting balance, is secretly woven anew from age to age. The tiny spiders that had worked through a single night amidst the lettered stones of the cemetery had reproduced in miniature the tight-strung gossamer that links us with the remains of our animal past.

We have noted that when Darwin intruded into the Galápagos and observed the biological rarities existing there, his growing suspicion about the reality of evolution was confirmed. By the time he wrote the *Origin of Species* and *The Descent of Man*, his attention had been occupied by his theory of how evolution had come about; namely, through natural selection. Natural selection was defined as the preservation of favorable genetic variations by means of the winnowing effect of the struggle for existence.

We need not here pursue the intricacies of modern genetic theory. In Darwin's time the emphasis upon intragroup struggle as a means of evolutionary advance sometimes took on exaggerated forms. It has tended so to re-emerge in simplistic interpretations of modern human problems. In the nineteenth century Moritz Wagner reminded Darwin of the bearing that isolation must exercise upon biological change. Darwin, in turn, responded, "it would have been a strange fact if I had overlooked the importance of isolation seeing that it was such cases as that of the Galápagos Archipelago, which chiefly led me to study the origin of species."

Nevertheless, by his own confession, he had "oscillated much" between what seemed to be the intense struggle present on the continents compared with the relaxed refuge area of the Galápagos. In reality, natural selection is a broad phrase, so broad that it implies many different kinds of change, or even repression of change, as when no opening exists in the constricting web we have examined. Islands are apt by their seclusion to offer doorways to the unexpected, rents in the living web, opportunities presented to stragglers who might be carrying concealed genetic novelty in their bodies—novelty that might have remained suppressed in a more drastic competitive environment.

Competition may simply suppress what exists only as potential. The first land-walking fish was, by modern standards, an ungainly and inefficient vertebrate. Figuratively, he was a water failure who had managed to climb ashore on a continent where no vertebrates existed. In a time of crisis he had escaped his enemies. For a long time he was in a position to perfect freely responses to his new mode of life. Adaptive radiation and struggle would only emerge later, as our fishy forerunner explored other land environments and began to create, in his turn, a new repressive web.

To have a genetic island there must be in the beginning an isolating barrier. It could be a genuine island, such as has occupied the speculative attention of voyagers since the days of Captain Cook. On the other hand, the "island" may be a mountaintop or the product of a glacial obstruction. An impassable stream may suffice, or the barrier of a season. The boundary may be also singularly present but invisible. Whatever else it may be, it must offer the opportunity of creative genetic change on a large scale if life is to advance. Struggle, of and by itself, does little but sharpen what exists to a superior efficiency. True, it plays an important role in evolution, but it is not necessarily the only, or even the primary, factor in the rare emergence of the completely novel. It must always be remembered that natural selection is one of those convenient magical phrases that can embrace both dramatic change and stultifying biological conservatism.

This morning, on my back lawn, giant mushrooms of a species unfamiliar have pushed up through the grass and leaves.

They have the appearance of distorted livers, or of some other unsightly organ. A single fog-filled night has sufficed to produce them. If the weather lasts, there will be more. Probably their spores have been merely waiting—waiting for who knows how long. The visitation of the fog represents a sudden rent in the living screen, an opportunity that fungi, as creatures of the night, are particularly suited to seize. The fact that the season is close onto winter makes their sudden upheaval intensely dramatic. It makes one ask what kind of word spores, what night-fog in the protohuman cranium, induced the emergence of that fantastic neurofungus which is man.

III

Islands can be regarded as something thrust up into recent time out of a primordial past. In a sense, they belong to different times: a crab time or a turtle time, or even a lemur time, as on Madagascar. It is possible to conceive an island that could contain a future time—something not quite in simultaneous relationship with the rest of the world. Perhaps in some obscure way everything living is on a different time plane. As for man, he is the most curious of all; he fits no plane, no visible island. He is bounded by no shore, except a shore of shadows. He has emerged almost as soundlessly as a mushroom in the night.

Islands are also places of extremes. They frequently produce opposites. On them may exist dwarf creations induced by lack of space or food. On the other hand, an open ecological niche, a lack of enemies, and some equally unopposed genetic drift may as readily produce giantism. The celebrated example of the monster Galápagos turtles comes immediately to mind. Man constitutes an even more unique spectacle for, beginning dwarfed and helpless within nature, he has become a Brocken specter as vast, murk-filled, and threatening as that described in old Germanic folk tales. Thomas De Quincey used to maintain that if one crossed oneself, this looming apparition of light and mist would do the same, but with reluctance and, sometimes, with an air of evasion. One may account this as natural in the case of an illusion moving against a cloud bank, but there is, in its delayed, uncertain gesture, a hint of ambiguity and terror projected from the original human climber on the mountain. To the discerning eye there is, thus, about both this creature and his reflection, something partaking of both microscopic and gigantic dimensions.

In some such way man arose upon an island—not on a visible oceanic island but in some hidden forest meadow. Man's selfhood, his future reality, was produced within the invisible island of his brain—the island clouded in a mist of sound. In this way the net of life was once more wrenched aside so that an impalpable shadow quickly wriggled through its strands into a new, unheard-of dimension of existence. Following this incredible event the natural world subsided once more into its place.

There was, of course, somewhere in the depths of time, a physical location where this episode took place. Unlike the sea barriers that Darwin had found constricting his island novelties, there had now appeared a single island whose shores seemed potentially limitless. This island, no matter where it had physically arisen, had been created by sound vibrating with meaning in the empty air. The island was based on man's most tremendous tool—the word. By degrees, the word would separate past from present, project the unseen future, contain the absent along with the real, and define them to human advantage. Man was no longer confined, like the animal, to what lay before his eyes or his own immediate attention. He could juxtapose, divide, and rearrange his world mentally. Upon the wilderness of the real, men came to project a phantom domain, the world of culture. In the end, their cities would lie congregated and gleaming like the nerve ganglions of an expanding brain.

Words would eventually raise specters so vast that man cowered and whimpered, where, as an animal, he had seen nothing either to reverence or to fear. Words expressed in substance would widen his powers; words, because he used them ill, would occasionally torture and imprison him. They would also lift him into regions of great light. Even the barbarian north, far from the white cities of the Mediterranean, would come to speak in wonder of the skald who could unlock the "word hoard." In our day this phantom island has embraced the planet, a world from which man has begun to eye the farther stars.

Isolation has produced man just as, on the Atlantic island of South Trinidad, there has emerged, as recorded by the late Apsley Cherry-Garrard, another and most eerie world, a world completely usurped by land crabs. Here, instead of a universe dominated by huge but harmless tortoises, the explorer encountered a nightmare. "The crabs," he recounted, "peep out at you from every nook and boulder. Their dead staring eyes follow your every step as if to say, 'If only you will drop down we will do the rest.' To lie down and sleep on any part of the island would be suicidal. . . . No matter how many are in sight they are all looking at you, and they follow step by step with a sickening deliberation."

Such remarkably confined but distinct worlds upon a single planet are infinitely informative. Because they are geologically younger than the planet itself, they tell us that life's power to create the new is still smoldering unseen about us in the living present. Darwin saw in the Galápagos Islands a reptilian world that had drowsed its antique way into the present. In a sense, Darwin was correct, but the islands are more than this. They are, in miniature, an alternative to the world we know, just as the nightmare crabs upon South Trinidad present an unthinkable alternative to ourselves. Each of these island worlds has taken through chance a different turning at some point in the past; each possesses an emergent novelty. In none has man twice appeared save as a wandering intruder from outside. Man belongs to his own island in nature, the invisible one of a growing genetic isolation about whose origins we know little.

Biological time never creates the same world twice, but out of its clefts and fissures creep, at long intervals, surprisingly original creatures whose destinies can never be anticipated before they arrive. It was so when the first Crossopterygian lungfish inched painfully into the new medium of the air. It was equally true when the first pre-ice-age man-apes drew an abstract line in speech between today and tomorrow. Out of just so much they would succeed in constructing a world.

Impelled by the rising flame of consciousness, they would ingest more than simple food. They would, instead, feed upon the contingent and the possible within their minds. They would step beyond the nature from which they had arisen and eventually turn upon her the level objective gaze of a stranger.

The rift so created is widening in our time. It is as though our self-created island were adrift on lawless currents like that same Galápagos Archipelago which seemed, to the old navigators, to set all their calculations at nought.

Over a century ago Thoreau, who had a sensitive ear for the tread of any overgrown Brocken specter in the shape of man, admitted that he would gladly fall "into some crevice along with leaves and acorns." The percipient philosopher felt the need for a renewed hermitage, a natural spot for hibernation. He was seeking a way back through the leafy curtain that has swung behind us, never to open again. Man had broken through that network of strangling vines by the magical utterance of the first word. In that guttural achievement he had created his destiny and taken leave of his kindred. There would be no way of return save perhaps one: through the power of imaginative insight, which has been manifested among a few great naturalists.

IV

Much has been spoken of oceanic islands as providing the haven, the sheltering crevice, the bleak and empty quarter in which potentially new forms of life might find the refuge necessary for immediate survival. In the words of one student of island faunas, Sherwin Carlquist, these creatures, far from being the ripe products of the war of nature, frequently represent, instead, the "inefficient, the unafraid and the obsolescent." How, then, can they be said to have any bearing upon the struggle for existence on the continents—the world where even the astute and perceptive Darwin had difficulty in explaining how the naked and physically inefficient protohumans had survived the claws of the great carnivores?

So puzzled had Darwin found himself that he had hesitated as to whether ancestral man might have been originally a genuine island product. Torn between viewing man as a weak creature needing protection in his early beginnings, or, on the other hand, attempting to visualize him as a fanged gorilloid monster fully equipped for the war of nature on the African land mass, the great biologist had failed to quite perceive the outlines of that invisibly expanding universe which man had unconsciously created out of airy nothing.

As we have seen, it is the wet fish gasping in the harsh air on the shore, the warm-blooded mammal roving unchecked through the torpor of the reptilian night, the lizard-bird launching into a moment of ill-aimed flight that shatter all purely competitive assumptions. These singular events reveal escapes through the living screen, penetrated, one would have to say in retrospect, by the "overspecialized" and the seemingly "inefficient," the creatures driven to the wall. Only after their triumphant planetary radiation is something new observed to have arisen in solitude and silence.

We begin in infancy with a universe that our minds constantly strive to subdue to the rational. But just as we seem to have achieved that triumph, some part of observed nature persists in breaking out once more into the unexpected. No greater surprise could have been anticipated than protoman's first stumbling venture into the hitherto unglimpsed ghostland of shifting symbols, that puzzling realm which Darwin vainly sought upon some real island in a more substantial sea. Reality has a way of hiding even from its most gifted observers.

In the Hall of Early Man in one of America's great archaeological museums there stands a reconstruction in the flesh of one of the early man-apes of preglacial times. He is reproduced as the bones tell us he must have basically appeared, small of brain, with just one foot, it could be said, across the human threshold. The whole appearance of the creature seems shrunken by modern standards—made more so by a fragment of bone clutched in a small uncertain hand. There remains with the observer an impression of fumbling weakness, as though the pygmy's dreams had depleted him—as though his genuine being had somehow been projected forward across millennia into the Herculean Grecian figure standing beyond him.

The dwarfed man-ape stands on the border of his invisible island; nor can there be any doubt of the nature of that island. It is Prospero's realm, whose first owner was Caliban. It is Shakespeare's island of sweet sounds and miraculous voices. It is the beginning rent in the curtain, the kingdom once confined to a single thicket, the isle of which it was once said, "this is no mortal business."

Some months ago, engaged in travel, I lay in a troubled sleep in the solitary freezing bed of a Canadian hotel. A blizzard had been raging. Beyond the raw new town stretched a stand of dark spruce forest such as one no longer sees in daylight. Somewhere toward dawn I dreamed.

The dream was of a great blurred bearlike shape emerging from the snow against the window. It pounded on the glass and beckoned importunately toward the forest. I caught the urgency of a message as uncouth and indecipherable as the shape of its huge bearer in the snow. In the immense terror of my dream I struggled against the import of that message as I struggled also to resist the impatient pounding of the frost-enveloped beast at the window.

Suddenly I lifted the telephone beside my bed, and through the receiver came a message as cryptic as the message from the snow, but far more miraculous in origin. For I knew intuitively, in the still snowfall of my dream, that the voice I heard, a long way off, was my own voice in childhood. Pure and sweet, incredibly refined and beautiful beyond the things of earth, yet somehow inexorable and not to be stayed, the voice was already terminating its message. "I am sorry to have troubled you," the clear faint syllables of the child persisted. They seemed to come across a thinning wire that lengthened far away into the years of my past. "I am sorry, I am sorry to have troubled you at all." The voice faded before I could speak. I was awake now, trembling in the cold. There was nothing to be seen at the window but the rising flurries of the snow.

Finally they ceased, and all around the little village, wrapped in an enormous eldritch winter, slumbered the dark forest. I looked below my window, but there were no tracks. I looked beside my bed, but there was, in reality, no phone. I lay back, huddled under the blanket, and thought briefly of the shrunken figure of the ape in the museum and of what had once been projected out of his living substance. My own body, in the freezing cold, felt wracked by a similar psychic effort that still persisted. Far off, as over an immense distance in my brain, I heard the echo of my own true voice, or perhaps it was mankind's accumulated voice, for the last time. It was hauntingly beautiful, but it was going. It would not be supplicated. As though in response to my thought, the incessant march of snow began again in the forest.

So this is the end, I thought, wrapping my shoulders closer against the increasing cold; we are, in truth, a failure. Beautiful and terrible, perhaps, but a failure, an island failure—an island whose origins are lost in time yet are still about us in such a way that we do not see them. To an invisible island we owe both triumph and disaster. At first, the island was a rent, a very small rent, in the curtain of life—perhaps no more than a few hairy creatures in a forest glade making experimental sounds that could be varied, one sound that defined the past and another that signified tomorrow. They were very small sounds with which to create an island, but the island, like those fostered by volcanoes on the sea floor, grew, and remained at the same time invisible. The sounds—passing from brain to brain, defining, measuring, remembering how stones could be broken —were responsible. If the island was invisible it was, unlike all other islands, shoreless as it grew. Nevertheless, man, like the other creatures on true islands, was isolated. He drifted insensibly from the heart of things. At first, he kept an uncertain memory of his origins in the animal world. He claimed descent from Grandfather Bear or Raven and on ritual occasions he talked to them or to his mother, the Earth.

The precarious thread that bound man to the living whole finally snapped. He had passed irrevocably into another dimension. His predicament is recognized in the myths of the Tree and of Pandora's box. He had learned to distinguish good from evil. Moreover, his capacity for evil increased as he discovered that the tiny sounds could be made to lie. This was an island within an island. It separated people into many islands. As man entered upon a wild new corridor of existence, some part of himself passed into a hypnotic slumber, but, in the diverse rooms of the mind, other sleepers awakened.

Man has indeed become a giant, but within him, growing at the pace of his own island, is locked the original minuscule dwarf who had stumbled out of the strangling grasp of the forest with a stone clutched in its hand. "This thing of Darknesse," speaks Shakespeare in the shape of the learned Prospero, "I acknowledge mine."

"There is no loneliness," once maintained the Egyptologist John Wilson, "like the loneliness of a mighty place fallen out of its proper service to man." Perhaps the same loneliness inevitably haunts modern man himself, that restless and vacant-eyed wanderer through the streets of cities, that man of ruinous countenance from whom the gods have hidden themselves. Above him, somewhere in the blue, like a hawk hovering over a deserted temple as if in expectation of some divine renewal, the spirit Ariel, dismissed and masterless, still lingers, soaring. Ariel had been long entrapped in the knotted pine that one suspects was human flesh. Does he now hesitate to leave his prison, grown used at last to the drunken mortal cry, "The sound is going. Hark, let's follow it"?

Perhaps it was just such music that I had heard traversing a phantom wire at midnight. Perhaps its purpose was to lead me to another doorway, another opening portal in the dark forest that is man. It is certain that the wooded shores that now confine us lie solely within ourselves. But they are the shores still frequented at midnight by a vengeful Caliban.

EIGHT

The Inner Galaxy

There is strong archaeological evidence to show that with the birth of human consciousness there was born, like a twin, the impulse to transcend it.

—ALAN MCGLASHAN

MANY YEARS AGO, I, with another youth of my own age whom I had persuaded to make the journey with me, walked throughout the day up a great mountain. There was a famous astronomical observatory upon the mountain. On certain nights, according to the guidebooks, the lay public might come to the observatory and look upon some remote planetary object. They could also hear a lecture.

The youth and I, who had much eager interest but no money, were unable to join one of the numerous tours organized from the tourist hotels in the valley. Instead, we had trudged for many hours in order to arrive before the crowds of visitors might frustrate our hopes for a glimpse of those far worlds about which we had read so avidly.

This was long ago, and we were naïve young men. We thought, though we were poor, that we would be welcome upon the mountain because of our desire to learn. There were reputed to dwell in the observatory men of wisdom who we hoped would receive us kindly since we, too, wished to gaze upon the wonders of outer space. We were, indeed, very unskilled in the ways of the adult world. As it turned out, we were never permitted to see the men of wisdom, or to gaze through the magic glass into outer space. I rather suspect that the eminent astronomers had not taken youths like us into their calculations. There was, it seemed, a relationship we had never suspected between the hotels in the valley and the men who inhabited the observatory upon the mountain.

Although by laborious effort we had succeeded in arriving before the busloads of tourists from the hotels, we were thrust forth and told to take our chances after the tourists had been accommodated. As busload after busload of people roared up before the observatory, we saw that this was an indirect dismissal. It would be dawn before our turn came, if, indeed, they chose to accept us at all.

The guard eyed us and our clothes with sullen distaste. Though it was freezing cold upon the mountain, it was plain we were not welcome in the inn that catered to the tourists. Reluctantly, with a few coins from our little store of change we purchased a bit of chocolate. We looked at each other. Wearily and without a word, we turned and began our long descent through the dark. It would take many hours; nor were we sustained by having seen the shining planet upon which our hopes were fixed.

This was my first experience of the commercial side of outer space, and though I now serve upon a committee that encourages the young in a direction once denied me, I feel that this youthful experience contributed to a certain growing introspection and curiosity about the relationship of science to the world about it.

Something was seriously wrong upon that mountain and among the wise men who flourished there. Knowledge, I had learned in the bleak wind by the shut door, was not free, and many to whom that observatory was only a passing curiosity had easier access to it than we who had climbed painfully for many hours. My memory is from the far days of the twenties, and I realize that we now beckon enticingly to the youth interested in space where before we ignored him. I still have an uncomfortable feeling, however, that it is the circumstances, and not the actors, which have changed. I remain oppressed by the thought that the venture into space is meaningless unless it coincides with a certain interior expansion, an ever growing universe within, to correspond with the far flight of the galaxies our telescopes follow from without.

Upon that desolate peak my mind had turned finally inward. It is from that domain, that inner sky, that I choose to speak—a world of dreams, of light and darkness, that we will never escape, even on the far edge of Arcturus. The inward skies of man will accompany him across any void upon which he ventures and will be with him to the end of time. There is just one way in which that inward world differs from outer space. It can be more volatile and mobile, more terrible and impoverished, yet withal more ennobling in its self-consciousness, than the universe that gave it birth. To the educators of this revolutionary generation, the transformations we may induce in that inner sky loom in at least equal importance with the work of those whose goals are set beyond the orbit of the moon.

No one needs to be told that different and private worlds exist in the heads of men. But in a day when some men are listening by radio telescope to the rustling of events at the ends of the universe, the universe of others consists of hopeless poverty amidst the filthy garbage of a city lot. A taxi man I know thinks the stars are just "up there," and that as soon as our vehicles are perfected we can all take off like crowds of summer tourists to Cape Cod. This man expects, and I fear has been encouraged to expect, that such flights will solve the population problem. Again, while I was sitting one night with a poet friend watching a great opera performed in a tent under arc lights, the poet took my arm and pointed silently. Far up, blundering out of the night, a huge Cecropia moth swept past from light to light over the posturings of the actors.

"He doesn't know," my friend whispered excitedly. "He's passing through an alien universe brightly lit but invisible to him. He's in another play; he doesn't see us. He doesn't know. Maybe it's happening right now to us. Where are we? Whose is the *real* play?"

Between the universe of the moth and the poet, I sat confounded. My mind went back to the heads of alabaster that the kings of the old Egyptian Empire sought to endow with eternal life, replacing thus against accident their own frail and perishable brains for the passage through eternity. The Pharaohs, like the moth among the arc lights, had been entranced by the flaming journey of the sun. Some had even constructed, hopefully, their own solar boats. Perhaps, I thought, those boats symbolized the frail vessel of which Plato was later to speak—that vessel on which to risk the voyage of life, or, rather, eternity, which was inevitably man's compulsive interest. As for me, I had come to seek wisdom no longer upon the improvised rafts of proud philosophies. I had seen the moth burn in its passage through the light. I had seen all the vessels fail but one—that word which Plato sought, and which none could long identify or hold.

There *was* a real play, but it was a play in which man was destined always to be a searcher, and it would be his true nature he would seek. The fragile vessel was himself, and not among the stars upon the mountain. Was not that what Plotinus had implied? Then if a man were to write further, I considered, he would write of that—of the last things.

II

Several years ago, a man in a small California town suffered an odd accident. The accident itself was commonplace. But the psychological episode accompanying it seems so strange that I recount it here. I had been long engaged upon a book I was eager to finish. As I walked, abstracted and alone, toward my office one late afternoon, I caught the toe of my shoe in an ill-placed drain. Some trick of mechanics brought me down over the curb with extraordinary violence. A tremendous crack echoed in my ears. When I next opened my eyes I was lying face down on the sidewalk. My nose was smashed over on one side. Blood from a gash on my forehead was cascading over my face.

Reluctantly I explored further, running my tongue cautiously about my mouth and over my teeth. Under my face a steady rivulet of blood was enlarging to a bright red pool on the sidewalk. It was then, as I peered nearsightedly at my ebbing substance there in the brilliant sunshine, that a surprising thing happened. Confusedly, painfully, indifferent to running feet and the anxious cries of witnesses about me, I lifted a wet hand out of this welter and murmured in compassionate concern, "Oh, don't go. I'm sorry, I've done for you."

The words were not addressed to the crowd gathering about me. They were inside and spoken to no one but a part of myself. I was quite sane, only it was an oddly detached sanity, for I was addressing blood cells, phagocytes, platelets, all the crawling, living, independent wonder that had been part of me and now, through my folly and lack of care, were dying like beached fish on the hot pavement. A great wave of passionate contrition, even of adoration, swept through my mind, a sensation of love on a cosmic scale, for mark that this experience was, in its way, as vast a catastrophe as would be that of a galaxy consciously suffering through the loss of its solar systems.

I was made up of millions of these tiny creatures, their toil, their sacrifices, as they hurried to seal and repair the rent fabric of this vast being whom they had unknowingly, but in love, compounded. And I, for the first time in my mortal existence, did not see these creatures as odd objects under the microscope. Instead, an echo of the force that moved them came up from the deep well of my being and flooded through the shaken circuits of my brain. I was they—their galaxy, their creation. For the first time, I loved them consciously, even as I was plucked up and away by willing hands. It seemed to me then, and does now in retrospect, that I had caused to the universe I inhabited as many deaths as the explosion of a supernova in the cosmos.

Weeks later, recovering, I paid a visit to the place of the accident. A faint discoloration still marked the sidewalk. I hovered over the spot, obscurely troubled. They were gone, utterly destroyed—those tiny beings—but the entity of which they had made a portion still persisted. I shook my head, conscious of the brooding mystery that the poet Dante impelled into his great line: "the love that moves the sun and other stars."

The phrase does not come handily to our lips today. For a century we have chosen to talk continuously about the struggle for existence, about man, the brawling half-ape and bestial fighter. We have explored with wavering candles the dark cellars of our subconscious and been appalled by the faces we have encountered there. It will do no harm, therefore, if we choose to examine the history of that great impulse—love, compassion, call it what one will—which, however discounted in our time, moved the dying Christ on Golgotha with a power that has reached across two thousand weary years.

"The conviction of wisdom," wrote Montaigne in the sixteenth century, "is the plague of man." Century after century, humanity studies itself in the mirror of fashion, and ever the mirror gives back distortions, which for the moment impose themselves upon man's real image. In one period we believe ourselves governed by immutable laws; in the next, by chance. In one period angels hover over our birth; in the following time we are planetary waifs, the product of a meaningless and ever altering chemistry. We exchange halos in one era for fangs in another. Our religious and philosophical conceptions change so rapidly that the theological and moral exhortations of one decade become the wastepaper of the next epoch. The ideas for which millions yielded up their lives produce only bored yawns in a later generation.

"We are, I know not how," Montaigne continued, "double in ourselves, so that what we believe we disbelieve, and cannot rid ourselves of what we condemn."

This complex, many-faceted, self-conscious creature now examines himself in the mirror held up to him by the modern students of prehistory. Increasingly he asks of the bony fragments recovered from pre-ice-age strata, not whether they are related to himself, but what manner of creature they proclaim us to be. Of the answer that may come up from underground we are all too evidently afraid. There are even those who have dared prematurely to announce the verdict. "Look," they say, "at the dark instincts that drive you. Look deep into your bloody, fossil, encrusted hearts. Then you will know man. You will know him from the caves to the Berlin wall. Thus he is and thus he will remain. It is written in his bones."

Yet the moment the words are said and documented, either the data are seized upon to give ourselves a fearsome picture to delight and excuse the black side of our natures or, strangely, even beautifully, the picture begins to waver and to change. St. Francis of the birds broods by the waters; Gilbert White of Selborne putters harmlessly with the old pet tortoise in his garden. Ishi, the primitive gentle philosopher, steps real as life from the Sierra forest—the idyllic man denounced as an invention of Rousseau's, yet the product of a world more primitive than black Africa today.

"Double in ourselves" we are, said Montaigne. Now with that doubleness in mind let us look once more into the fossil past, full into the hollow sockets of the halfmen from whom we sprang. Their bones are known; their remains have been turning up for over a century in almost every area of the Old World land mass. They have been found in the caves and gravels of ice age Europe, in the cemented breccias of deposits near Peking, in Asian coastal isles like Java, shaken at intervals by turbulent volcanoes. They have been found, as well, in the high uplands of eastern Africa and in the grottoes of the Holy Land.

Nevertheless, the faces of our ancestors remain forever unknown to us even as they stare from the illustrations of the poorest and most obscure textbook. The color of their skins is lost, the texture of their hair unknown, the expression of their once living features is as masked as those of the anonymous cadaver that represents collective humanity in the pages of medical textbooks. It is the same gray anonymity in which man's formidable enemy, the saber-toothed tiger, is lost, or even the dinosaur.

In the case of man, the representations are particularly ungratifying. Man is a creature volatile of expression, and across his features in a day may flow happiness and remorse, rage and charity. Individually, as on a modern street, one should be able to sight the sly, the brutal, and the benignant. If, in the world of fossils, however, we seek the soul of man himself, we are forced to draw it from the empty sockets of skulls or the representations of artists quick to project their own conceptions of the past upon the indifferent dead.

It is man's folly, as it is perhaps a sign of his spiritual aspirations, that he is forever scrutinizing and redefining himself. A mole, so far as we can determine, is content with its dim world below the grass roots, a snow leopard with being what he is—a drifting ghost in a blizzard. Man, by contrast, is marked by a restless inner eye, which, in periods of social violence, such as characterize our age, grows clouded with anxiety. There are times when our bodies seem to waver from within and bulge lumpishly with the shape of contending forces.

There is danger as well as wisdom, however, in such self-scrutiny. Man, unlike the lower creatures locked safely within their particular endowed natures, possesses freedom. He can define and redefine his own humanity, his own conception of himself. In so doing, he may give wings to the spirit or reshape himself into something more genuinely bestial than any beast of prey obeying its own nature. In this ability to take on the shape of his own dreams, man extends beyond visible nature into another and stranger realm. It is part of each person's individual evolutionary status that he possesses this power in unequal degrees.

Few of us can be saints; few of us are total monsters. To the degree that we let others project upon us erroneous or unbalanced conceptions of our natures, we may unconsciously reshape our own image to less pleasing forms. It is one thing to be "realistic," as many are fond of saying, about human nature. It is another thing entirely to let that consideration set limits to our spiritual aspirations or to precipitate us into cynicism and despair. We are protean in many things, and stand between extremes. There is still great room for the observation of John Donne, made over three centuries ago, however, that "no man doth refine and exalt Nature to the heighth it would beare."

As one surveys the artistic conceptions of the past, whether sculptured or drawn, one frequently encounters an adenoidal, open-mouthed brute with a club representing Neanderthal man. Then, by contrast, we encounter a neatly groomed model of Peking man, looking as clear-eyed and intelligent as a broker on his way to the Stock Exchange. Something is obviously wrong here. The well-groomed Peking specimen belongs on the same anatomical level as Pithecanthropus, sometimes represented in older illustrations as possessing snarling fangs. The fangs are a figment of the artist's imagination. They have been stolen from our living relative, the gorilla. The mispictured adenoidal moron with the club is known to have buried his dead with offerings, and to have cared for the injured and maimed among his kind.

Men are subjects of society. It is true that they carry bits and pieces of their past about with them, but they also covertly examine in the social mirror of their minds the way they look. Thus there is a quality of illusion about all of us. Emerson knew this well when he asked, in one of his more profound moments, "Why do men feel that the natural history of man has never been written, but he is always leaving behind what you have said of him, and it becomes old and books of metaphysics worthless?"

This comment of Emerson's is perhaps one of the most difficult pieces of wisdom that man has to learn. We are inclined to visualize our psychological makeup as fixed—as something bestowed upon the first man. In pre-evolutionary times, the human mind, with its reason, its conscience, its free will, was regarded as divinely and immediately created in the human organism just as it stands today.

With the rise of Darwinian evolution in the mid-nineteenth century, the concept of the stably endowed species correctly gave way to the notion of man and other animal forms as transient, imperfect, forever moving from one set of conditions to another. "Cosmic nature," wrote Thomas Huxley, Darwin's colleague and defender, "is no school of virtue. . . . For his successful progress as far as the savage state, man has been largely indebted to those qualities which he shares with the ape and the tiger."

No intelligent person today, surveying the low skull vault and heavy brow ridges of fossil man, can deny that man has changed through the aeons of prehistory, however difficult may seem the road he has traveled. Natural selection has undoubtedly played a leading role in that process. Here we must proceed with care, if we are not to fall into fallacious reasoning. Otherwise we will emerge from our survey of the past with another set of stereotypes as to the nature of man, which may well prove to be just as rigid and dogmatic as those developed in pre-evolutionary thought—stereotypes that have been thrust forward even today as evidence of man's bestial nature.

Man's altruistic and innately co-operative character has brought him along the road to civilization far more than the qualities of the ape and tiger of Huxley's analysis. These are bad metaphors at best. The ape is a largely inoffensive social animal, the tiger a solitary, carnivorous hunter. To lump them in a comparison with man is spectacular but confusing. As for the fearful war of nature painted by the early evolutionists and symbolized by the tiger, we know today that even the great carnivores exist, normally, in balance with their prey. When satiated and not involved in the hunt, they may stroll scarcely noticed among the herd creatures they stalk.

Some members of the Darwinian circle could only conceive of man achieving his high intellect through the heavy selection of incessant war. Today we know that early man was small and scant in numbers and that most of his efforts must have been given over to food-getting rather than conflict. This is not to minimize his destructive qualities, but his long-drawn-out, helpless childhood, during which his growing brain matured, could only have flourished in the safety of a stable family organization—groups marked by altruistic and long-continued care of the young.

The nineteenth-century evolutionists, and many philosophers still today, are obsessed by struggle. They try to define natural selection in one sense only—something that Darwin himself avoided. They ignore all man's finer qualities—generosity, self-sacrifice, universe-searching wisdom—in the attempt to enclose him in the small capsule that contained the brain of protoman. Such writers often fail to explore man's growing sense of beauty, the language that has opened and defined his world, the little gifts he came to lay beside his dead.

None of these acts could have been prophesied before man came. They reveal something other than what the pure materialist would be able to draw out of the dark concourse of matter before the genuine emergence of these novel human phenomena into time. There is no definition or description of man possible by reducing him to ape or tree shrew. Once, it is true, the shrew contained him, but he is gone. He has broken from the opened seed pod of the prehominid brain, a thistledown now drifting toward the empty spaces of the universe. He is full of the lights and visions—yes, and the fearful darknesses—of the next age of man.

The world we now know is open-ended, unpredictable. Man has partially domesticated himself; in this lies the story of his strange nature, of that love which transcends the small Darwinian matters of tribal cooperation and safety. For man, be it noted, can love the music of Ariel's isle, or, in his heart, that ideal city of the Greeks which is not and yet is forever.

The law of selection that acts upon living creatures in the wild is frequently repressive. A coat color a little off tone and visible, a variation in instinct, may make for death. The powerful creative surge from the under-darkness of nature is held in check, awaiting, perhaps, a season that never comes; the white stag is struck down by the hunter. It is this unending struggle that those who would picture man from the beginning as a monster of terror would delineate—the man with the stone striking down in barbaric rage, not only his game, but his brother and his son.

Natural selection is real but at the same time it is a shifting chimera, less a "law" than making its own law from age to age. Let us see, before we approach what I shall call domesticated man, what mutual aid can mean in the life of a European sea bird, the common tern. This bird lacks the careful concealing coloration of some of nature's species. It is variable in matters of egg form and nest shape. Capricious deviation in all these features prevails among the terns. The conformist pressures of natural selection have here given way to the creative forces of random mutation. The potential hidden in nature has flowered into a greater variety of behavior. Thus, what we call natural selection, "the war of nature," can either enclose living creatures in specialized prisons or, on occasion, open amazing doorways into unsuspected worlds. Even such a lowly relative of man as the existing lemur Propithecus, which lives in groups, may exhibit marked individual variation, because these animals recognize and behave differently toward each other. Conformity has here given way to selective pressure for at least limited physical diversity and corresponding individuality of behavior.

Though the case of man is complicated, it seems evident that just such a remarkable doorway opened when man, as a social animal, fell under selective forces that no longer severely channeled the nature of his mind or the minds of his aberrant offspring. Through language, this creature could communicate his dreams around the cave fires. Inevitably, a great wealth of intellectual diversity, and consequent selective mating, based upon mutual attraction, would emerge from the dark storehouse of nature. The cruel and the gentle would sit at the same fireside, dreaming already in the Stone Age the different dreams they dream today.

The visionary was already awaiting the eternal city; the gifted musician sat hearing in his brain sounds that did not yet exist. All waited upon and yet possessed, in some dim way, the future in their heads. Abysmal darkness and great light lay invisibly about their camps. The phantom cities of the far future awaited latent talents for which, in that unspecialized time, there was no name.

Above all, some of them, a mere handful in any generation perhaps, loved—they loved the animals about them, the song of the wind, the soft voices of women. On the flat surfaces of cave walls the three dimensions of the outside world took animal shape and form. Here—not with the ax, not with the bow—man fumbled at the door of his true kingdom. Here, hidden in times of trouble behind silent brows, against the man with the flint, waited St. Francis of the birds—the lovers, the men who are still forced to walk warily among their kind.

III

I am middle-aged now, and like the Egyptian heads of buried stone, or like the gentle ones who came before me, I am resigned to wait out man's lingering barbarity. I have walked much to the sea, not knowing what I seek. The west headland I visit is always boiling, even on calm days. Spume leaps up from the sea caverns of buried reefs and the blue and purple of the turbulent waters are roiled and twisted with clashing and opposed currents. I go there frequently and sit for hours on an old whiskey crate half-buried in the sand.

Staring into those uncertain and treacherous waters with their unexpected and lifting apparitions is like looking into the future. You can see its forces constantly gathering, expending themselves, streaming away and streaming back, contorting or violently lifting into huge and grotesque shapes. The meaning escapes one, but day after day the harpy gulls scream and mew over it and the crabs scuttle like spiders along its edge, waving threatening pincers.

But I wander.

On one occasion, there was just this broken crate in the sand, myself, and the sea—and then this other. I only became aware of him after several days had passed. I first encountered him when I had ventured at low tide up to the verge of the reef beyond which burst that leaping, spouting thunder, which, in my isolated wanderings, I had come to conceive of as containing the future. As I reached the flat, slippery stones over which passed a constant surf, I saw a gray wing tilt upward and move a few feet farther on. It was a big gray-backed gull, who slid quietly down again amidst the encrusted sea growth. He moved just enough, out of old and wise judgment, to keep me at arm's length, no more. He was no longer with his kind, hovering and mewing over the outer rock masses of a dubious future. He had a space of his own on the last edge of the present. He fed there upon such things as the sea brought. He was old and he rested, if one could be said to rest amidst such waters.

I disturbed him once by coming closer, whereupon he rose and tilted slightly in the blast from over the reef. If I did not move, neither did he. Since I am not one to go rushing over dangerous crevices, we achieved, after some days, a dignified relationship. We were both gray, and disinclined toward a future that had come to have little meaning to either of us. We stood or sat a little apart and ignored each other, being, after all, creatures diverse.

Every morning when I came he was there. He was growing thinner, but he still rose at my coming and hovered low upon his great seagoing wings. Then I would seek my box and he would swoop back to the little space that contained his last of life. I came to look for this bird as though we shared some sane, enormously simple secret amidst a little shingle of hard stones and broken beach.

After several days he was gone. A sector of my own life had been sheared away with his going. I shied a stone uncertainly toward the still-spouting future. Nothing came of it; no hand reached out, no shape emerged. The only rational shape had been that aged gull, too wise to venture more than a tilting wing's length upward in such air. Finally, the extremest edge of his space had hesitantly touched mine. Neither of us had much farther to go, and the harsh simplicity of it was somehow appropriate and gratifying. A little salt-washed rock had contained us both.

Here, I thought, is where I shall abide my ending, in the mind at least. Here where the sea grinds coral and bone alike to pebbles, and the crabs come in the night for the recent dead. Here where everything is transmuted and transmutes, but all is living or about to live.

It was here that I came to know the final phase of love in the mind of man—the phase beyond the evolutionists' meager concentration upon survival. Here I no longer cared about survival—I merely loved. And the love was meaningless, as the harsh Victorian Darwinists would have understood it or even, equally, those harsh modern materialists of whom Lord Dunsany once said: "It is very seldom that the same man knows much of science, and about the things that were known before ever science came."

I felt, sitting in that desolate spot upon my whiskey crate, a love without issue, tenuous, almost disembodied. It was a love for an old gull, for wild dogs playing in the surf, for a hermit crab in an abandoned shell.

It was a love that had been growing through the unthinking demands of childhood, through the pains and rapture of adult desire. Now it was breaking free, at last, of my worn body, still containing but passing beyond those other loves. Now, at last, it was truly "the bright stranger, the foreign self," of which Emerson had once written.

Through shattered and receding skulls, growing ever smaller behind us in the crannies of a broken earth, a stranger had crept and made his way. But precisely how he came, and what might be his destiny, except that it is not wholly of our time or this our star, we do not know.

Perhaps it is always the destined role of the compassionate to be strangers among men. To fail and pass, to fail and come again. For the seed of man is thistledown, and a puff of breath may govern it, or a word from a poet torment it into greatness. There are few among us who can notice the passage of a moth's wing across an opera tent at midnight and ask ourselves, "Whose is the real play?"

I had turned to the young man who spoke those words as to one whose eye reached farther than the giant lens upon the mountain in my youth. Before us had seemed to stretch the infinite pathways of space down which, like the questing moth, it was henceforth man's doom to wander. But the void had become to me equally an interior void—the void of our own minds—a sea as infinite as the one before which I had been meditating.

Amidst the fall of waters on that desolate shore I watched briefly an exquisitely shaped jellyfish pumping its little umbrella sturdily along only to subside with the next wave on the strand. "Love makyth the lover and the living matters not," an old phrase came hesitantly to my lips. We would win, I thought steadily, if not in human guise then in another, for love was something that life in its infinite prodigality could afford. It was the failures who had always won, but by the time they won they had come to be called successes. This is the final paradox, which men call evolution.

NINE

The Innocent Fox

*Only to a magician is the world forever fluid, infinitely mutable
and eternally new. Only he knows the secret of change, only he
knows truly that all things are crouched in eagerness to become
something else, and it is from this universal tension that he draws
his power.*

—PETER BEAGLE

SINCE MAN FIRST saw an impossible visage staring upward
from a still pool, he has been haunted by meanings—
meanings felt even in the wood, where the trees leaned over
him, manifesting a vast and living presence. The image in the
pool vanished at the touch of his finger, but he went home and
created a legend. The great trees never spoke, but man knew
that dryads slipped among their boles. Since the red morning
of time it has been so, and the compulsive reading of such
manuscripts will continue to occupy man's attention long after
the books that contain his inmost thoughts have been sealed
away by the indefatigable spider.

Some men are daylight readers, who peruse the ambiguous
wording of clouds or the individual letter shapes of wandering
birds. Some, like myself, are librarians of the night, whose
ephemeral documents consist of root-inscribed bones or whatever rustles in thickets upon solitary walks. Man, for all his
daylight activities, is, at best, an evening creature. Our very
addiction to the day and our compulsion, manifest through
the ages, to invent and use illuminating devices, to contest
with midnight, to cast off sleep as we would death, suggest
that we know more of the shadows than we are willing to recognize. We have come from the dark wood of the past, and
our bodies carry the scars and unhealed wounds of that transition. Our minds are haunted by night terrors that arise from
the subterranean domain of racial and private memories.

Lastly, we inhabit a spiritual twilight on this planet. It is
perhaps the most poignant of all the deprivations to which

man has been exposed by nature. I have said *deprivation*, but perhaps I should, rather, maintain that this feeling of loss is an unrealized anticipation. We imagine we are day creatures, but we grope in a lawless and smoky realm toward an exit that eludes us. We appear to know instinctively that such an exit exists.

I am not the first man to have lost his way only to find, if not a gate, a mysterious hole in a hedge that a child would know at once led to some other dimension at the world's end. Such passageways exist, or man would not be here. Not for nothing did Santayana once contend that life is a movement from the forgotten into the unexpected.

As adults, we are preoccupied with living. As a consequence, we see little. At the approach of age some men look about them at last and discover the hole in the hedge leading to the unforeseen. By then, there is frequently no child companion to lead them safely through. After one or two experiences of getting impaled on thorns, the most persistent individual is apt to withdraw and to assert angrily that no such opening exists.

My experience has been quite the opposite, but I have been fortunate. After several unsuccessful but tantalizing trials, which I intend to disclose, I had the help, not of a child, but of a creature—a creature who, appropriately, came out of a quite unremarkable and prosaic den. There was nothing, in retrospect, at all mysterious or unreal about him. Nevertheless, the creature was baffling, just as, I suppose, to animals, man himself is baffling.

II

An autumn midnight in 1967 caught me staring idly from my study window at the attic cupola of an old Victorian house that loomed far above a neighboring grove of trees. I suppose the episode happened just as I had grown dimly aware, amidst my encasing cocoon of books and papers, that something was missing from my life. This feeling had brought me from my desk to peer hopelessly upon the relentless advance of suburban housing. For years, I had not seen anything from that particular window that did not spell the death of something I loved.

Finally, in blundering, good-natured confidence, the last land tortoise had fallen a victim to the new expressway. None of his kind any longer came to replace him. A chipmunk that had held out valiantly in a drainpipe on the lawn had been forced to flee from the usurping rats that had come with the new supermarket. A parking lot now occupied most of the view from the window. I was a man trapped in the despair once alluded to as the utterly hopeless fear confined to moderns—that no miracles can ever happen. I considered, as I tried to will myself away into the attic room far above the trees, the wisdom of a search, a search unlikely to yield tangible results.

Since boyhood I have been charmed by the unexpected and the beautiful. This was what had led me originally into science, but now I felt instinctively that something more was needed—though what I needed verged on a miracle. As a scientist, I did not believe in miracles, though I willingly granted the word broad latitudes of definition.

My whole life had been unconsciously a search, and the search had not been restricted to the bones and stones of my visible profession. Moreover, my age could allow me folly; indeed, it demanded a boldness that the young frequently cannot afford. All I needed to do was to set forth either mentally or physically, but to where escaped me.

At that instant the high dormer window beyond the trees blazed as blue as a lightning flash. As I have remarked, it was midnight. There was no possibility of reflection from a street lamp. A giant bolt of artificial lightning was playing from a condenser, leaping at intervals across the interior of the black pane in the distance. It was the artificial lightning that only one or several engineers with unusual equipment could produce.

Now the old house was plebeian enough. Rooms were rented. People of modest middle-class means lived there, as I was to learn later. But still, in the midmost of the night, somebody or some group was engaged in that attic room upon a fantastic experiment. For, you see, I spied. I spied for nights in succession. I was bored, I was sleepless, and it pleased me to think that the mad scientists, as I came to call them, were engaged, in their hidden room, upon some remarkable and unheard-of adventure.

Why else would they be active at midnight, why else would they be engaged for a brief hour and then extinguish the spark? In the next few days I trained high-powered field glasses upon the window, but the blue bolt defeated me, as did the wavering of autumn boughs across the distant roof. I could only believe that science still possessed some of its old, mad fascination for a mind outside the professional circle of the great laboratories. Perhaps, I thought eagerly, there was a fresh intelligence groping after some secret beyond pure technology. I thought of the dreams of Emerson and others when evolution was first anticipated but its mechanisms remained a mystery entangled with the first galvanic batteries. Night after night, while the leaves thinned and the bolt leaped at its appointed hour, I dreamed, staring from my window, of that coruscating arc revivifying flesh or leaping sentient beyond it into some unguessed state of being. Only for such purposes, I thought, would a man toil in an attic room at midnight.

I began unconsciously to hang more and more upon that work of which, in reality, I knew nothing. It sustained me in my waking hours when the old house, amidst its yellowing leaves, assumed a sleepy and inconsequential air. For me, it had restored wonder and lifted my dreams to the height they had once had when, as a young student, I had peeped through the glass door of a famous experimenter's laboratory. I no longer read. I sat in the darkened study and watched and waited for the unforeseen. It came in a way I had not expected.

One night the window remained dark. My powerful glasses revealed only birds flying across the face of the moon. A bat fluttered about the tessellated chimney. A few remaining leaves fell into the dark below the roofs.

I waited expectantly for the experiment to be resumed. It was not. The next night it rained violently. The window did not glow. Leaves yellowed the wet walks below the street lamps. It was the same the next night and the next. The episode, I came to feel, peering ruefully from my window, was altogether too much like science itself—science with its lightning bolts, its bubbling retorts, its elusive promises of perfection. All too frequently the dream ended in a downpour of rain and leaves upon wet walks. The men involved had a way, like my mysterious neighbors, of vanishing silently and leaving, if anything at all, corroding bits of metal out of which no one could make sense.

I had once stood in a graveyard that was a great fallen city. It was not hard to imagine another. After watching fruitlessly at intervals until winter was imminent, I promised myself a journey. After all, there was nothing to explain my disappointment. I had not known for what I was searching.

Or perhaps I did know, secretly, and would not admit it to myself: I wanted a miracle. Miracles, by definition, are without continuity, and perhaps my rooftop scientist had nudged me in that direction by the uncertainty of his departure. The only thing that characterizes a miracle, to my mind, is its sudden appearance and disappearance within the natural order, although, strangely, this loose definition would include each individual person. Miracles, in fact, momentarily dissolve the natural order or place themselves in opposition to it. My first experience had been only a tantalizing expectation, a hint that I must look elsewhere than in retorts or coiled wire, however formidable the powers that could be coerced to inhabit them. There was magic, but it was an autumnal, sad magic. I had a growing feeling that miracles were particularly concerned with life, with the animal aspect of things.

Just at this time, and with my thoughts in a receptive mood, a summons came that made it necessary for me to make a long night drive over poor roads through a dense forest. As a subjective experience, which it turned out to be, I would call it a near approach to what I was seeking. There was no doubt I was working further toward the heart of the problem. The common man thinks a miracle can just be "seen" to be reported. Quite the contrary. One has to be, I was discovering, reasonably sophisticated even to *perceive* the miraculous. It takes experience; otherwise, more miracles would be encountered.

One has, in short, to refine one's perceptions. Lightning bolts observed in attics, I now knew, were simply raw material, a lurking extravagant potential in the cosmos. In themselves, they were merely powers summoned up and released by the human mind. Wishing would never make them anything else and might make them worse. Nuclear fission was a ready example. No, a miracle was definitely something else, but that I would have to discover in my own good time.

Preoccupied with such thoughts, I started my journey of descent through the mountains. For a long time I was alone. I followed a road of unexpectedly twisting curves and abrupt descents. I bumped over ruts, where I occasionally caught the earthly starshine of eyes under leaves. Or I plunged at intervals into an impenetrable gloom buttressed by the trunks of huge pines.

After hours of arduous concentration and the sudden crimping of the wheel, my eyes were playing tricks with me. It was time to stop, but I could not afford to stop. I shook my head to clear it and blundered on. For a long time, in this confined glen among the mountains, I had been dimly aware that something beyond the reach of my headlights, but at times momentarily caught in their flicker, was accompanying me.

Whatever the creature might be, it was amazingly fleet. I never really saw its true outline. It seemed, at times, to my weary and much-rubbed eyes, to be running upright like a man, or, again, its color appeared to shift in a multiform illusion. Sometimes it seemed to be bounding forward. Sometimes it seemed to present a face to me and dance backward. From weary consciousness of an animal I grew slowly aware that the being caught momentarily in my flickering headlights was as much a shapeshifter as the wolf in a folk tale. It was not an animal; it was a gliding, leaping mythology. I felt the skin crawl on the back of my neck, for this was still the forest of the windigo and the floating heads commemorated so vividly in the masks of the Iroquois. I was lost, but I understood the forest. The blood that ran in me was not urban. I almost said not human. It had come from other times and a far place.

I slowed the car and silently fought to contain the horror that even animals feel before the disruption of the natural order. But was there a natural order? As I coaxed my lights to a fuller blaze I suddenly realized the absurdity of the question. Why should life tremble before the unexpected if it had not already anticipated the answer? There was no order. Or, better, what order there might be was far wilder and more formidable than that conjured up by human effort.

It did not help in the least to make out finally that the creature who had assigned himself to me was an absurdly spotted dog of dubious affinities—nor did it help that his coat had the curious properties generally attributable to a magician. For how, after all, could I assert with surety what shape this dog had originally possessed a half mile down the road? There was no way of securing his word for it.

The dog was, in actuality, an illusory succession of forms finally, but momentarily, frozen into the shape "dog" by me. A word, no more. But as it turned away into the night how was I to know it would remain "dog"? By experience? No, it had been picked by me out of a running weave of colors and faces into which it would lapse once more as it bounded silently into the inhuman, unpopulated wood. We deceive ourselves if we think our self-drawn categories exist there. The dog would simply become once more an endless running series of forms, which would not, the instant I might vanish, any longer know themselves as "dog."

By a mental effort peculiar to man, I had wrenched a leaping phantom into the flesh "dog," but the shape could not be held, neither his nor my own. We were contradictions and unreal. A nerve net and the lens of an eye had created us. Like the dog, I was destined to leap away at last into the unknown wood. My flesh, my own seemingly unique individuality, was already slipping like flying mist, like the colors of the dog, away from the little parcel of my bones. If there was order in us, it was the order of change. I started the car again, but I drove on chastened and unsure. Somewhere something was running and changing in the haunted wood. I knew no more than that. In a similar way, my mind was leaping and also changing as it sped. That was how the true miracle, my own miracle, came to me in its own time and fashion.

III

The episode occurred upon an unengaging and unfrequented shore. It began in the late afternoon of a day devoted at the start to ordinary scientific purposes. There was the broken prow of a beached boat subsiding in heavy sand, left by the whim of ancient currents a long way distant from the shifting coast. Somewhere on the horizon wavered the tenuous outlines of a misplaced building, growing increasingly insubstantial in the autumn light.

After my companions had taken their photographs and departed, their persistent voices were immediately seized upon and absorbed by the extending immensity of an incoming fog. The fog trailed in wisps over the upthrust ribs of the boat. For a time I could see it fingering the tracks of some small animal, as though engaged in a belated dialogue with the creature's mind. The tracks crisscrossed a dune, and there the fog hesitated, as though puzzled. Finally, it approached and enwrapped me, as though to peer into my face. I was not frightened, but I also realized with a slight shock that I was not intended immediately to leave.

I sat down then and rested with my back against the overturned boat. All around me the stillness intensified and the wandering tendrils of the fog continued their search. Nothing escaped them.

The broken cup of a wild bird's egg was touched tentatively, as if with meaning, for the first time. I saw a sand-colored ghost crab, hitherto hidden and immobile, begin to sidle amidst the beach grass as though imbued suddenly with a will derived ultimately from the fog. A gull passed high overhead, but its cry took on the plaint of something other than itself.

I began dimly to remember a primitive dialogue as to whether God is a mist or merely a mist maker. Since a great deal of my thought has been spent amidst such early human and, to my mind, not outworn speculations, the idea did not seem particularly irrational or blasphemous. How else would so great a being, assuming his existence, be able thoroughly to investigate his world, or, perhaps, merely a world that he had come upon, than as he was now proceeding to do?

I closed my eyes and let the tiny diffused droplets of the fog gently palpate my face. At the same time, by some unexplained affinity, I felt my mind drawn inland, to pour, smoking and gigantic as the fog itself, through the gorges of a neighboring mountain range.

In a little shaft of falling light my consciousness swirled dimly over the tombstones of a fallen cemetery. Something within me touched half-obliterated names and dates before sliding imperceptibly onward toward an errand in the city. That errand, whatever its purpose, perhaps because I was mercifully guided away from the future, was denied me.

As suddenly as I had been dispersed I found myself back among the boat timbers and the broken shell of something that had not achieved existence. "I am the thing that lives in the midst of the bones"—a line from the dead poet Charles Williams persisted obstinately in my head. It was true. I was merely condensed from that greater fog to a smaller congelation of droplets. Vague and smoky wisplets of thought were my extensions.

From a rack of bone no more substantial than the broken boat ribs on the beach, I was moving like that larger, all-investigating fog through the doorways of the past. Somewhere far away in an inland city the fog was transformed into a blizzard. Nineteen twenty-nine was a meaningless date that whipped by upon a flying newspaper. The blizzard was beating upon a great gate marked St. Elizabeth's. I was no longer the blizzard. I was hurrying, a small dark shadow, up a stairway beyond which came a labored and importunate breathing.

The man lay back among the pillows, wracked, yellow, and cadaverous. Though I was his son he knew me only as one lamp is briefly lit from another in the windy night. He was beyond speech, but a question was there, occupying the dying mind, excluding the living, something before which all remaining thought had to be mustered. At the time I was too young to understand. Only now could the hurrying shadow drawn from the wrecked boat interpret and relive the question. The starving figure on the bed was held back from death only by a magnificent heart that would not die.

I, the insubstantial substance of memory, the dispersed droplets of the ranging fog, saw the man lift his hands for the last time. Strangely, in all that ravished body, they alone had remained unchanged. They were strong hands, the hands of a craftsman who had played many roles in his life: actor, laborer, professional runner. They were the hands of a man, indirectly of all men, for such had been the nature of his life. Now, in a last lucid moment, he had lifted them up and, curiously, as though they belonged to another being, he had turned and flexed them, gazed upon them unbelievingly, and dropped them once more.

He, too, the shadow, the mist in the gaping bones, had seen these seemingly untouched deathless instruments rally as though with one last purpose before the demanding will. And I, also a shadow, come back across forty years, could hear the question at last. "Why are you, my hands, so separate from me at death, yet still to be commanded? Why have you served me, you who are alive and ingeniously clever?" For here he turned and contemplated them with his old superb steadiness. "What has been our partnership, for I, the shadow, am going, yet you of all of me are alive and persist?"

I could have sworn that his last thought was not of himself but of the fate of the instruments. He was outside, he was trying to look into the secret purposes of things, and the hands, the masterful hands, were the only purpose remaining, while he, increasingly without center, was vanishing. It was the hands that contained his last conscious act. They had been formidable in life. In death they had become strangers who had denied their master's last question.

Suddenly I was back under the overhang of the foundered boat. I had sat there stiff with cold for many hours. I was no longer the extension of a blizzard beating against immovable gates. The year of the locusts was done. It was, instead, the year of the mist maker that some obscure Macusi witch doctor had chosen to call god. But the mist maker had gone over the long-abandoned beach, touching for his inscrutable purposes only the broken shell of the nonexistent, only the tracks of a wayward fox, only a man who, serving the mist maker, could be made to stream wispily through the interstices of time.

I was a biologist, but I chose not to examine my hands. The fog and the night were lifting. I had been far away for hours. Crouched in my heavy sheepskin I waited without thought as the witch doctor might have waited for the morning dispersion of his god. Finally, the dawn began to touch the sea, and then the worn timbers of the hulk beside which I sheltered reddened just a little. It was then I began to glimpse the world from a different perspective.

I had watched for nights the great bolts leaping across the pane of an attic window, the bolts Emerson had dreamed in the first scientific days might be the force that hurled reptile into mammal. I had watched at midnight the mad scientists intent upon their own creation. But in the end, those fantastic flashes of the lightning had ceased without issue, at least for me. The pane, the inscrutable pane, had darkened at last; the scientists, if scientists they were, had departed, carrying their secret with them. I sighed, remembering. It was then I saw the miracle. I saw it because I was hunched at ground level, smelling rank of fox, and no longer gazing with upright human arrogance upon the things of this world.

I did not realize at first what it was that I looked upon. As my wandering attention centered, I saw nothing but two small projecting ears lit by the morning sun. Beneath them, a small neat face looked shyly up at me. The ears moved at every sound, drank in a gull's cry and the far horn of a ship. They crinkled, I began to realize, only with curiosity; they had not learned to fear. The creature was very young. He was alone in a dread universe. I crept on my knees around the prow and crouched beside him. It was a small fox pup from a den under the timbers who looked up at me. God knows what had become of his brothers and sisters. His parent must not have been home from hunting.

He innocently selected what I think was a chicken bone from an untidy pile of splintered rubbish and shook it at me invitingly. There was a vast and playful humor in his face. "If there was only one fox in the world and I could kill him, I would do." The words of a British poacher in a pub rasped in my ears. I dropped even further and painfully away from human stature. It has been said repeatedly that one can never, try as he will, get around to the front of the universe. Man is destined to see only its far side, to realize nature only in retreat.

Yet here was the thing in the midst of the bones, the wide-eyed, innocent fox inviting me to play, with the innate courtesy of its two forepaws placed appealingly together, along with a mock shake of the head. The universe was swinging in some fantastic fashion around to present its face, and the face was so small that the universe itself was laughing.

It was not a time for human dignity. It was a time only for the careful observance of amenities written behind the stars. Gravely I arranged my forepaws while the puppy whimpered with ill-concealed excitement. I drew the breath of a fox's den into my nostrils. On impulse, I picked up clumsily a whiter bone and shook it in teeth that had not entirely forgotten their original purpose. Round and round we tumbled for one ecstatic moment. We were the innocent thing in the midst of the bones, born in the egg, born in the den, born in the dark cave with the stone ax close to hand, born at last in human guise to grow coldly remote in the room with the rifle rack upon the wall.

But I had seen my miracle. I had seen the universe as it begins for all things. It was, in reality, a child's universe, a tiny and laughing universe. I rolled the pup on his back and ran, literally ran for the nearest ridge. The sun was half out of the sea, and the world was swinging back to normal. The adult foxes would be already trotting home.

A little farther on, I passed one on a ridge who knew well I had no gun, for it swung by quite close, stepping delicately with brush and head held high. Its face was watchful but averted. It did not matter. It was what I had experienced and the fox had experienced, what we had all experienced in adulthood. We passed carefully on our separate ways into the morning, eyes not meeting.

But to me the mist had come, and the mere chance of two lifted sunlit ears at morning. I knew at last why the man on the bed had smiled finally before he dropped his hands. He, too, had worked around to the front of things in his death agony. The hands were playthings and had to be cast aside at last like a little cherished toy. There was a meaning and there was not a meaning, and therein lay the agony.

The meaning was all in the beginning, as though time was awry. It was a little beautiful meaning that did not stay, and the sixty-year-old man on the hospital bed had traveled briefly toward it through the dark at the end of the universe. There was something in the desperate nature of the world that had to be reversed, but he had been too weak to tell me, and the hands had dropped helplessly away.

After forty years I had been just his own age when the fog had come groping for my face. I think I can safely put it down that I had been allowed my miracle. It was very small, as is the way of great things. I had been permitted to correct time's arrow for a space of perhaps five minutes—and that is a boon not granted to all men. If I were to render a report upon this episode, I would say that men must find a way to run the arrow backward. Doubtless it is impossible in the physical world, but in the memory and the will man might achieve the deed if he would try.

For just a moment I had held the universe at bay by the simple expedient of sitting on my haunches before a fox den and tumbling about with a chicken bone. It is the gravest, most meaningful act I shall ever accomplish, but, as Thoreau once remarked of some peculiar errand of his own, there is no use reporting it to the Royal Society.

TEN

The Last Neanderthal

For thou shalt be in league with the stones of the field: and the beasts of the field shall be at peace with thee.

—JOB 5:23

IT HAS LONG been the thought of science, particularly in evolutionary biology, that nature does not make extended leaps, that her creatures slip in slow disguise from one shape to another. A simple observation will reveal, however, that there are rocks in deserts that glow with heat for a time after sundown. Similar emanations may come from the writer or the scientist. The creative individual is someone upon whom mysterious rays have converged and are again reflected, not necessarily immediately, but in the course of years. That all of this wispy geometry of dreams and memories should be the product of a kind of slow-burning oxidation carried on in an equally diffuse and mediating web of nerve and sense cells is surprising enough, but that the emanations from the same particulate organ, the brain, should be so strikingly different as to disobey the old truism of an unleaping nature is quite surprising, once one comes to contemplate the reality.

The same incident may stand as a simple fact to some, an intangible hint of the nature of the universe to others, a useful myth to a savage, or any number of other things. The receptive mind makes all the difference, shadowing or lighting the original object. I was an observer, intent upon my own solitary hieroglyphics.

It happened a long time ago at Curaçao, in the Netherlands Antilles, on a shore marked by the exposed ribs of a wrecked freighter. The place was one where only a student of desolation would find cause to linger. Pelicans perched awkwardly on what remained of a rusted prow. On the edge of the littered beach beyond the port I had come upon a dead dog wrapped in burlap, obviously buried at sea and drifted in by the waves. The dog was little more than a skeleton but still articulated, one delicate bony paw laid gracefully—as though its owner merely slept, and would presently awaken—across a stone at the water's edge. Around his throat was a waterlogged black strap that showed he had once belonged to someone. This dog was a mongrel whose life had been spent among the island fishermen. He had known only the small sea-beaten boats that come across the strait from Venezuela. He had romped briefly on shores like this to which he had been returned by the indifferent sea.

I stepped back a little hesitantly from the smell of death, but still I paused reluctantly. Why, in this cove littered with tin cans, bottles, and cast-off garments, did I find it difficult, if not a sacrilege, to turn away? Because, the thought finally came to me, this particular tattered garment had once lived. Scenes on the living sea that would never in all eternity recur again had streamed through the sockets of those vanished eyes. The dog was young, the teeth in its jaws still perfect. It was of that type of loving creature who had gamboled happily about the legs of men and striven to partake of their endeavors.

Someone had seen crudely to his sea burial, but not well enough to prevent his lying now where came everything abandoned. Nevertheless, vast natural forces had intervened to clothe him with a pathetic dignity. The tide had brought him quietly at night and placed what remained of him asleep upon the stones. Here at sunrise I had stood above him in a light he would never any longer see. Even if I had had a shovel the stones would have prevented his burial. He would wait for a second tide to spirit him away or lay him higher to bleach starkly upon coral and conch shells, mingling the little lime of his bones with all else that had once stood upright on these shores.

As I turned upward into the hills beyond the beach I was faintly aware of a tracery of lizard tails amidst the sand and the semidesert shrubbery. The lizards were so numerous on the desert floor that their swift movement in the bright sun left a dizzying impression, like spots dancing before one's eyes. The creatures had a tangential way of darting off to the side like inconsequential thoughts that never paused long enough to be fully apprehended. One's eyesight was oppressed by subtly moving points where all should have been quiet. Similar darting specks seemed to be invading my mind. Offshore I could hear the sea wheezing and suspiring in long gasps among the caverns of the coral. The equatorial sun blazed on my un-protected head and hummingbirds flashed like little green flames in the underbrush. I sought quick shelter under a manzanillo tree, whose poisoned apples had tempted the sail-ors of Columbus.

I suppose the apples really made the connection. Or perhaps merely the interior rustling of the lizards, as I passed some cardboard boxes beside a fence, brought the thing to mind. Or again, it may have been the tropic sun, lending its flames to life with a kind of dreadful indifference as to the result. At any rate, as I shielded my head under the leaves of the poison tree, the darting lizard points began to run together into a pattern.

Before me passed a broken old horse plodding before a cart laden with bags of cast-off clothing, discarded furniture, and abandoned metal. The horse's harness was a makeshift combi-nation of straps mended with rope. The bearded man perched high in the driver's seat looked as though he had been com-pounded from the assorted junk heap in the wagon bed. What finally occupied the center of my attention, however, was a street sign and a year—a year that scurried into shape with the flickering alacrity of the lizards. "R Street," they spelled, and the year was 1923.

By now the man on the wagon is dead, his cargo dispersed, never to be reassembled. The plodding beast has been over-taken by whatever fate comes upon a junkman's horse. Their significance upon that particular day in 1923 had been resolved to this, just this: The wagon had been passing the intersection between R and Fourteenth streets when I had leaned from a high-school window a block away, absorbed as only a sixteen-year-old may sometimes be with the sudden discovery of time. It is all going, I thought with the bitter desperation of the young confronting history. No one can hold us. Each and all, we are riding into the dark. Even living, we cannot remember half the events of our own days.

At that moment my eye had fallen upon the junk dealer passing his fateful corner. Now, I had thought instantly, now, save him, immortalize the unseizable moment. The junkman is the symbol of all that is going or is gone. He is passing the intersection into nothingness. Say to the mind, "Hold him, do not forget."

The darting lizard points beyond the manzanillo tree converged and tightened. The phantom horse and the heaped, chaotic wagon were still jouncing across the intersection upon R Street. They had never crossed it; they would not. Forty-five years had fled away. I was not wrong about the powers latent in the brain. The scene was still in process.

I estimated the lowering of the sun with one eye while at the back of my mind the lizard rustling continued. The blistering apples of the manzanillo reminded me of an inconsequential wild-plum fall far away in Nebraska. They were not edible but they contained the same, if a simpler, version of the mystery hidden in our heads. They were hoarding and dispersing energy while the inanimate universe was running down around us.

"We must regard the organism as a configuration contrived to evade the tendency of the universal laws of nature," John Joly the geologist had once remarked. Unlike the fire in a thicket, life burned cunningly and hoarded its resources. Energy provisions in the seed provided against individual death. Of all the unexpected qualities of an unexpected universe, the sheer organizing power of animal and plant metabolism is one of the most remarkable, but, as in the case of most everyday marvels, we take it for granted. Where it reaches its highest development, in the human mind, we forget it completely. Yet out of it history is made—the junkman on R Street is prevented from departing. Growing increasingly archaic, that phantom would be held at the R Street intersection while all around him new houses arose and the years passed unremembered. He would not be released until my own mind began to crumble.

The power to free him is not mine. He is held enchanted because long ago I willed a miniature of history, confined to a single brain. That brain is devouring oxygen at a rate out of all proportion to the rest of the body. It is involved in burning, evoking, and transposing visions, whether of lizard tails, alphabets from the sea, or the realms beyond the galaxy. So important does nature regard this unseen combustion, this smoke of the planet's autumn, that a starving man's brain will be protected to the last while his body is steadily consumed. It is a part of unexpected nature.

In the rational universe of the physical laboratory this sullen and obstinate burning might not, save for our habit of taking the existent for granted, have been expected. Nonetheless, it is here, and man is its most tremendous manifestation. One might ask, Would it be possible to understand humanity a little better if one could follow along just a step of the evolutionary pathway in person? Suppose that there still lived . . . but let me tell the tale, make of it what you will.

II

Years after the experience of which I am about to speak, I came upon a recent but Neanderthaloid skull in the dissecting room—a rare enough occurrence, one that the far-out flitting of forgotten genes struggles occasionally to produce, as if life sometimes hesitated and were inclined to turn back upon its pathway. For a time, remembering an episode of my youth, I kept the indices of cranial measurement by me.

Today, thinking of that experience, I have searched vainly for my old notebook. It is gone. The years have a way of caring for things that do not seek the safety of print. The earlier event remains, however, because it was not a matter of measurements or anthropological indices but of a living person whom I once knew. Now, in my autumn, the face of that girl and the strange season I spent in her neighborhood return in a kind of hazy lesson that I was too young to understand.

It happened in the West, somewhere in that wide drought-ridden land of empty coulees that carry in sudden spates of flood the boulders of the Rockies toward the sea. I suppose that, with the outward flight of population, the region is as wild now as it was then, some forty years ago. It would be useless to search for the place upon a map, though I have tried. Too many years and too many uncertain miles lie behind all bone hunters. There was no town to fix upon a road map. There was only a sod house tucked behind a butte, out of the prevailing wind. And there was a little spring-fed pond in a grassy meadow—that I remember.

Bone hunting is not really a very romantic occupation. One walks day after day along miles of frequently unrewarding outcrop. One grows browner, leaner, and tougher, it is true, but one is far from the bright lights, and the prospect, barring a big strike, like a mammoth, is always to abandon camp and go on. It was really a gypsy profession, then, for those who did the field collecting.

In this case, we did not go on. There was an eroding hill in the vicinity, and on top of that hill, just below sod cover, were the foot bones, hundreds of them, of some lost Tertiary species of American rhinoceros. It is useless to ask why we found only foot bones or why we gathered the mineralized things in such fantastic quantities that they must still lie stacked in some museum storeroom. Maybe the creatures had been immured standing up in a waterhole and in the millions of succeeding years the rest of the carcasses had eroded away from the hilltop stratum. But there were the foot bones, and the orders had come down, so we dug carpals and metacarpals till we cursed like an army platoon that headquarters has forgotten.

There was just one diversion: the spring, and the pond in the meadow. There, under the bank, we cooled our milk and butter purchased from the soddy inhabitants. There we swam and splashed after work. The country people were reserved and kept mostly to themselves. They were uninterested in the dull bones on the hilltop unenlivened by skulls or treasure. After all, there was reason for their reserve. We must have appeared, by their rural standards, harmless but undoubtedly touched in the head. The barrier of reserve was never broken. The surly farmer kept to his parched acres and estimated to his profit our damage to his uncultivated hilltop. The slatternly wife tended a few scrawny chickens. In that ever blowing landscape their windmill largely ran itself.

Only a stocky barefoot girl of twenty sometimes came hesitantly down the path to our camp to deliver eggs. Some sixty days had drifted by upon that hillside. I began to remember the remark of an old fossil hunter who in his time had known the Gold Coast and the African veldt. "When calico begins to look like silk," he had once warned over a fire in the Sierras, "it is time to go home."

But enough of that. We were not bad young people. The girl shyly brought us the eggs, the butter, and the bacon, and then withdrew. Only after some little time did her appearance begin to strike me as odd. Men are accustomed to men in their

various color variations around the world. When the past intrudes into a modern setting, however, it is less apt to be visible, because to see it demands knowledge of the past, and the
past is always camouflaged when it wears the clothes of the
present.

The girl came slowly down the trail one evening, and it
struck me suddenly how alone she looked and how, well, *alien*,
she also appeared. Our cook was stoking up the evening fire,
and as the shadows leaped and flickered I, leaning invisibly
against a rock, was suddenly transported one hundred thousand years into the past. The shadows and their dancing high-
lights were the cause of it. They had swept the present out of
sight. That girl coming reluctantly down the pathway to the
fire was removed from us in time, and subconsciously she knew
it as I did. By modern standards she was not pretty, and the
gingham dress she wore, if anything, defined the difference.

Short, thickset, and massive, her body was still not the body
of a typical peasant woman. Her head, thrust a little forward
against the light, was massive-boned. Along the eye orbits at
the edge of the frontal bone I could see outlined in the flames
an armored protuberance that, particularly in women, had
vanished before the close of the Würmian ice. She swung her
head almost like a great muzzle beneath its curls, and I was
struck by the low bun-shaped breadth at the back. Along her
exposed arms one could see a flash of golden hair.

No, we are out of time, I thought quickly. We are each and
every one displaced. She is the last Neanderthal, and she does
not know what to do. We are those who eliminated her long
ago. It is like an old scene endlessly re-enacted. Only the
chipped stones and the dead game are lacking.

I came out of the shadow then and spoke gently to her,
taking the packages. It was the most one could do across that
waste of infinite years. She spoke almost inaudibly, drawing an
unconscious circle in the dust with a splayed bare foot. I saw,
through the thin dress, the powerful thighs, the yearning fer-
tility going unmated in this lonesome spot. She looked up, and
a trick of the fire accentuated the cavernous eye sockets so that
I saw only darkness within. I accompanied her a short distance
along the trail. "What is it you are digging for?" she managed
to ask.

"It has to do with time," I said slowly. "Something that happened a long time ago."

She listened incuriously, as one at the morning of creation might do.

"Do you like this?" she persisted. "Do you always just go from one place to another digging these things? And who pays for it, and what comes of it in the end? Do you have a home?" The soddy and her burly father were looming in the dusk. I paused, but questions flung across the centuries are hard to answer.

"I am a student," I said, but with no confidence. How could I say that suddenly she herself and her ulnar-bowed and golden-haired forearms were a part of a long reach backward into time?

"Of what has been, and what will come of it we are trying to find out. I am afraid it will not help one to find a home," I said, more to myself than her. "Quite the reverse, in fact. You see—"

The dark sockets under the tumbled hair seemed somehow sadly vacant. "Thank you for bringing the things," I said, knowing the customs of that land. "Your father is waiting. I will go back to camp now." With this I strode off toward our fire but went, on impulse, beyond it into the full-starred night.

This was the way of things along the Wild Cat escarpment. There was sand blowing and the past mingling with the present in more ways than professional science chose to see. There were eroded farms no longer running cattle and a diminishing population waiting, as this girl was waiting, for something they would never possess. They were, without realizing it, huntsmen without game, women without warriors. Obsolescence was upon their way of life.

But about the girl lingered a curious gentleness that we know now had long ago touched the vanished Neanderthals she so resembled. It would be her fate to marry eventually one of the illiterate hard-eyed uplanders of my own kind. Whatever the subtle genes had resurrected in her body would be buried once more and hidden in the creature called *sapiens*. Perhaps in the end his last woman would stand unwanted before some fiercer, brighter version of himself. It would be no more than justice. I was farther out in the deep spaces than I knew, and the fire was embers when I returned.

The season was waning. There came, inevitably, a time when the trees began to talk of winter in the crags above the camp. I have repeated all that can be said about so fragile an episode. I had exchanged in the course of weeks a few wistful, scarcely understood remarks. I had waved to her a time or so from the quarry hilltop. As the time of our departure neared I had once glimpsed her shyly surveying from a rise beyond the pond our youthful plungings and naked wallowings in the spring-fed water. Then suddenly the leaves were down or turning yellow. It was time to go. The fossil quarry and its interminable foot bones were at last exhausted.

But something never intended had arisen for me there by the darkening water—some agonizing, lifelong nostalgia, both personal and, in another sense, transcending the personal. It was—how shall I say it?—the endurance in a single mind of two stages of man's climb up the energy ladder that may be both his triumph and his doom.

Our battered equipment was assembled in the Model T's, which, in that time, were the only penetrators of deep-rutted upland roads. Morose good-byes were expressed; money was passed over the broken sod cover on the hilltop. Hundreds of once galloping rhinoceros foot bones were stowed safely away. And that was it. I stood by the running board and slowly, very slowly, let my eyes wander toward that massive, archaic, and yet tragically noble head—of a creature so far back in time it did not know it represented tragedy. I made, I think, some kind of little personal gesture of farewell. Her head raised in recognition and then dropped. The motors started. *Homo sapiens*, the energy devourer, was on his way once more.

What was it she had said, I thought desperately as I swung aboard. Home, she had questioned, "Do you have a home?" Perhaps I once did, I was to think many times in the years that followed, but I, too, was a mental atavism. I, like that lost creature, would never find the place called home. It lay somewhere in the past down that hundred-thousand-year road on which travel was impossible. Only ghosts with uncertain eyes and abashed gestures would meet there. Upon a surging tide of power first conceived in the hearth fires of dead caverns mankind was plunging into an uncontrolled future beyond anything the people of the Ice had known.

The cell that had somehow mastered the secret of controlled energy, of surreptitious burning to a purpose, had finally produced the mind, judiciously, in its turn, controlling the inconstant fire at the cave mouth. Beyond anything that lost girl could imagine, words in the mouth or immured in libraries would cause substance to vanish and the earth itself to tremble. The little increments of individual energy dissolving at death had been coded and passed through the centuries by human ingenuity. A climbing juggernaut of power was leaping from triumph to triumph. It threatened to be more than man and all his words could master. It was more and less than man.

I remembered those cavernous eye sockets whose depths were forever hidden from me in the firelight. Did they contain a premonition of the end we had invited, or was it only that I was young and hungry for all that was untouchable? I have searched once more for the old notebooks but, again, in vain. They would tell me, at best, only how living phantoms can be anatomically compared with those of the past. They would tell nothing of that season of the falling leaves or how I learned under the night sky of the utter homelessness of man.

III

I have seen a tree root burst a rock face on a mountain or slowly wrench aside the gateway of a forgotten city. This is a very cunning feat, which men take too readily for granted. Life, unlike the inanimate, will take the long way round to circumvent barrenness. A kind of desperate will resides even in a root. It will perform the evasive tactics of an army, slowly inching its way through crevices and hoarding energy until someday it swells and a living tree upheaves the heaviest mausoleum. This covert struggle is part of the lifelong battle waged against the Second Law of Thermodynamics, the heat death that has been frequently assumed to rule the universe. At the hands of man that hoarded energy takes strange forms, both in the methods of its accumulation and in the diverse ways of its expenditure.

For hundreds of thousands of years, a time longer than all we know of recorded history, the kin of that phantom girl had lived without cities along the Italian Mediterranean or below the northern tentacles of the groping ice. The low archaic skull vault had been as capacious as ours. Neanderthal man had, we now know after long digging, his own small dreams and kindnesses. He had buried his dead with offerings—there were even evidences that they had been laid, in some instances, upon beds of wild flowers. Beyond the chipped flints and the fires in the cavern darkness his mind had not involved itself with what was to come upon him with our kind—the first bowmen, the great artists, the terrible creatures of his blood who were never still.

It was a time of autumn driftage that might have lasted and been well forever. Whether it was his own heavy brow that changed in the chill nights or that somewhere his line had mingled with a changeling cuckoo brood who multiplied at his expense we do not know with certainty. We know only that he vanished, though sometimes, as in the case of my upland girl, a chance assemblage of archaic genes struggles to reemerge from the loins of *sapiens*.

But the plucked flint had flown; the heavy sad girls had borne the children of the conquerors. Rain and leaves washed over the cave shelters of the past. Bronze replaced flint, iron replaced bronze, while the killing never ceased. The Neanderthals were forgotten; their grottoes housed the oracles of later religions. Marble cities gleamed along the Mediterranean. The ice and the cave bear had vanished. White-robed philosophers discoursed in Athens. Armed galleys moved upon the waters. Agriculture had brought wealth and diversification of labor. It had also brought professional soldiery. The armored ones were growing and, with them, slavery, torture, and death upon all the seas of the world.

The energy that had once sufficed only to take man from one camping place to another, the harsh but innocent world glimpsed by Cook in the eighteenth century on the shores of Australia, century by century was driving toward a climax. The warriors with the tall foreheads given increasingly to fanatic religions and monumental art had finally grown to doubt the creations of their own minds.

The remnants of what had once been talked about in Athens and been consumed in the flames of Alexandria hesitantly crept forth once more. Early in the seventeenth century Sir

Francis Bacon asserted that "by the agency of man a new aspect of things, a new universe, comes into view." In those words he was laying the basis of what he came to call "the second world," that world which could be drawn out of the natural by the sheer power of the human mind. Man had, of course, unwittingly been doing something of the sort since he came to speak. Bacon, however, was dreaming of the new world of invention, of toleration, of escape from irrational custom. He was the herald of the scientific method itself. Yet that method demands history also—the history I as an eager student had long ago beheld symbolically upon a corner in the shape of a junkman's cart. Without knowledge of the past, the way into the thickets of the future is desperate and unclear.

Bacon's second world is now so much with us that it rocks our conception of what the natural order was, or is, or in what sense it can be restored. A mathematical formula traveling weakly along the fibers of the neopallium may serve to wreck the planet. It is a kind of metabolic energy never envisaged by the lichen attacking a rock face or dreamed of in the flickering shadows of a cave fire. Yet from these ancient sources man's hunger has been drawn. Its potential is to be found in the life of the world we call natural, just as its terrifying intricacy is the product of the second visionary world evoked in the brain of man.

The two exist on the planet in an increasingly uneven balance. Into one or the other or into a terrifying nothing one of these two worlds must finally subside. Man, whose strange metabolism has passed beyond the search for food to the naked ingestion of power, is scarcely aware that the energy whose limited planetary store lies at the root of the struggle for existence has passed by way of his mind into another dimension. There the giant shadows of the past continue to contend. They do so because life is a furnace of concealed flame.

Some pages back I spoke of a wild-plum thicket. I did so because I had a youthful memory of visiting it in autumn. All the hoarded juices of summer had fallen with that lush untasted fruit upon the grass. The tiny engines of the plant had painstakingly gathered throughout the summer rich stores of sugar and syrup from the ground. Seed had been produced; birds had flown away with fruit that would give rise to plum trees miles away. The energy dispersion was so beneficent on that autumn afternoon that earth itself seemed anxious to promote the process against the downward guttering of the stars. Even I, tasting the fruit, was in my animal way scooping up some of it into thoughts and dreams.

Long after the Antillean adventure I chanced on an autumn walk to revisit the plum thicket. I was older, much older, and I had come largely because I wondered if the thicket was still there and because this strange hoarding and burning at the heart of life still puzzled me. I have spoken figuratively of fire as an animal, as being perhaps the very *essence* of animal. Oxidation, I mean, as it enters into life and consciousness.

Fire, as we have learned to our cost, has an insatiable hunger to be fed. It is a nonliving force that can even locomote itself. What if now—and I half closed my eyes against the blue plums and the smoke drifting along the draw—what if now it is only concealed and grown slyly conscious of its own burning in this little house of sticks and clay that I inhabit? What if I am, in some way, only a sophisticated fire that has acquired an ability to regulate its rate of combustion and to hoard its fuel in order to see and walk?

The plums, like some gift given from no one to no one visible, continued to fall about me. I was old now, I thought suddenly, glancing at a vein on my hand. I would have to hoard what remained of the embers. I thought of the junkman's horse and tried to release him so that he might be gone.

Perhaps I had finally succeeded. I do not know. I remembered that star-filled night years ago on the escarpment and the heavy-headed dreaming girl drawing a circle in the dust. Perhaps it was time itself she drew, for my own head was growing heavy and the smoke from the autumn fields seemed to be penetrating my mind. I wanted to drop them at last, these carefully hoarded memories. I wanted to strew them like the blue plums in some gesture of love toward the universe all outward on a mat of leaves. Rich, rich and not to be hoarded, only to be laid down for someone, anyone, no longer to be carried and remembered in pain like the delicate paw lying forever on the beach at Curaçao.

I leaned farther back, relaxing in the leaves. It was a feeling I had never had before, and it was strangely soothing. Perhaps

I was no longer *Homo sapiens*, and perhaps that girl, the last Neanderthal, had known as much from the first. Perhaps all I was, really, was a pile of autumn leaves seeing smoke wraiths through the haze of my own burning. Things get odder on this planet, not less so. I dropped my head finally and gazed straight up through the branches at the sun. It was all going, I felt, memories dropping away in that high indifferent blaze that tolerated no other light. I let it be so for a little, but then I felt in my pocket the flint blade that I had carried all those years from the gravels on the escarpment. It reminded me of a journey I would not complete and the circle in the dust around which I had magically traveled for so long.

I arose then and, biting a plum that tasted bitter, I limped off down the ravine. One hundred thousand years had made little difference—at least, to me. The secret was to travel always in the first world, not the second; or, at least, to know at each crossroad which world was which. I went on, clutching for stability the flint knife in my pocket. A blue smoke like some final conflagration swept out of the draw and preceded me. I could feel its heat. I coughed, and my eyes watered. I tried as best I could to keep pace with it as it swirled on. There was a crackling behind me as though I myself were burning, but the smoke was what I followed. I held the sharp flint like a dowser's twig, cold and steady in my hand.

Bibliography

Barlow, Nora. "Charles Darwin and the Galápagos Islands," *Nature*, Vol. 136 (1935), p. 391.

Beagle, Peter. *The Last Unicorn*, New York, Viking Press, 1968.

Beaglehole, J. C. *The Journals of Captain James Cook on His Voyages of Discovery*, 3 vols. to date and a portfolio, London, Cambridge University Press, 1955–67.

Brown, Lloyd A. *The Story of Maps*, New York, Crown Publishers, 1949.

Cameron, Hector Charles. *Sir Joseph Banks: The Autocrat of the Philosophers*, London, The Batchworth Press, 1952.

Campbell, Joseph. *The Hero with a Thousand Faces*, New York, Meridian Books, 1956.

———. *The Masks of God: Primitive Mythology*, New York, Viking Press, 1959.

Carlquist, Sherwin. *Island Life: A Natural History of the Islands of the World*, Garden City, N.Y., The Natural History Press, 1965.

Carozzi, Albert V. "Agassiz's Amazing Geological Speculation: The Ice Age," *Studies in Romanticism*, Vol. 5 (1966), pp. 57–83.

Carrington, Hugh. *Life of Captain Cook*, London, Sidgwick and Jackson, 1939.

Chaning-Pearce. Melville, *The Terrible Crystal*, London, Kegan Paul, 1940.

Channing, William Ellery. *Thoreau, the Poet Naturalist*, New York, Biblo and Tannen, 1966.

Chapman, Walker. *The Loneliest Continent*, Greenwich, Conn., New York Graphic Society Publishers Ltd., 1964.

Charlesworth, J. K. *The Quaternary Era: With Special Reference to Its Glaciation*, 2 vols., London, Edward Arnold, 1957.

Cherry-Garrard, Apsley. *The Worst Journey in the World*, London, Penguin Books, 1948.

Christie, John Aldrich. *Thoreau as World Traveler*, New York, Columbia University Press, 1965.

Clarke, Howard W. *The Art of the Odyssey*, Englewood Cliffs, N.J., Prentice-Hall, 1967.

Coburn, Kathleen (ed.). *Inquiring Spirit: A New Presentation of Coleridge from His Published and Unpublished Writings*, New York, Pantheon Books, 1951.

Creed, John Martin, and Boys Smith, John Sandwith (eds.). *Religious Thought in the Eighteenth Century*, London, Cambridge University Press, 1934.

D'Ancona, Umberto. *The Struggle for Existence*, Leiden, Holland, E. J. Brill, 1954.

Darwin, Charles. *Journal of Researches*, London, Henry Colburn, 1839.

———. *On the Origin of Species*, reprint of 2nd rev. ed., London, Oxford University Press, 1860.

Darwin, Francis (ed.). *Life and Letters of Charles Darwin*, 3 vols., London, John Murray, 1888.

———. and Seward, A. C. (eds.). *More Letters of Charles Darwin*, London, John Murray, 1903.

Dyson, James L. *The World of Ice*, New York, Alfred Knopf, 1962.

Eiseley, Loren. *Francis Bacon and the Modern Dilemma*, Lincoln, University of Nebraska Press, 1963.

———. "Neanderthal Man and the Dawn of Human Paleontology," *The Quarterly Review of Biology*, Vol. 32 (December 1957), pp. 323–329.

Eydoux, Henri-Paul. *The Buried Past: A Survey of Great Archaeological Discoveries*, New York, Frederick A. Praeger, 1966.

Gelli, Giovanni Battista. *Circe*, trans. by Thomas Brown, ed. by Robert M. Adams, Ithaca, N.Y., Cornell University Press, 1963.

Gibbs, Frederic Andrews. "The Most Important Thing," *American Journal of Public Health*, Vol. 41 (1951), pp. 1503–1508.

Harding, Walter (ed.). *Thoreau, Man of Concord*, New York, Holt, Rinehart and Winston, 1960.

Hooker, Sir Joseph Dalton (ed.). *Journal of the Right Honorable Sir Joseph Banks*, London, Macmillan, 1896.

Hyman, Stanley Edgar. "Descent, Fall and Sex—Darwin's Victorianism," *The Carleton Miscellany*, Vol. 2, No. 4 (1961), pp. 11–25.

————. *The Tangled Bank: Darwin, Marx, Frazer and Freud as Imaginative Writers*, New York, Atheneum, 1962.

Ivanova, I. K. "The Significance of Fossil Hominids and Their Culture for the Stratigraphy of the Quaternary Period," *Arctic Anthropology*, Vol. 4 (1967), pp. 212–223.

Jolly, Alison. *Lemur Behavior: A Madagascar Field Study*, Chicago, University of Chicago Press, 1966.

Joly, John. *The Birth-Time of the World, and Other Scientific Essays*, New York, E. P. Dutton, 1915.

Kazantzakis, Nikos. *The Odyssey: A Modern Sequel*, New York, Simon and Schuster, 1958.

Lack, David. *Darwin's Finches*, London, Cambridge University Press, 1947.

Lattimore, Richmond. *The Odyssey of Homer: A Modern Translation*, New York, Harper and Row, 1965.

MacArthur, Robert H. and Wilson, E. O. *The Theory of Island Biogeography*, Princeton, Princeton University Press, 1967.

McGlashan, Alan. *The Savage and Beautiful Country*, London, Chatto and Windus, 1966.

Mayr, Ernst. "Changes of Genetic Environment and Evolution," in *Evolution as a Process*, ed. by Julian Huxley et al. New York, Collier Books, 1963.

Mellersh, H. E. L. *Fitzroy of the Beagle*, London, Rupert Hart-Davis, 1968.

Partridge, R. B. "The Primeval Fireball Today," *The American Scientist*, Vol. 57 (1969), pp. 37–74.

Pascoli, Giovanni. *Poesie*, Verona, Italy, Arnoldo Mondadori Editore, 1965.

Price, A. Grenfell (ed.). *The Explorations of Captain James Cook in the Pacific, as Told by Selections of His Own Journals, 1768–1779*, New York, Heritage Press, 1958.

Ritchie, James. "The Edinburgh Explorers," *University of Edinburgh Journal*, Vol. 12 (1943), pp. 155–159.

————. "Evolution and the Galápagos Islands," *University of Edinburgh Journal*, Vol. 12 (1943), pp. 97–105.

Santayana, George. *The Birth of Reason*, New York, Columbia University Press, 1968.

Schultz, Gwen. *Glaciers and the Ice Age*, New York, Holt, Rinehart and Winston, 1963.

Sillman, Leonard. "The Genesis of Man," *International Journal of Psychoanalysis*, Vol. 34 (1953), pp. 146–152.

Simpson, George Gaylord. *The Geography of Evolution*, New York and Philadelphia, Chilton Books, 1965.

Smith, Edward. *The Life of Sir Joseph Banks: President of the Royal Society, with Some Notices of His Friends and Contemporaries*, London, John Lane, The Bodley Head, 1911.

Smith, Joseph Lindon. *Tombs, Temples and Ancient Art*, Norman, Okla., University of Oklahoma Press, 1956.

Stanford, W. B. *The Ulysses Theme: A Study in the Adaptability of a Traditional Hero*, 2nd rev. ed., Oxford, England, Basil Blackwell, 1963.

Sullivan, Walter. "The Neanderthal Man Liked Flowers," *The New York Times*, June 13, 1968.

Thoreau, Henry David. *The Journal of Henry David Thoreau*, ed. by Bradford Torrey and Francis H. Allen, 14 vols., Boston, Houghton Mifflin, 1949.

Villiers, Alan. *Captain James Cook*, New York, Charles Scribner's Sons, 1967.

———. *The Coral Sea*, New York, Whittlesey House, McGraw-Hill, 1949.

Wallace, Alfred Russel. *Island Life*, 2nd rev. ed., London, Macmillan & Co., 1892.

Wheeler, John Archibald. "Our Universe: The Known and the Unknown," *The American Scholar*, Vol. 37 (1968), pp. 248–274.

UNCOLLECTED PROSE

Autumn—A Memory

There is a voice that speaks to men through the rose and yellow of dying autumn leaves . . .

I REMEMBER STILLNESS and the faint flutter of red and golden leaves down long dim shafts of sunlight through the trees.

I asked a cow-puncher lounging in the shade where to find the Aztec ruins.

"Up the road and across the river. But the place'll be closed now," he cogitated.

I went, hoping he was mistaken. He wasn't, but there was no one around. I stole a hasty glance about and climbed over the fence.

The round stone pits of the kiva were deserted and the walls were crumbling. It is a lonely thing to look on men's broken handiwork and muse wide-eyed over their disappearance . . . Still, perhaps it was their dust that floated in a slight breeze over the ruined wall . . .

Red and sombre, with many lights on purple stone the afternoon was fading.

The place was weird with stillness in which one caught oneself straining to catch the undernote of long dead activity, of something that lingered, that would linger till the last stone had fallen, something that would not go away.

One listened as at the edge of a void through which no sound might penetrate. And yet one listened oddly sure—as though a voice *were* there that could not pierce the blackness . . . One wondered . . .

There were visions of the laboring copper bodies that built this place under the blazing sun. Was the land so grim then? From where did they come—and why—and what fate overtook them?

Did they die fiercely, breast to breast and knife to thong and stone? (The twilight air was peaceful.)

There were generations, there were brave little friendships, hatred and feasts and there was love . . . A coyote laughed till the stone-piled gullies rang . . .

There may have come a time when the offended god turned his face away, when prayers and the gay-colored prayer stones and the holy medicine lost power and men died in gasping heaps . . . Or maybe the harvests burned. Did any live to go . . . ?

A star burned with a steady silver flame. I was a shadow among shadows brooding over the fate of other shadows that I alone strove to summon up out of the all-pervading dusk.

They had built, stone upon heavy stone. Women and men and children—children and women and men . . . To the great watching god men's faces pass as a blur . . .

Starlight and dust in starlight . . . Does it matter now at all . . . ?

Prairie Schooner, October 1927

Riding the Peddlers

I F YOU ARE young and there is speed in your blood you begin
as a passenger stiff. In the jungles you study time tables,
speak familiarly as a plutocrat of the Golden State Limited, and
carve speed records on watertanks a thousand miles apart. If
you ride a freight it must be a hot-shot or a manifest. It does
not matter that there is no place for you to go and only yard
bulls to greet you upon your arrival. Going is the business of
your class, and as a young novitiate you grit your teeth on the
decks of passengers hurtling through mile-long tunnels and
strive to equal the time records of paid fares. Later, after sun-
dry encounters with the railroad detectives who lurk with vigilance about such trains, and after hearing from companions
tales of this or that tramp who left his anatomy strewn along
the right-of-way, you experience a tremor of doubt. You notice
that the older and still healthy members of the profession show
no great concern in this matter of speed; in fact, many of them
can be found aboard the peddlers.

When this fact was first brought home to me I considered it,
with the usual cockiness of youth, to be a loss of nerve due to
advancing age. I pitied these elderly stiffs and was moved to
sighs over the most forlorn object of all—the bindle stiff who
had abandoned the trains for the road. But my conversion was
soon to come. It chanced that most of my hoboing career up
to that time had been carried on along the main trunk lines
where fast trains were plentiful; hence there had been no reason to patronize the slow freights making all the little local
stops. These trains we referred to scornfully as "peddlers."

I was making fast time down through the middle states one
autumn when the news spread that some hoboes had murdered a Union Pacific yard bull and tossed his body from a
moving train near a town some fifty miles ahead of us. The
police were arresting every vagrant in sight and since the town
had an evil reputation among the brotherhood in the best of
times, I had no desire to pass through it at that moment. But
also I had no desire to walk. Something had to be done—and
quickly. In another ten minutes we would be pulling out. At

this moment my eyes chanced to fall upon a Burlington time-table in the hands of another vagrant.

"Say, Jack, can I see that?" I asked.

"Sure." He passed it over.

I turned hastily to a map of the system. The gods, it began to appear, were with me. The map showed a side fork of the Burlington angling up from their main line into the very town where we now sat. So far so good. But how often did the trains run? I had no intention of sitting in that town a week waiting for enough freight to accumulate to make a train. I stood up on our boxcar and glanced about. There was a spur running off to the south that must be the branch in question. A few coal cars rested on a siding. As far as I could see, cornfields wavered to the south. My heart sank.

I climbed off and pursued our braky, who had gone past a couple of minutes before on his way toward the head of the train. A dozen cars away I overtook him. Yes, there was a spur here. No, he didn't know when the trains ran. "You'd better go down to their depot and ask," he said. "That little yellow building way down there is it." I looked. It was a good mile beyond the end of our train. That meant about two miles alto-gether. If there was no train, I would never make it back before this one had gone. I set out in a stumbling run down the tracks, accompanied by another bum of the same mind.

When we had passed our caboose, more track came in sight. From behind the little depot came an engine about as big and just as busy as an overgrown tumblebug, pushing a couple of cars. Our hopes soared. They might be making up a train. We vaulted a U. P. fence, thumbed a mental nose at their No Tres-passing signs, and dashed into the Burlington yard office.

"Who's in charge here?" asked my companion with the air of an official inspector. I grew conscious of our rags and waited to be thrown out.

"I am," said a pugnacious looking individual. "Whatta yuh want?"

"Is that a freight making up out there?"

"Yep."

"When does she pull out?"

"In about thirty minutes, boys," said the yard master grinning.

"Imagine pulling a stunt like that on the main line," I said with a sigh of relief as we escaped into the outer air. "Let's go eat. The engineer's eating his lunch."

"Hamburger—but for God's sake hurry it," I pleaded with the waitress. "Five times on this trip I've ordered hamburger and smelled it cooking and then had to leave it and run for a train."

We chafed impatiently at the counter, one eye cocked on the train. Sure enough, as soon as the hamburgers were half cooked, the engine bell set up a clamor. We tore them smoking from the hands of the waitress and fled away across the tracks. Dangling our feet from a side-door pullman, we waited for the train to pull out. Nothing happened. One or two idlers rolled over in the grass of the Square. We ate our sandwiches in melancholy silence. Thirty minutes later the train pulled out.

We rattled away over a dozen little bridges and across innumerable fields. Some of them were wild, with coarse grass springing up around the old glacial boulders. A few rabbits loped along unhurriedly. We stopped at sleepy little towns, and people who chanced to pass talked idly with us. One of the train crew suggested a more comfortable car. It wasn't long until we realized that we had stepped, momentarily, at least, out of a world of class hatred and struggle. It was what I had imagined the West of the old days to be.

It was not that these people had not lived elsewhere—many of them—or that they had not partaken of the struggle. But here along this little backwater, it did not matter. There were not enough of us to count. So there was time for us to speak drowsily and courteously to each other. Room for the prairie grass to grow between the stones, for the rabbits to hop less furtively. We knew how the latter felt. We too had hopped, skipped, and run through the city jungle in fear of the fist and the blackjack. Here we thrust out tentacles of friendliness again from under a hardened armor of toughness and cynicism. Released from fear, our small horizon expanded to take in the curving brown back of the train and ourselves as personalities.

"Y'know, Jack," said my companion, who never in a thousand miles had vouchsafed a word about himself, "I like this. I was born in Indiana on a farm. I never seen a train till I was sixteen. Funny, for a bum."

"Yeah," I said. "I've been in Indiana. The hills are swell."

"Say, have you?" He warmed visibly. "We used to have a spring back of the house. Colder than ice. My old man used to cool the raisin wine in it. And in the fall things were swell." He groped for words. "You know, the leaves and everything. I might be there yet, but I went to live with an uncle. He had a big voice and all he did was work and go to church and beller to God because his crops weren't good enough.

"There was an atheist fellow lived down the road a ways. He was a nice fellow but he always had good crops. My uncle couldn't stand it. He'd complain to God on Sundays. I hated it. You know. It was sort of like bein' a tattle-tale in school. And when the atheist took sick one summer and didn't get well I sorta blamed Uncle Carl. I was just a kid and I liked that fellow. I sorta felt if Uncle Carl hadn't kept shootin' his mouth off that way God wouldn't a noticed.

"The railroad ran through one of Uncle Carl's pastures and I'd been used to watchin' it. I used to see the 'boes ridin' and wonder about 'em. You know? Like a kid will. Where they're goin' and what they'll be seein' and all? Well, right after this happened I was down in the pasture tendin' the cows and thinkin' about Jake—he was the atheist—and rememberin' how nice he'd been to me—showin' me pictures out of books of city things an' all—when along comes a freight, slow like this, and she stops right in the pasture account of somethin' bein' wrong.

"They was some 'boes on her and one of 'em gets off and comes down and asks me for a drink. I took him up to the house, askin' him questions all the time. He answers me. You know, like we string these kids along? Adventure and all that. Well, everything woulda still been all right, but Uncle Carl was at the house with a bender on. He musta been feelin' proud at cookin' Jake's goose. Made him feel sorta like the Lord's appointed. Anyway he starts tellin' this 'bo that by the sweat of your brow you had to earn your drink and bread. He goes on like that a while and then says no, he can't have any drink.

"I can still see that 'bo, a sorta tired lookin' fellow. He don't say anything. He just sort of sinks into himself and walks quiet down into the pasture again. Not like he's afraid, but just tired

from it happenin' so many times. Like he's sick of hearin'
words. I never said a thing to Uncle Carl. I don't think he even
noticed me. But you know how kids are. I 'spose you ran away
from home, too, sometime. All sick and twisted inside. Yeah.
That's the way I felt. I caught up with that fellow down in the
pasture. 'I'm old enough,' I said. 'I'm goin' too.'

"'All right, kid,' he says. Not coaxin'-like, but just as though
it had to be done. If he'd coaxed, I wouldn't have gone, but
that's the way I felt too—like it had to be done. I was scared,
but it had to be done. Like a sort of justice or somethin'. Like
savin' yourself from bein' damned."

His face looked proud for a moment under the hard lines
and the dirt. "I never went back," he said.

I sat silent. There was nothing to say. Here was the end of
his forlorn odyssey and revolt. In less than a hundred miles the
freight wilderness had separated us. In many thousand miles I
never saw his face again.

II

Months later in a little town below the Sierras I missed the last
passenger. I missed the manifest that followed it. There was
nothing left but a peddler. It was a long train laboring forward
under the combined efforts of three engines. Bums sat along
the roof in scores. It was a question whether one could make
faster time riding or walking beside it. I chose to ride.

We ran for miles through the fierce heat of the Sacramento
Valley. The drowsy clicking of the wheels lulled me to sleep.
Later there were tunnels. Toward evening I awoke.

Sleepily, and then with surprise I glanced back along the car
roofs. The scene that presented itself might have been some
modern evangelist's dream of the last train arriving in Paradise.
There were dreary faces, mad faces, happy faces, evil faces. The
wise alone were not with us. It was just as well. They would
have been as much interlopers as the snake in Eden. Through
the slow, beautiful sunset along the mountain the train labored
grotesquely. On either side great orchards stood at harvest.
Everywhere hungry men were climbing down from the cars
and squeezing through fences. Men already in the trees shook

frantically. Apples rained down and found their way into hats and ragged pockets. Their owners ran and swung aboard the train. This process continued for miles along the grade.

Someone squatting on the roof beside me hummed a stanza of "The Big Rock Candy Mountains":

> "Oh, the railroad bulls they all are blind
> In the big rock candy mountains."

The Limited might be due in Reno, but one of her passengers was not. I was aboard a peddler and no one expects the peddlers. They come when they arrive. In that hour, in the late sun with the pine scent of evening already in the air it was not possible that we were crawling toward the high Sierras and the deserts of Utah. No. We had slipped tunnel-like through time. We had taken a phantom track into the hazy country—the Big Rock Candy Mountains—the restful haven of all bums. Stretched on the car roof I thought, fleetingly, of "Indiana." I wished he had made this train. They would wait for us in vain in Reno.

Prairie Schooner, Winter 1933

The Mop to K. C.

I STOLE an overcoat and caught the K. C. drag at the crossing. I had a thousand miles to go, and there was already frost in the fields. I thought I'd be the last 'bo left in that country. Worse yet, there wasn't an empty in sight. But a gondola came by with a couple of heads watching me. I'd have company anyhow. So I climbed in. There were three 'boes huddled in some straw in the forward end and a little guy all black with coal dust standing looking at me.

"Hello, kid," I said. "Headin' South? We got the snow right on our tail."

"We sure have." He had a voice with a girlish quaver in it, and every now and then he shook like a leaf.

It wasn't much to wonder at. He had on a pair of overalls, and an old suit coat pinned around his neck with a safety pin. And that was all he had, too, except the shoes he stood in.

The Mop to K. C. is a rough road. We began to bounce like hell. I was glad of it. It shook all the shivers out of you and you hung on to your guts and the effort made you feel warmer.

"I been riding passengers all the way from Rochester," the kid yelled between whistle blasts. I sat down and he sat beside me. "Do you think we could catch another passenger?"

It was funny sitting there listening to him talk about passengers in that girlish voice. Every now and then it would sort of rise and break. You know—like the way a guy talks when he's scared. I didn't believe he was that good a stiff. It takes guts to ride passengers.

"You'd better leave 'em alone, kid," I said. "But if you want to fly there's a manifest through Falls City, and we wait over there for her."

He asked me if I wanted to make it with him. I didn't like this "we" stuff.

"You'll have a better chance alone," I said. "Where ya goin'?"

"Leavenworth."

"Then what the hell's the hurry? You're damn near home."

"Yeah," he said. "That's right. That's right." He kept getting up and looking out of the car till I figured he had the bull horrors. There were hollows under his eyes and he looked sick. He talked sometimes like he was a little off.

Once he went up and talked to the other bums, but I noticed they sort of looked at each other and didn't say much. Pretty soon he was back. I don't know why I didn't give him the air. But there was something in his eyes. I couldn't do it. And he looked so damned miserable sitting there and shaking.

Pretty soon he saw the ring on my hand. "What's that?" he asked me. It was a college ring, but I lied and said it was a high school ring. A guy hasn't got any right to ask you things like that on the road.

His eyes brightened with that funny fever look again, and he said: "Gee, you musta had a swell home to get edjacated like that."

Yeah! That was all right, but not for me. I had a thousand miles to go, and I didn't want to talk about that.

"How's the bulls up North?" I asked him.

He started shivering again. "They're mean. When I was ridin' passengers a couple of 'em stopped right beside the tender, but I laid low, quiet as a mouse, and they went away." He laughed that cracked laugh and looked at me sort of admiringly. "You're big. You're darn near big as a bull. I bet they don't bother you none."

"Listen!" I said. "I'm not fightin' no bulls. I look big 'cause I got on two suits of underwear. See?"

"But the bums is what I'm scared of. I'm scared to get into box cars with 'em. Don't the wolves ever bother you?"

"Christ, no," I said. But I felt sick inside. He ought never to have been born.

We were easing into a town. "I'm gonna hunt an empty," I said. "It'll be cold as hell by midnight."

I dropped off and went back along the cars watching for a broken seal.

Away down the track I heard him yell, and pretty soon he pattered up. I'd found an empty and was opening it. He helped me push it open, and I climbed in. He stood outside and asked if he could get in. He wouldn't look at me. I didn't want him, but what was a guy to do? The engine kicked and we started.

"Get in," I said, and lent him a hand. To make it worse, the door jammed and wouldn't shut.

I tried to sleep, but it was no use; so finally I just lay quiet while it got darker and darker. The kid curled up beside me and went to sleep. Dead tired. He moaned and shivered and once he talked, but I couldn't make it out.

Once I lit a cigarette and looked at him. He was a road kid, all right. The dirt was ground into him until not even an undertaker could get it off. His cap had fallen sideways and a few straggly curls stuck out. Asleep—when he wasn't scared—he had a friendly sort of mouth.

We rolled on to a side-track in Falls City. I looked out. Sure enough the manifest was coming in. The engine steamed by. I shook the kid.

"Wake up; here's the manifest." He jumped back from me into a corner. Then he woke up. "I'm not gonna hurt ya. Here's the manifest."

He looked out with me. "You just got time to make her," I said. "Good luck."

"Ain't you coming?" he asked. "No," I said. "I'm going to K. C. I don't want to get in there at midnight and walk the streets to keep warm. But you'll make Leavenworth all right. You live there, don't you?"

"I ain't got no home," he said, that sick tremble coming back to his mouth. "I'm just goin' there. I ain't never had no home."

"I'm sorry, kid," I said. "I thought—" He was running toward the head of the train.

I went back to the corner for a smoke. A thing like that's hard on your sleep. Besides, the little fool'd freeze unless he found an empty reefer. What the hell was he going to do at Leavenworth?

I heard the manifest pull out. By and by they tested our air. It was clear dark now. I could see the lights of the town twinkle up on the hill.

She high-balled just as a head blocked the doorway. Somebody scrambled in. We started to move. Whoever it was blundered toward the corner. I ran into him in the dark.

"Jack! Is it you, Jack?" That voice was unmistakable. I was sore. "S'matter?" I asked. "Did they ditch you?"

"I couldn't make it."

I knew he was lying and he knew I knew it. He didn't answer any more, but shoved something into my hands in the dark. It was a paper sack that sagged like there was meat to it.

"I bummed a house over the tracks," he said.

I'd already got a sandwich into my mouth by that time; so I didn't say any more about the manifest. I shoved the sack at him and struck another light. "Aren't you gonna eat?" I asked.

He was leaning against the wall, and cold as it was, I noticed beads of sweat on his face. His eyes had a funny far-away look. I put the match on the floor, stepped on it, and said, "You'd better eat."

We were running through a bunch of cornfields, and I could see the frost and the cold light on them. I gave the sack another shove.

"No thanks," he said. "I—I don't want anything. I can't eat. There's—there's somethin' wrong with my guts—inside. That's why I went to Rochester. A doc told me I ought to."

He stopped. I could hear him fumbling in the dark, and waited. Pretty soon he handed me something that felt like cardboard. I struck another match.

"It's the doc's report," he explained faintly, from somewhere in the dark.

The match went out. I sat there in the dark, holding the envelope. I didn't read it. I knew already what was in it. I waited as long as I could and then I said, "What did they say?"

"They said I—I'm gonna die." He got it out without his voice breaking, but I knew what it cost him. I could feel him over there in the blackness pressing his hands against the floor.

"I—it's tough," I blundered, but saying it soft, and trying to ease him down. "Didn't they say you could do something —anything?"

"Well, they said quiet and rest 'ud help, and they said a lot of things I didn't understand, but I knew what they meant."

"And they kicked you back out on the road?"

"No." He was almost apologetic. "They was nice. They told me they'd take care of me, but I was afraid. It was all cold and white and bare and I was afraid. It was like the Home. I was in the Home till I was thirteen. I don't know who my folks was. I guess my folks wasn't married. That was the way with most of the kids there. The place they wanted to put me was like the Home. I couldn't stand it—those rows of beds and things. I told 'em I knew somebody at Leavenworth."

"Do you?"

He had snudged over until he was against me. Then he went on more steadily. "I wasn't gonna die in one of those places— all cold and shiny where you're a number or something."

I wanted to get out of that car. I patted his shoulder and said: "Listen, Bud, you shouldn't of done that. They might 'a fixed you up. You shouldn't be doin' that. Do you know somebody at Leavenworth? Shall I take you to a hospital?"

Oh, it was swell, all right, sitting there in the dark, getting colder and colder, with that voice like a sick girl's going on and on in the darkness. And the wheels going faster and faster and that damned hoghead hooting away off over the river like he wanted a clear track to hell.

I hardly knew what he was saying, at first. I just kept willing inside, "Don't let him cry, for Christ's sake, don't let him cry." You can't keep a tension like that. You'd go mad. There wasn't anything I could do. It was all in the cards. I said that to myself and eased down and tried to listen to him.

He didn't cry, either. He just kept leaning against me and babbling. It was doing him good. He kept talking about a long string of places he'd been, and at first I couldn't figure it out. Every bum knows towns from New York to Frisco, but why talk about that? There's the same bulls and the same people in all of them.

First it was Muncie, Indiana.

"I lived in a big box in Muncie for a week," he told me. It was like a kid talking of his playhouse time, like the things a girl might dream about. "I ain't never had a home—I guess I told you that. But I pretty near did in Muncie. This box was on the ends of town, and I was gonna borrow a saw and fix real windows in it. Some bums broke it up for kindling wood." He stopped and thought and then said he guessed it really hadn't mattered, the police wouldn't have let him stay, anyway.

The kid knew every town from New York to Tia Juana. I doubt if he had spent a month in one spot since he left the orphanage.

"And once outside of Sacramento a guy gave me a real cabin. Anyhow he said I could come and live with him. But the other guys didn't like him. A fellow beat him up, and I ran away."

"Yeah," I said. That sort of thing used to make me feel crawly. Now I just felt sort of empty like nothing mattered. I don't believe I'd have blinked if we'd run off a trestle.

"An' once in K. C.—you know the yards in K. C.?—I slept for three days in a corn-shock out by Argentine. I was getting to like it. It was pretty, and quiet-like around there and nobody'd found me out. The next night a mouse scared me and I crawled out, but it was so cold I crawled back in again. I was going to Cal., but I stayed there three days with the mice. The last day I gave them some crumbs from a handout. They was gettin' to know me. Mice ain't so bad when you get to know them. Mostly I'm afraid of them, though. I guess I was lonesome."

He drew a long breath, and I stood up in the dark and took the overcoat off and tossed it over him as he lay shivering.

"You sleep now," I said. "I'll tell you when we get to Leavenworth." I'd have given him my shirt to get him quiet.

"Ya won't leave, will ya Jack?" he asked me under the coat.

"No," I said.

"Jack—"

"Yeah?"

"I was ridin' passengers 'cause I was lonesome and they went fast." I waited awhile and pretty soon he said again, half asleep, "'Cause they went fast."

Poor devil—running from himself, I suppose. I walked up and down in the other end of the car to keep warm. It was quiet except for the steady roar of the wheels. We were running alongside the Missouri. The driftwood in the moonlight looked sometimes like bodies floating.

Once we passed a hoboes' night camp on the line. They were burning cornstalks and huddling around the fire to keep warm. Every time the fire blazed they bobbed up and down like big shadows. There wasn't another thing alive in the fields.

We made Leavenworth at midnight. The train stops in a sort of cut, with the river on the left and a high bank on the right. Somewhere along that bank is a spring with a pipe in it, and a little jungle at one side. I didn't expect anybody there, it was so late, but sure enough there was a little red blaze way down the tracks. When we got closer, I could see somebody hunched over beside it. I shook the kid and yelled in his ear, "We're comin' in."

He got up, bumped into the wall, then staggered over to the door and almost fell out before I could grab him.

I shook him and pointed over to the fire. "For Christ's sake get over there and warm up."

He started, and then looked at me. "Ain't you comin'?"

"No," I said. "She pulls out in a minute. Do you know anybody here?" I thought maybe he'd answer me this time.

He just stood there stupid a minute, looking at his toes.

"I did."

"Is he here now?"

"I don't know. He had a place down there in the bottoms. I—I worked for him a few weeks. He said for me to come back some winter. I could trap with him, he said. I thought maybe I could fix up a little shack. I've always wanted to."

"What if you don't find him?"

"I don't know. I guess it won't matter much anyhow."

"There's a hospital at the prison," I said, looking the other way. "I don't know much about such things. Maybe they'd take you in, in a pinch."

He didn't say anything. A jerk traveled down the cars and she whistled. I felt sore; I don't know what at.

"You're a damned fool," I said.

He didn't say a word.

I swung into the car and stood and watched from the doorway as we backed up and then went forward. The crouching old man by the fire never moved. The track runs straight for a long way. I watched until there was just a red eye of light in the dark and a little shape jumping and gesticulating like a shadow, to keep warm. And that other shape that never moved. Then we rounded a curve and the light was gone.

Neanderthal Man and the
Dawn of Human Paleontology

IF THE RECORD of the rocks had never been, if the stones had remained closed, if the dead bones had never spoken, still man would have wondered. He would have wondered every time a black ape chattered from the trees as they do in the Celebes where, of old, simple forest people had called them ancestors of the tribe. He would have wondered when he saw the huge orangs pass in the forest, their bodies festooned with reddish hair like moss and on their faces the sad expression of a lost humanity. He would have seen, even in Europe, the mischievous fingers and half-human ways of performing monkeys. He would have felt, aloof in religious pride and the surety of revelation though he was, a vague feeling of unease. It is a troubling thing to be a man, with a very special and assured position in the cosmos, and still to feel those amused little eyes in the bush—eyes so maddeningly like our own.

Wherever man has been aware of monkeys, they have demanded explanation; without them we might regard ourselves as unique in the universe. With them appears the sole convincing evidence of man's relationship to the lowly world about him. The rumor of apes had passed far and wide even before the days of the great voyagers. Nor is it surprising to find, even in a Europe which, for over a thousand years after the Greek philosophers, eschewed the idea of organic change, that folklore and speculation arose around the living primates, so that parts of the Christian world evoked a second, a "black" creation to account for them.

In this view the Devil and his demon associates decided to attempt a creation in imitation of the work of God. When the diabolical rites were completed and the devil gazed upon the being which he had created it was found not to be human, but an ape. Great though Satan's powers were, he could not implant a soul in his abortive offspring. The ape remains, therefore, in the words of H. W. Janson (1952), a "kind of indistinct echo or reflection of man."

But whether we emphasize the tale of the black creation, or turn to various other Christian or pagan legends of differing character, we can observe one fact eternally present wherever apes are known: they are recognized as in some manner related to man, or as being deviant men or transformed men. *Naturae degenerantis homo, figura diaboli, homo sylvestris*—whatever they are named, whatever the legend of their relationship to man, they travel down the ages with him in a companionship which is inseparable. It is as though man had a shadowy companion, a psychological Doppelgänger who had remained in the forest and yet lingered along its edge to mock his civilized brother.

His shape is confused and often seen unclearly. Sometimes it is the ape, sometimes it is an untamed wild man with a club, a lurker in woody dells and caves. Sometimes the creature moves across the modern stage in a traditional harlequin patchwork coat of leaves and tatters. Or he is the hermit with strange wisdom in a cave. He is the Abominable Snow Man whose footprints were pictured as late as yesterday's newspaper. Western civilization has never quite forgotten this hairy and primordial shadow. Compounded, perhaps, of man's suppressed subconscious longings, he is ape and man and dancing bear. He is all that remains of the lost world of the trees; of the time before the cities.

We shall see that something of this creature emerges in the early interpretations of Neanderthal man, and we shall also observe that the skull in the little cave by the Neander served as a kind of prism or lens through which passed at mid-century, a hundred years ago, a curious conglomeration of pre-Darwinian beliefs which, transmogrified and altered, entered the world of the Darwinists. Nothing, in short, so clearly reveals the transitional nature of Darwinian thinking as the treatment accorded this single human fossil. To define and evaluate the currents of thought which swirled around that heavy-browed calvarium one must examine at some little length the beliefs about man existing in the first half of the century, and then compare those beliefs with those which obtained after the publication of the *Origin of Species* and the *Descent of Man*. Strangely enough the alterations will not be as great as we commonly assume.

Space will not permit us an exhaustive pursuit of each

thought element to sources in some cases more remote than the eighteenth century. For this reason we shall begin by an improvisation which may enable us to grasp quickly a key distinction between two approaches to the evolutionary problem. The one involves the comparative morphology of the living, the other the historical morphology of the dead. Both are based upon the comparative anatomy which was brought to such heights of perfection under the guidance of Cuvier.

The first used of these two methods, or ladders into the past, as we might call them, is the taxonomical approach to life which in its eighteenth century origins arose largely out of natural theology—the Scale of Nature concept which has been so extensively examined by Professor Lovejoy in his well-known book, *The Great Chain of Being* (1936).

Theologically, it was largely assumed in the eighteenth century that creation had been instantaneous, but that organisms were arranged in a continuous chain which anatomically led to man as the crowning glory of earthly life. Such views led to the search for connections, for the links between living forms. Taxonomy and comparative anatomy were, in other words, being unconsciously promoted by the religious philosophy of the times. Seeming gaps in the living scale of life were often assumed to be filled by creatures in unknown portions of the world. It was not yet clear, either geologically or paleontologically, that the world was old, or that many of its life-forms had vanished long ago; the doctrine of Special Creation held the field.

If one examines the Scale of Nature concept carefully, however, it can be observed that this linked chain of development implies a position for monkeys and anthropoid apes as standing next to man on the anatomical ladder. The only thing needed to transform the Scale of Nature idea into an evolutionary scheme is to introduce within it the conception of time in vast quantities, and the further notion that one form of life can give rise to another, that the links, in other words, are not fixed but can move their positions on the scale. This theory, the dawn of a truly evolutionary philosophy, began to come in toward the close of the eighteenth century, but its supporters constituted at that time a small and largely ignored minority.

As one examines the writings of the early evolutionists, in so far as they touch upon man, it is necessary to remember that human fossils of any sort were undiscovered, and that even the possibility that forms of life *could* become totally extinct was still under debate. It is not surprising under such circumstances that the early evolutionists tended to regard the orang as a possible existing human "ancestor" or genuine *Homo alalus* who needed only speech and the refining influence of civilization to become a man.

The higher primates, actually as yet little studied but reported upon by the voyagers with much anthropomorphic embellishment, come close to complete humanization, only a grade below the Hottentot. African kings were said to have guards who resembled monkeys (Ritson, 1802) and, in many of the early descriptions, it is apparent that the distinction between man and ape is extremely blurred. Amongst the Australians, writes one chronicler, "there was one man who, but for the gift of speech, might very well have passed for an orang-outang. He was remarkably hairy; his arms appeared of an uncommon length; in his gait he was not perfectly upright; and his whole manner seemed to have more of the brute, and less of the human species about him than any of his countrymen. The gift of speech, however, which he must, if at all, have acquired in his infancy, will not alone prevent his actually being what he might very well have passed for." (Cited by Ritson, p. 23.)

The writer, in other words, implies in this last remark that speech is an invention which could be acquired by an ape. Natives strikingly different in appearance from the white man were being arranged on a kind of evolutionary scale as intermediate links between the great apes and man. Thus, though between the first decade of the nineteenth century and the discovery of Neanderthal man in 1856, there had been a vast increase in our knowledge of the fossil past, nothing had been learned of man save the likelihood of an antiquity extending into the glacial period.

The comparative anatomy of the existing primates pointed to some close relationship with man, but it was still only possible for the evolutionist to compare living man with living ape and the living races. So far as man was concerned, the ladder into the past was still the taxonomic ladder all of whose rungs were still in existence. Important though this approach has been in revealing the secret of organic change, it is bound to be, without the checks provided by paleontology, in some degree mythological and figurative.

Lacking adequate human materials, the attention of the Darwinists was necessarily confined to the depths of the Bornean and other tropical rain forests, partly because of the known great apes that haunted their depths, but also partly because of the feeling that other more human "anthropoids," true "missing links" might still be lurking thereabouts.

A typical expression of this view actually precedes the publication of the *Origin of Species*. It occurs in Chambers's Journal in 1856 and reveals once again the popular interest in the orang (Anon., 1856). The anonymous author, after commenting that the animal may possibly be taught to speak, ventures a remark which reveals his concern with living missing links in the human phylogeny. "The Mias papan," he ventures, "may form only the external link of a chain, the other extremity of which lies hidden in the wild solitudes of Borneo." "We would therefore," he urges, "suggest to philosophers the desirableness of giving a new direction to their researches, and trying what may be done in the regions of the further east." Toward the close of the nineteenth century those regions would yield Pithecanthropus, but one hundred years ago it was still possible to seek for a living connection between man and the great anthropoids.

It may thus be seen that prior to the discovery of the Neanderthal cranium, and prior, also, to the publication of the *Origin of Species*, the rising interest in the geological history of the planet had done nothing but extend suspicions of man's antiquity and lead some individuals to anticipate a relationship between man and the great primates. If man was really an educated ape who had learned to speak and wear clothes, some degree of time must be assumed in the process, but this time element still did not seem to demand the assumption of a long chain of extinct ancestral hominids.

As evolutionary and geological speculation intensified past mid-century, Fühlrott and Schaaffhausen announced the discovery whose hundredth anniversary we are celebrating, namely, that of the first human cranium recognizable as lying outside the known limits of human variation. Schaaffhausen must be accredited with a complete recognition of the skull as "due to a natural conformation, hitherto not known to exist even in the most barbarous races" (Schaaffhausen, 1860). By a fortunate chance the discovery only slightly anticipated the publication of Darwin's and Wallace's evolutionary theories, and thus it drew more attention than would otherwise have been the case. Before long, argument would rage as to whether it was a genuine early link in the human phylogeny, yet at the time of Schaaffhausen's first report it is interesting to note that he merely expressed a suspicion that the skull might represent an individual member of one of those wild northern races spoken of by the Latin writers.

So far we have spoken of the hints and intimations of human primitiveness derived from the examination of living apes and their arrangement in a sequence of stages with existing man. It is not, therefore, without interest that when a cast of the Neanderthal calvarium reached London it was exhibited between the skull of a gorilla and two negro crania. "Should this Neanderthal man prove to be an intermediate species between the Papuan and the gorilla, a great point of controversy would be gained by the transmutationists . . ." commented one observer (Blake, 1864). Another anthropologist, Pruner-Bey, did not weight the negro resemblance so highly, but was fascinated by the close resemblance of the skull to that of existing Irishmen (Blake, 1865).

"Part at least of the ancient memorials of Nature were written in a living language," Sir Charles Lyell once wisely observed. That key, that Rosetta stone to which Lyell refers is, of course, comparative anatomy. Without it we would be at a loss for a clue by which to understand the life of the past, or its relationship to ourselves. It is for this reason that Neanderthal man and his interpretation bulk so large in the mid-nineteenth century. He was, to pursue Sir Charles's simile, the first archaic human syllable caught echoing from the remoter past of man, yet still recognizably human. It was inevitable that his discovery should raise a storm of contention. Crania of similar rugosity were thought to be observed among the living peasantry of Europe, preferably among nationalities other than one's own.

There was ample justification for Virchow's remark that

after the Neanderthal discovery, skulls from peat bogs, from time levels known today to be Neolithic, began to be regarded as primitive. "They smelt out," he wryly remarks, "the scent of the ape." And, further, if it was pointed out that these crania from the old pile villages were capacious and modern, some of the more rabid evolutionists contended that these people "had more interstitial tissue than is now usual, and that in spite of the size of the brain, their nerve substance may have remained at a lower stage of development" (Virchow, 1878). It was, to put the matter bluntly, a way of arguing for evolution by saying that even when ancient skulls had the appearance of *Homo sapiens* they had, to all intents and purposes, been stuffed with connective tissue, a polite euphemism for sawdust. Those who would castigate Virchow today for his opposition to the acceptance of the Neanderthal skull as a valid human fossil must remember that such excesses as represented by the above statements did nothing to increase the tolerance of the old master scientist.

Turning from the Continent to the British Isles, we may ask what views were entertained by Darwin and Huxley. Darwin, always cautious, did not express himself early. Huxley, writing to Lyell in 1862, commented that "the Neanderthal skull may be described as a slightly exaggerated modification of one of the two types (and the lower) of Australian skulls" (Huxley, 1913). Darwin, laboring upon the *Descent of Man* (1871), and drawing from Carter Blake before him, spoke of the enormous projecting canines of the La Naulette specimen. Nevertheless, he is on the whole wary, particularly because he was impressed with the capacity of the Neanderthal cranium. The Darwinians who had eyed microcephalic idiots as possible reversions to primitive man were actually not very happy or enthusiastic about the first Neanderthal discovery.

The massive supraorbital development attracted them, but the face was missing, a factor which made it impossible to be quite certain that the Neander skull was not within the variant range of the modern races. Some, like Carl Vogt (1864, p. 304), labelled the Neanderthal cranium as that of an idiot, but one has to realize that in the eyes of certain of these workers, and particularly Vogt, this does not destroy the value of the specimen as representing an early human stage of development. Instead, this theory rather cleverly evades the issue of time. It is equivalent to saying: if the specimen is old it is "normal" for its period. If it should prove to be a comparatively recent burial it is "abnormal," but a remarkable case of reversion, nevertheless, to a primitive ancestor (Eiseley, 1954).

The western world throughout the last few centuries has tended to see itself either in terms of steps going up, or steps coming down. Progress or degeneration has been a debate often influenced by the individual temperament of philosophers and perennially renewed under the impact of new systems of thought. The role played by the discovery of Neanderthal man cannot be entirely grasped without some reference to this aspect of the western mind. Under the influence of Archbishop Whately, a strong belief in the conception that modern primitives and even archeological remains of past primitives represented a state of degradation from more civilized conditions was variously expressed by numerous writers at the mid-century and beyond.

"We have found nothing yet," comments H. B. Tristram, "to prove that the barbarous dwellers on the kitchen middens were not the wandering outcasts from the pre-existent civilization of the valleys of the Euphrates or the Nile, nor is there any chronological argument against it. Nor have we yet seen the traces of the barbaric epoch underlying the vestiges of the earliest civilization in its sites. Nor in the face of the relics of the Mississippi valley, of Central America, or of Mesopotamia, can we admit that there is no evidence before us of man relapsing from civilization." (Tristram, 1866.)

It would be impossible in a short space of time to explore fully the degeneration theory which, if pursued, could be shown to be related to the "Decay of Nature" concept of the Elizabethan era. As an argument to explain certain archeological facts, however, it reemerges a little prior to the publication of the *Origin of Species* and at a time when the bone caves of Europe are just beginning to be probed. The revelations of the caverns were beginning to trouble religious orthodoxy. There can be no doubt that the emergence of the "degeneration" idea was the direct attempt of the spokesman for the traditional viewpoint to explain away the crude artifacts which, when found in Europe, hinted that remote and bestial beginnings might lie at the root of western culture.

Though the argument is no longer attractive, it was a persuasive alternative to evolutionism in the mid-century decades. There were serious debates, and some of the leading scholars of the period—men like the Duke of Argyll and Sir John Lubbock—arrayed themselves on opposite sides of the question. Archeology could supply the material for debate, but the truth was that its materials lent themselves readily to either the degeneration or the development hypothesis. Rudolf Schmid (1883), a contemporary student of the period, confessed frankly that: "Archeology, as a whole, seems to do no more than admit that its results can be incorporated into the theory of an origin of the human race through gradual development, *if this theory can be shown to be correct in some other way . . .*" Schmid, in other words, recognized that something other than the approach through artifacts was needed to establish the reality of human evolution. That something was provided by the discovery of the Neanderthal skull. The skull and the La Naulette mandible of 1866 focussed attention on the fact that the bone caves might contain more than simple artifacts and evidence of extended human antiquity; they might provide conclusive proof as well that the human body itself had undergone the transfiguring touch of time.

It is true that the Darwinian passion for present-day atavisms obscured, for a time, this revelation, and that the skulls of idiots occupied on the demonstrator's platform a place between the skull of the orang and the Hottentot, but as the evidence from the caves slowly accumulated, this fashion passed. The Neanderthal discovery, coming as it did, early enough so that translations and accounts appeared in England almost simultaneously with the opening of the evolutionary debate, ensured that the skull would not be ignored nor forgotten.

For a striking new fact to receive attention, it must fit into a theory, and that theory was immediately at hand. The earlier Neanderthal discovery at Gibraltar had had no such good fortune. Schaaffhausen's remarks about the wild tribes mentioned by the Romans would be quickly forgotten by their author, nor would it matter that the Neanderthal specimen would be challenged from every conceivable point of view. The real point lay in the world-wide, if hesitant reception of the fossil, in the interest it had stirred. Out of the bones would grow a new and specialized science, and as a consequence there would occur a slow fading of the degeneration hypothesis. The discovery would lead on, also, to the truly historical morphology which Darwin and his colleagues had not yet succeeded in extricating from the comparative morphology of the living which continued in some degree to dominate their epoch and their thinking. Nor did it dominate biologists alone. As late as 1905, a prominent American anthropologist, W. J. McGee, who had charge of the anthropological exhibits at the Louisiana Purchase Exposition in St. Louis, spoke as follows: "the next physical type chosen was the Ainu . . . their small stature . . . their use of the feet as manual adjuncts, their elongated arms and incurved hands, and their facility in climbing, approximate them to the quadrumanes and betoken a tree-climbing ancestry." (McGee, 1905.)

Alfred Russel Wallace would, in his emphasis upon the shift of evolution to the mind, with its enormous latent possibilities, achieve, in the end, a partially effective compromise between the degenerationists and the evolutionists, for Wallace had a foot in both camps. He would explain the apparent slowness of human evolution as then understood, and, at the same time, in his meditations over the mysterious powers of the human mind, he would go far to abandon the linear biology which placed living races in a series of subordinate "fossil" levels, and which revolted many Christian thinkers.

If we now attempt to summarize briefly the separate elements of human thinking which were in some degree altered by the discovery at the Neanderthal cave, we may list them as follows:

First: The discovery aroused and maintained interest for a number of reasons. Anatomically, the massive character of the orbital ridges and the unusual conformation of the vault led to suspicions that its type was extinct in Europe. Also, and most importantly from the speculative standpoint, the skull came to serious attention almost simultaneously with the developing Darwinian controversy. Seized upon by some of the more avid evolutionists as a proof of man's descent from the great anthropoids, it was interpreted in terms of the taxonomical ladder which had been utilized by the transformationists of the first part of the century. Its capacious brain-case did not prevent it from being labelled as a brute, and its characters were so transformed that, without the slightest basis in fact, it was described as possessing huge projecting canines and "an appearance in the highest degree hideous and ferocious." This ferocity, of course, represents a distorted Victorian emphasis upon the struggle for existence in its more lurid forms.

Second: As the number of big-brained primitives in the upper Pleistocene became slowly established, the Darwinists were forced to reassess the time involved in the human transformation, and to extend it.

Third: The most astute researchers were led to consider the possibility that the earth, rather than the jungle or the insane hospital, might contain the clue to man's evolutionary past.

Fourth: The degeneration theory so successfully advocated by Archbishop Whately and his followers, as long as they had only the archeologist to face, began to lose its efficacy once it could be shown that the caves of Europe contained actual organic traces of men more primitive in physical type than men of the present day.

Finally, if we were to ask what lesson this discovery has to teach us today—one hundred years after the quarrymen sank their shovels in that little enclosed grotto in the gorge of the Neander—we might venture to say this: that man, irrespective of whether he is a theologian or a scientist, has a strong tendency to see what he hopes to see. The nineteenth century evolutionists—those who accepted the Neanderthal specimen —saw a savage animal whose skull contours reminded them, paradoxically, of the Negro, the Mongol, the Hottentot, or the Irishman—all peoples at that time economically depressed. The scholars tended to cling to a linear phylogenetic line in which each living race represented a frozen step on the way to the emergence of the civilized Caucasian. We smile at this today, yet it may be wondered whether our present frequent tendency to arrange our varied assemblage of human fossils in a single unilinear phylogeny may not threaten to lead to a historical simplification as rigid and potentially wrong as we now know the unilinear taxonomical approach through the living races to have been. After all, we know paleontologically that families have proliferated and later contracted as a single type has achieved ecological dominance.

In conclusion, when we consider this creature of "brute be-nightedness" and "gorilloid ferocity," as most of those who peered into that dark skull vault chose to interpret what they saw there, let us remember what was finally revealed at the little French cave near La Chapelle aux Saints in 1908. Here, across millennia of time, we can observe a very moving spectacle. For these men whose brains were locked in a skull reminiscent of the ape, these men whom serious scientists had contended to possess no thoughts beyond those of the brute had laid down their dead in grief.

Massive flint-hardened hands had shaped a sepulchre and placed flat stones to guard the dead man's head. A haunch of meat had been left to aid the dead man's journey. Worked flints, a little treasure of the human dawn, had been poured lovingly into the grave. And down the untold centuries the message has come without words: "We too were human, we too suffered, we too believed that the grave is not the end. We too, whose faces affright you now, knew human agony and human love."

It is important to consider that across fifty thousand years nothing has changed or altered in that act. It is the human gesture by which we know a man though he looks out upon us under a brow suggestive of the ape. If, in another fifty thousand years, man can still weep, we will know humanity is safe. This is all we need to ask about the onrush of the scientific age.

LIST OF LITERATURE

Anonymous. 1856. "The Wild Man of the Woods," *Chambers' J.*, ser. 3, 6: 129–131.

Blake, C. C. 1864. "On the alleged peculiar characters and assumed antiquity of the human cranium from Neanderthal," *Anthrop. Rev.*, 2: cxli.

———. 1865. "On certain simious skulls," *Rep. Brit. Ass.*, 114.

Darwin, Charles. 1871. *The Descent of Man and Selection in Relation to Sex.* John Murray, London.

Eiseley, L. C. 1954. "The reception of the first missing links," *Proc. Amer. phil. Soc.*, 98: 453–465.

Huxley, Leonard. 1913. *Life and Letters of Thomas Huxley*, 3 vol. (Vol. I, p. 286.) Macmillan, London.

Janson, H. W. 1952. *Apes and Ape Lore in the Middle Ages and the Renaissance*. Warburg Institute, Univ. London, London.

Lovejoy, Arthur O. 1936. *The Great Chain of Being*. Harvard Univ. Press, Cambridge.

McGee, W. J. 1905. "Anthropology at the Louisiana Purchase Exposition," *Science*, n. s., 22: 811–826.

Ritson, Joseph. 1802. *An Essay on Abstinence From Animal Food*. R. Phillips, London.

Schaaffhausen, D. 1861. "On the crania of the most ancient races of man," *Nat. Hist. Rev.*, 1: 155–172.

Schmid, Rudolf. 1883. *The Theories of Darwin and Their Relation to Philosophy, Religion and Morality* (pp. 90–91). Janson, McClurg & Co., Chicago.

Tristram, H. B. 1866. "Recent geographical and historical progress in zoology," *Contemp. Rev.*, 2: 124.

Virchow, Rudolph. 1878. *Freedom of Science In The Modern State*, (pp. 58–61). London.

Vogt, Carl. 1864. *Lectures on Man*. London.

Whately, Richard. 1861. *Miscellaneous Lectures and Reviews*. London.

The Quarterly Review of Biology, December 1957

FROM *The Lost Notebooks*

"THERE is a gate," he said, "under the sand there in the western desert." We were speaking around that table of a buried Egyptian city. "The temple mound lies here"—he gestured at the maps that lay before us. "Petrie sunk a shaft there in the nineties but found nothing. I cannot recommend it because of the cost, but the gate"—a flush suffused his face, he looked at us importunately—"the gate may lead somewhere."

"It is a horizontal stripping job that would cost a fortune," someone said practically. "You would have a wall and a gate. Which way would you go?"

"Toward the city and away from the temple," he said, his accent growing greater in his excitement.

"Why?" someone asked gently. "Isn't it true that this would lead toward the city and that to follow such a roadway might basically demand the uncovering of a whole city, something we are not equipped to do financially."

"It goes somewhere," stubbornly reiterated the aging archaeologist. "I think three seasons more—." I thought of the way the wind comes in across that desert, and the flies. The man was old and there was a family to consider.

"In three seasons," said the director gently, "you might be following a road. Then you would have to ask for incalculable sums to follow it further. It is impossible."

"There is a gate and a road," said the old Egyptologist, speaking painfully in broken English. "It should be followed." He made unconscious digging motions with his hands.

A slow doubt went from eye to eye around the committee at the director's table. The man grew conscious of it, arose, and excused himself.

"He won't retire," someone said affectionately. "We will have to make the decision for him." Another sighed. "If we send him back he will never see the end of his road."

"It is the road we are all digging in," someone more perceptive ventured. "Find him another part of it here at home.

Explain to him the road runs into a multiplicity of roads. None of us here will see the end of them."

* * *

THE blizzards by the Pole are not a likely place to search for the secrets of life. Only this, I think, can explain the amazing indifference which the public of two generations has shown toward the contents of one of the sledges drawn painfully southward over the Antarctic ice barrier by a trio of starving men. Captain Scott and that immortal company who perished with him are well known to science. Nevertheless, in the chronicles of their passing, in the records of those who found them huddled in their last bivouac, a line at best is devoted to that single sledge. They dragged it with their dying breath, yet its contents, by the curious whims that sometimes afflict the final reports of ill-fated expeditions, have been little mentioned since. Its significance is therefore uncertain, but, this we know, it was not gold. It was a sledload of fossils. There are men, I think, who would have understood this effort had it been made in other continents. But the Antarctic waste is a dead world and the drama of man's attempts to enter it is so spectacular. . . .

* * *

I WOULD never again make a profession of time. My walls are lined with books expounding its mysteries. My hands have been split and raw with grubbing into its waste bins and hidden crevices. I have stared so much at death that I can recognize the lingering personalities in the faces of skulls and feel accompanying affinities and revulsions. I am the last in an Ice Age leaf fall.

* * *

THE early brain, shielded with a warrior's helmet of bone as though nature itself was dubious of the survival of this strange instrument and yet had taken steps to protect it, is, in itself, a minor mystery. If we go back to the beginnings of life, not man, he has taken three billion years to produce and perhaps in a few more years thirty minutes to destroy. I like to think that with the invention of a brain capable of symbolic thought and, perhaps, as an unsought corollary, philosophy, nature rejoiced to look out upon itself. But man has dragged out of the Ice Age with him an old and lower brain that reemerges in the mists of alcohol or which I have seen snarling on the bed of madness. There are claws in it by now fantastically extended.

* * *

Few laymen realize that every bone that one holds in one's hands is a fallen kingdom, a veritable ruined world, a totally unique object which will never return through time.

What of it, a chain is displaced.

* * *

Just now in the midst of a cold, driving rain and the turbulent winter boiling of clouds over his favorite perch on a wind-whipped sapling, I found him singing again. The sound came up into my study and from the study into my heart, the melody hanging there like something from the timeless world of Plato's forms, so that I was distracted from my tasks and began to think of springs long ago and a girl I remembered.

I even put up the window in desperation and laid out some seeds, though I knew it was useless, since that outer world was his, and he eyed me, perhaps with justice, as being tinged with evil, otherwise why would I be seeking to purchase favors with good seed?

"But you have only two or three years more to learn the meaning of this," I said to him darkly. "I am forty-five and determined to know. With that song and what I have found in books maybe the secret could be learned by us together. There is not much time, you know, and less for you than me; it is important to know why one is singing."

He did not fly up to the window. He sang on, those beautiful notes falling carelessly as raindrops across the house roofs, falling into my heart, falling in the lanes of memory, bothering my pen in its slow way across the paper. I tore up the sheet, but I began again. I will start at the very beginning, I told him severely, though he was fainter now and had, as I judged, perched farther away. I will begin at the beginning and find the way of it— to that song, I mean, and why I, who am not a bird, can be so troubled by it. I will begin at the very beginning.

The pure voice did not mock me, it sang on as I wrote, but it does not help me. Out of the eternal garden of the forms it does not know me, but as I listen I know mankind should find its way thither; music reaches down to us, that at least I know.

* * *

NEARLY every week when I pick up the literary reviews I find that a new book on space travel has appeared. Even the navigational problems are being considered in serious and erudite tones. Needless to say, I am glad to see this. After all, one doesn't want to leave one's own little dust-fleck universe without some idea of where one is heading and how one is to get back. Space is lonely and vast beyond our conception, but we pretty well understand that now. It is not this that troubles me as I read the increasing literature—pulp or otherwise—upon space travel; it is, instead, the appalling poverty of imagination manifested in our descriptions of the world outside—whether within our universe or far beyond in other galaxies and remote from us by light-years.

The lack of grasp of biological principles is exhibited by both readers and writers in this field. I have read about cabbage men and bird men; I have read the loves of the lizard men and the tree men, but in every case I have labored under no illusion. I have been reading about a man, *Homo sapiens*, that common earthling clapped into an ill-fitting coat of feathers and retaining all his basic human attributes, including, generally, an eye for the pretty girl who has just emerged from the spaceship. His lechery and miscegenating proclivities have an oddly human ring, and if this is all we are going to find on other planets, I for one am going to be content to stay at home. There is quite enough of that sort of thing down here, without encouraging it throughout the starry systems.

* * *

May 21, 1953

EARLY in this month Mabel and I took an evening stroll around the Wynnewood Park Apts. and in three separate instances observed along the walks in the twilight, dark collections of a small ant so thickly clustered as to appear like freshly excavated dirt—Mabel, in fact, first assuming that it was a newly opened nest. The ants seemed quiescent, certainly not moving rapidly, and thus enhancing their similarity, in the half darkness, to fresh, dark earth. There were no signs of winged forms, only these curious masses as though some kind of meeting were going on. There appeared to be no substance along the walk which was attracting them. I assume, because of the three instances observed, that the event was not unusual but was somehow an appropriate action of this form at this season. Doubtless if it had been feasible to observe, other gatherings would have been seen in the surrounding fields. What they were occupied in doing I have not the slightest idea. Does some physiological impulse hit them all at the same time? Certainly the colonies we observed were too remote from each other to be mutually influenced by contact.

A few days ago I observed, after a night's rain, cicadas (the seventeen-year form?) emerging from their larval cases around the tree near the Wynnewood station. The grubs coming to the surface were creeping toward the nearest tree. Those I observed were creeping in the proper direction to reach it. Was this chance, or do they have a way of sensing or seeing the tree? The grubs were a yellowish earth-stained brown, the emerging adults white with red eyes.

I was greatly impressed with the number who failed to achieve their passage into the world of light. Some failed to wriggle free of their old garb and died only partially out of the grub form. Some worked clear, but their wings failed to develop properly; others crept painfully about with stiff legs or other defects which seemed largely due to hormonal failures or low viability. The wastage is tremendous, even leaving aside the birds that attack the emergents in their helpless stages. Nevertheless, the wood, next day, was full of the successful. They appeared, in contrast to the middlewestern two-year variety, very sluggish and unwilling fliers. They were easy to pick up and seemed to prefer to cling rather than fly. The color is dark, the red eyes remaining.

A high wind on the day of emergence caused much damage to wings still soft and forming, as well as hurling many to the ground. This event certainly added to the casualties. I noticed an attacking sparrow striking the wings from one individual, and I found other wings in pairs with the body gone, mute testimony to the attack of birds. Later I saw an ant carrying away the wings.

This great loss is reminiscent of Richard Jefferies' account of young mice of the last brood freezing in the English winter. I tried to help a few—such is one's eagerness to further life against death.

* * *

Sunday, June 7, 1953

WALKED along the edge of the daisy field. Interesting to see nature going on here at the verge of the Penn Fruit Company. Abandoned cups, match cases, whiskey bottles slowly being covered and made "natural" by red clover and daisies. How fast life flows from one form to another. *The illusion of things.* To the generations of cabbage butterflies I must be as everlasting as the Sphinx. I turned round sharply to make sure the Wynnewood shopping center had not disappeared while I gazed. I half expected to see a little plaster and a few bricks disappearing into the insatiable maw of the clover. Matter disappears or modifies its appearances so fast that everything takes on an aspect of illusion—a momentary fizzing and boiling with smoke rings, like pouring dissident chemicals into a retort. To a time-foreshortened eye all form would become bubblings or momentary imaginative shapes such as one sees when dreaming before a winter fire. One thinks one sees a shape and it is gone. Perhaps it is so with God half-asleep over the dying embers of his universe, and dreaming the shapes that come and go through the coals.

Got to thinking that Ernest Borek's *Man the Chemical Machine* is a further extension of the Cartesian world view. First, man is a machine of pulleys and levers, then when this fails to provide the complete answer he becomes the *chemical* machine; what, after this—the crystal-lattice machine, the electron machine? We must be near to reaching the end of this road.

* * *

Sunday, February 14, 1954

A LONG abandonment of this record, but I shall now try to get back into the spirit of it. The thought struck me yesterday that one queer thing about mammals is that they have never developed footless land forms. Weasels and certain of their relatives are sinuous and long-bodied but none of them have actually paralleled the snake in adaptation. Perhaps the latter requires belly plates to be successful, since whales have certainly lost the hind limbs. It is interesting to speculate on the amount of correlation involved in (1) loss of legs, (2) development of locomotion by plates which requires muscular adjustments, etc.

A good essay could also be written on animals living in holes, etc., and the dangers of going out into the world.

* * *

August 12, 1955

MANY months since my last entry. I am a poor diary keeper—no doubt of that! This morning I saw a very remarkable sight. It was a rainy, windy day and the huge water lily leaves in the museum pond were all projecting above the surface and blowing in the wind. There are possibly two species represented—one lying flat to the water—the other standing a little above as in this drawing. The lily leaves have

an adaptation (probably of fine waxy hairs) which causes waterdrops to roll over them without penetrating or soaking the leaf. As the cuplike big leaves swayed in the wind, sizable drops of water rolled around and around in them like drops of quicksilver, flashing and quivering in the grey light. There were many leaves and much water and the whole pond seemed to quiver and flash as the drops rolled in the unsoakable leaves. Darwin was right that this phenomenon deserves investigation.

* * *

September 1, 1955

(a)   It occurs to me that there is a very clear analogy between the way in which an apartment house (or another building, for that matter) acquires its biota, and the way an oceanic island acquires its plant and animal population. An apartment house newly built (a recent volcanic island upthrust from the waves) is destitute at first of a fauna. If it is remote from neighborhoods where such a fauna may be acquired (islands far at sea), it may be destitute of insects, silverfish, etc., for a longer period. As time runs on, however, the chance of immigrants arriving intensifies. A pair of roaches may arrive in a box (floating timber) and escape into the basement, and soon the house is populated so extensively that even the professional exterminators can only keep the population reduced. The closer the apartment house may lie to other older ones or to neighborhood groceries, as in, say, New York, will play a part in the time involved before population is acquired. Now, to give this a figurative evolutionary twist, we might imagine each house more self-contained than it actually is and lasting over more than one geological period. Let us conceive that this genetic isolation is further played upon by different methods of extermination used by the human inhabitants. Let us say that in one house roaches have grown adjusted to a given poison, in another they have developed clever adaptations for evading the traps set for them by people, or perhaps other insects have been introduced to combat them. Say, spiders. All of this would mean a series of self-contained apartment (island) worlds in which intense but quite different forms of selection were going on. In the end, after the passage of some millions of years, the once-similar biotas might be quite different. The heating system (climate) in the different apartments might also be hot, cool, humid, dry, etc. Eggs, larvae, etc., will thus be struggling against quite different natural environments.

(b) I could not help being amused today. An exterminator came and let loose a gas that gave me unpleasant physical symptoms. After it was all over, I went into the bathroom and found a vigorous, fast-moving young roach perched on my toothbrush. I was showing more effects from the "exterminator" than he, and he easily evaded my lunge at him.

(c) Engels, I believe, says somewhere that Darwin failed to distinguish between struggle to survive in a single environment and the adaptation leading into new environments. Critics of Malthus have pointed out that he created a false situation in that he claimed population was increasing faster than the food supply, when what was really happening was that population was increasing faster than a given type of economic system could make use of people. In other words, there was no shortage of food in a natural sense, there was only an ecological failure. Perhaps an analogy to human society exists here in the animal world. Cities allow more niches in which diverse talents can be manifest. Similarly, a biota which has arranged itself in ecological communities has reduced the struggle for existence so that more types can take advantage of a given region (law of divergence). Thus diverse adaptive mutation is important, more important for life in general than straight-line evolution. Man's mental variability, which is not wholly cultural, has, in his cities, replaced physical evolution.

***

September 20, 1955

Saw a pigeon on the campus today with feet carrying little feathered ruffles, or teddy bear shoes. Obviously wild but sporting some remnant of high breeding in a fancier's cote— a fallen aristocrat.

Also one with an injured leg who used his wings to move when the other walked. He used to rest in the sun with his wings spread out, lying on his side in an ungainly fashion but looking nevertheless well fed and managing to cope with his infirmity. Perhaps in winter he would suffer more. The college campus was a better refuge for him than the city streets would have been and hurt his remaining foot less. Where he roosted and how, I had, of course, no idea. Sooner or later a cat. . . .

There is something intensely pathetic about harmless animals dying alone. I shall never forget coming across the University of Kansas campus one summer evening and finding a dying turtle dove on the walk. It had not been visibly injured, but had been perhaps poisoned. Its mate, in very obvious emotional distress, walked nervously about it and paid absolutely no attention to me as I squatted down to see if I could help. I lifted up the dying bird, whose eyes were already glazing, and placed it on the flat roof of a nearby shed, where it might at least be safe from dogs or cats. All the time, the female bird stood fearlessly at my feet and fluttered up to the shed roof to be beside her mate. It was hopeless. In the last light of evening I hurried away asking myself questions I did not want to ask. Was there a nest? Were there young? That affectionate life-indifferent bird by nightfall would be alone in the vast

universe feared by us all. I did not go that way again in the
morning. I knew too well what I would find.

***

January 28, 1956

M<small>AN</small> is always marveling at what he has blown apart, never at
what the universe has put together, and this is his limitation.
He still has something of the destructive primate mind within
him. He is at heart a "fragmenting" creature. He resents even
Bergson's indeterminism and would see evolution as the response of the dead to the dead—not indeterminate at all. Man,
the creation of the universe, after all created the bomb.

***

April 14, 1956

T<small>HERE</small> must be dogs barking at the bottom of chaos—great,
hoarse hounds whose voices bounce eternally against falling rock
and echo and reecho in the crevices of eternity. Only a dog's voice
out of the deep abysses carries the proper menace and at the same
time preserves the weird objectivity and indifference which is
part of the hunting pack. A lion's voice is great but personal, a
dog's bark by contrast contains the maniacal essence of chaos,
dumb matter come alive in the dark and howling its voice endlessly and stupidly against the sleeping quiet of nonexistence—but
I overelaborate—perhaps you have not heard as I have hounds
beneath you as you cling desperately to a cliff wall.

When I get to the bottom they turn warm and wagging and
friendly—again with the total irrationality that obtains over
the great cliff of chaos. Did they take me finally, because of my
successful descent, as a demon like themselves—for if I had
fallen, they had given every indication of devouring me—or
are the dogs of Cerberus, the hoarse-voiced, much-feared
guardians of Nothing, actually abysmally friendly and lonely
creatures? Since that long, agonizing descent before I reached
the city on the plain, I have never been quite sure. When I
come to the Final Pit in which they howl, I shall, without too
great a show of confidence, put out my hand once more and
speak. Perhaps the great hounds of fear may wait with wagging
tails for a voice who knows them. It may as well be mine. For
who is to know one demon from another in the dark. . . .

***

December 22, 1956

READING in Thoreau's journals today. It strikes me that Thoreau's writing is like his own landscape—a vast expanse of weeds, brush, thickets, and just occasionally a singing bird with a soft note hidden in some unexceptionable underbrush. Thoreau, in other words, is as chaotic as the real world of nature and just as full of trivia, with here and there some remarkable observational nugget.

***

"DEVIATION" IN PHYSICAL ANTHROPOLOGY

IN Russia, when the state for political reasons desires to destroy the public reputation of a writer, scientist, or official, one of the several epithets which may be used to convey the man's deviation from the party line is the word "mystic." It was used, for example, in connection with the fall from grace of the Polish leader Gomulka before his recent rehabilitation under Khrushchev. As an occasional student of the history of thought, I cannot resist the observation that this name-calling device occasionally emerges here in some few scientific quarters where there is an unconscious attachment to an extremely materialistic world view similar to that which broods with such intensity over the Russian landscape. One may write, for example, a nature essay in the purely literary tradition, expressing some feeling for the marvelous, or the wonder of life—things perfectly acceptable when perused in such old classics as Thoreau or Hudson, and then awake to discover that a certain element in the "union" regards one's activities in this totally separate field as "mystical" and "alien to the spirit of science." It was not so among the great Renaissance thinkers, but the growing compartmentalization of thought has contributed to a trade feeling that the shoemaker should stick to his last. The feeling is more evident in some sciences than in others; in fact, some of the older sciences, whose members, perhaps, are more secure and with a longer tradition behind them, are manifestly less nervous in this regard than younger ones.

***

July 23, 1957

A GOOD idea for an essay: the trigger, simplifying mechanisms of life—all the way up to man, who, culturally, however, has a far more varied approach to reality. The dream animal again. And is it not true that life, as manifested through its instincts, demands a security guarantee from nature that is largely forthcoming? The inorganic world could and does really exist in a kind of chaos, but before life can peep forth, even as a flower, or a stick insect, or a beetle, it has to have some kind of unofficial assurance of nature's stability, just as we can read that stability in the ripple marks impressed in stone, or the rain marks on a long-vanished beach, or in the eye of a vanished hundred-million-year-old trilobite. The nineteenth century was amazed when it discovered these things, but wasps and migratory birds were not. They had an old contract, an old promise, never broken till man began to interfere with things, that nature, in degree, is steadfast and continuous. Her laws do not change, nor moons shift, nor the seasons come and go too violently. There is change, but throughout the past, life alters with the slow pace of inorganic change. Calcium, iron, phosphorus, I suppose, could exist in the jumbled world of the inorganic without the certainties we know. Taken up into a living system, however, *being* that system, they must, in a sense, have knowledge of the future; tomorrow's rain may be important, tomorrow's wind, tomorrow's sun—their own or another's. Life in contrast to the inorganic is historic in a new way. It reflects the past but must also expect something of the future. It has nature's promise—a guarantee that has not been broken in four billion years that the universe has a queer kind of rationality and expectedness about it. Lord Dunsany says, "If we change too much we may no longer fit into the scheme of things; but the glow-worm shows no signs of making any change." (*Patches of Sunlight*, p. 25)

***

October 12, 1957

I AM not nearly so interested in what monkey man was derived from as I am in what kind of monkey he is to become.

***

November 3, 1957

A DAY or two before coming to Lawrence, troubled by the onerous burdens laid upon me by your learned committee, I took the trouble to walk through the Hall of Man in one of the large eastern museums. A person of great learning had been instrumental in erecting those exhibits, and I hoped to find there some clue as to human destiny, some key that might unlock in a few succinct sentences the nature of man. The exhibit ended in a question mark before an atomic machine and a graph showing the almost incredible energy that now lay open to the hand of man. Needless to say, I agreed with the question mark which ended the history of humanity within that hall. But as I turned and went in the other direction, step by step, eon by eon, back into the past, I came to a scarcely human thing crouched over a little fire of sticks and peering up at me under shaggy brows. The caption read: "Man begins his technological climb up the energy ladder. He discovers fire." I walked a short ways backward and forward. I read the captions. I looked again at the creatures huddled over a fire of sticks—at the woman clutching a child to her breast. Again I searched the hall. Here was the sum total of all that history and anthropology had seen fit to emphasize as important in the human story. The hunter's tools were there, the economic revolution effected by agriculture was ably presented. Summarized before my eyes populations grew, cities and empires rose and fell, and still man's energy accumulated. One saw another thing. One saw the armored legions grow and grow until at last continent confronted continent and the powers of death to a world lay in the hands of the descendants of that naked woman and her consort by the fire of sticks.

But as I stood and gazed at that learned presentation of the past, a disturbing doubt crossed my mind. Was the story true? Of course it is true, my reason told me. It is the verdict of science. It is a story that you yourself have helped, in a small way, to make; there is the arrow buried in the bone, there is the fire, there is the trail of man across five continents. There is his first timid reaching out into the seas, and now let us get on to the skies. Forget the cities in the sand. Mars will be next.

I hesitated again before those engines of the past and still my doubts continued—for it seemed to me that there was lacking here some clues, some vital essence of the creature man—that I was looking upon stone and potsherd, sword and catapult, from someplace just a little remote and distorted. "This is the story of man," the caption ran through my head, and at that moment, finally, I knew I was looking at the past through the eyes of a modern twentieth-century American, or, for that matter, a Russian. There was no basic difference. In that whole exhibit was ranged the energies of wheat and fire and oil, but of what man had dreamed of his relations with men there was no trace. Yet it is only on paper, or in human heads, we might say in paraphrase of Shaw, that man has sought successfully to transcend himself, his appetites and his desires. In that great room was no slightest hint of the most remarkable story of all—the rise of a value-creating animal and the way in which his values had been modified and transformed to bring him to the world he faces today. It is of this I would speak this evening.

***

October 11, 1958

A FEW days ago, passing through one of the less frequented corridors of Pennsylvania Station, I heard the wonderful loud trilling of a single cricket apparently all alone in this great rumbling place. Was he calling for a female in his loneliness or merely singing an autumn song to himself? Why cannot any man do this in his machine-raucous society—trill out anyhow, sing to himself if to no one else, purr in the autumn sun? If he cannot love one thing, then love all, leaf, brick, and autumn spider.

***

THE BLUE UNIVERSE

I FOUND him making little marks and calculations on an old envelope beside his plate in the Faculty Club. "What is it?" I asked curiously, for though I am one of those professors who is commonly overawed by deans and presidents, their computations and thoughts never cease to interest me. At a faculty table one can never tell what will turn up in the way of likely information. Ours always reminds me of one of those foreign restaurants in the Swiss Alps where the agents of great powers meet and each looks out of a separate window at some mountain or other and speaks abstractedly into the air about the politics of mean little states to whom someone has loaned a tank or two. "It's finished," he said. "The new laboratory," I said. I was about to ask about the budget, in which I had a peculiar interest, but I thought better of it. "No, no," he said impatiently, "the universe."

I maintained the aplomb upon which we pride ourselves in that particular club, even waited a moment before answering. "How long have we got?" I asked quietly. He looked at me across the table and made another scratch on the pad. "Fifty billion years," he said. "In that case," I said, and hesitated, for I was afraid he would know what I was thinking, "why bother with commencement, why bother with another of those dreadful little speeches, 'By virtue of the authority vested in me . . . ,' why bother with another student?"

I made a slight movement to rise. "In that case," I said, "you will excuse my failure to appear this afternoon. I feel one should make some preparation, take stock of things. You know how it is." He nodded, vaguely immersed again with the accuracy of his figures. I left him there calculating by the cosine of Cygnus or whatever it is these cosmologists swear by. It wasn't through mathematics that I saw the doom was close in. When a university official of such eminence as the man opposite me begins to neglect commencement, the case is probably worse than his public utterance. I wasn't going to be fooled by this soothing pronouncement of fifty billion years. No, anytime the man worked like that you could be sure of it—the doom was close in.

I went all the way home on the local and got safely into my study. I had something there that I wanted beside me in my final hour—something I had faith in. This will take a moment of explaining on my part because, you see, I had a universe in the study—a universe replete with suns and passing shadows and remote blue distances. All mine, the most beautiful thing I have ever owned—believe me, I know at last how a god outside creation might stare into his creation without any power except to smash it, and that he could never do. He could stare forever into the smoky dark at the hot blue flame of suns or roll the great ball of space around in his fingers, but never, never reach in to alter its fate a hair's breadth. That was the way with the universe on my desk. It was always different, something was always changing color—and it was about the size, I would estimate, of the universe that collects in a human head when you train a telescope on the Milky Way. How many light-year distances can be hung there in the grey folds of the brain? How many pinpoint stars around which circle the black planets we will never see? And which is real, the macrocosm outside or the microcosm inside? I did not have the answer, but I had a universe there on the desk, a blue universe that someone had brought home from an obscure junk shop years before and presented to me. A solid glass ball paperweight into which some forgotten craftsman had poured the stars and night of a century before I was born.

***

EVOLUTION CAN BE SEEN

When I climb I almost always carry seeds with me in my pocket. Often I like to carry sunflower seeds, or an acorn, or any queer "sticktight" that has a way of gripping fur or boot tops as if it had a deliberate eye on the Himalayas and meant to use the intelligence of others to arrive at them. More than one lost mountaineer lying dead at the bottom of a crevasse has proved that his sole achievement in life was to inch some plant a half-mile further toward the moon. His body may have been scarcely cold before that illicit transported seed had been getting a foothold beneath him on a patch of stony ground or writhing its way into a firm engagement with the elements on the moisture of his life's blood. I have carried such seeds up the sheer walls of mesas and I have never had illusions that I was any different to them than a grizzly's back or a puma's paw.

They had no interest in any of us—bear, panther, or man—but they had a preternatural knowledge that at some point we would lie down and there they would start to grow. All the same, I have carried them on their journeys and even deliberately aided their machinations, though not quite in the way they intended. I have dropped sunflower seed on stony mesa tops and planted cactus in Alpine meadows amidst bluebells and the sound of water. I have sowed northern seeds south and southern seeds north and crammed acorns into the most unlikely places. You can call it a hobby if you like. In a small way I am a world changer and leaving my mark on the future. Most of it will come to nothing, but some of it may not. Anyhow, I like to see life spreading and not receding. Blake once said that you could not pluck a flower without troubling a star. It can similarly be observed that by planting a seed in a new place you may be running a long shadow out into the future and tampering with the world's axis—it may even have happened the night my pet blacksnake took a notion to live in the neighbor's basement. Life is never fixed and stable. It is always mercurial, rolling and splitting, disappearing and reemerging in a quite bewildering fashion. It is constantly changing, and now it has affected me to the extent that I never make a journey to a wood or a mountain without the temptation to explode a puffball in a new clearing or encourage some sleepy monster that is just cracking out of the earth mold. It is, of course, an irresponsible attitude, since I cannot tell what will come of it. No doubt man himself may have been the indirect product of a tumbleweed blowing past the eyes of a curious primate hanging poised from a bough to which he forgot to climb back after he had chased the weed out into the grass. Naturally this is a simplification, but if the world hangs on such matters it may be as well to act boldly and realize all immanent possibilities at once. Shake the seeds out of their pods, launch the milkweed down, and set the lizards scuttling. We are in a creative universe. Let us then create. In the spring when a breath of air sets the propellers of the maple seeds to whirring, I always say to myself hopefully, "After us the dragons. . . ."

* * *

EVERY one of us is a hidden child. We are hidden in our beginnings—in the vagaries of the genetic cards that have been dealt out to us by nature; we are hidden in the episodes of our individual lives, environments, events unduplicable that have aided or ineradicably scarred us. We may have come from poverty or riches, but ironically this tells us nothing either, for

on occasion it may be the rich child who has suffered trauma
and the poor one who has achieved insight, and balance. It is
not always so, nor do I give this last as a nostrum or a desirable
recipe for success. I merely say to begin with that life is infinitely complex and hidden. We share together some elements
of a complicated culture and a language, which, while common
to all of us, whispers to each of us secret meanings, depending
upon the lives we have led. Our diverging roads begin before
birth and intensify afterwards.

* * *

In the old days of the New England transcendentalists, it used
to be stated that the cosmos was a reflection of man, that his
shadow ran a long way out through nature. Though the idea
may be, in some sense, out of fashion, I would venture to remark that men like Emerson and Thoreau, whose interior
thoughts contained a place for muskrats, bean fields, and uninhabited peaks, were closer to an analysis of man's original nature, his soul, if you will, than much that has gone on in
laboratories since. A wilderness exists in man which refuses to
be studied. "There has been but the sun and the eye since the
beginning," Thoreau once wrote, and some of us prefer to
have that eye round, open, and as undomesticated as an owl's
in a primeval forest—a forest which invisibly surrounds us still.

* * *

THE OTHER PLAYER

Sometime on the verge of adolescence I took to entering
empty houses. "Breaking and entering," I suppose it would be
called on the police blotters of today. The gang of boys who
accompanied and initiated me into these exploits, however,
did so for no earthly gain. At heart we were bored metaphysi-
cians. We sought to meet ghosts. Ghosts were to be found in
old and empty houses. Empty houses were locked, and frequently their gloom was made more appalling by boarded
windows. Therefore it is true that in the police phrase we
broke and entered. We went down coal chutes into dark base-
ments. We passed through unlatched second-story windows

by way of tree limbs. We estimated the size of chimneys, the scalability of porch roofs, the possibility that a drainpipe might sustain one of our lighter members until he could reach the roof and descend through an attic window. We had among us those who excelled in specialities of strength or light agility and who did not hesitate to make their way alone through the looming terrors of abandoned rooms and stairwells.

We were scientists only in our study of the material means necessary to escape the world of the ordinary. It was our unexpressed purpose to cross out of the ordinary world and challenge the dark powers in their own domain. The laboratory was not for us. We demanded immediate confrontation before every dark tower known to us. Looking back at these episodes I can only say, with a slight lifting of the scalp, that it is remarkable that one among us did not have the final premature confrontation that he sought. From drainpipe to skylight we had asked for terror and had been refused. But the powers were older and far more subtle than we. Now in maturity I am convinced they merely shrank into the shadows as we passed. For on one occasion alone I had my confrontation, though except for an inexplicable sense of foreboding no shape confronted me. I say foreboding because why else for forty years would I remember two dice rolling on an empty floor—two dice cast only by myself.

I was alone in the great ballroom of an abandoned mansion. No cheerful gang of swaggerers waited at the window. I had come by myself on impulse, and something, before I left, had appeared to meet my challenge. That I was to know—but I was too young to grasp its full intent. My years were few and not sufficient for the game I started. I would not know by growing degrees until the end.

For example, I remembered for the first time the ruined farmhouse I stumbled upon at sunset—no one will believe this—that had school papers labeled Eiseley from a generation before lying amidst the collapsed plaster. I remember I found two dice on the plank floor at evening while something seemed to draw back in the shadows. I did not know how adults played dice. I merely cast and recast the pretty cubes in the growing sunset. Sometimes I thought that I won, sometimes that I lost. One of the dice survives in my desk drawer. I liked their sound, but I never understood their spots, and the road was growing dark. I thought it strange. I knew no other Eiseleys. They were certainly gone and the house was ruined, the plaster fallen. I hurried away down the road, never to return.

***

September 18, 1966

LAST night I encountered an amazing little creature in a windy corner of the Wynnewood shopping area. It seemed at first glance some long-limbed feathery spider teetering rapidly down the edge of a storefront. Then it swung into the air and hesitantly, like a spider on a thread, blew away into the parking lot. Then it returned on a gust of wind and ran toward me once more with amazing rapidity on its spindly silken legs. It was only with great difficulty and trial that I discovered it was actually a silken filamentous seed seeking a hiding place and scurrying with the uncanny surety of a conscious animal. In fact, it did escape me before I could secure it. Its limbs were stiffer than milkweed down, and propelled by the wind it ran rapidly and evasively over the pavement. It unsettled me as though I had come upon some part of nature that I had not anticipated, something only recently emerged from the crevices of evening.

***

NO, it is not because I am filled with obscure guilt that I step gently over, and not upon, an autumn cricket. It is not because of guilt that I refuse to shoot the last osprey from her nest in the tide marsh. I possess empathy; I have grown with man in his mind's growing. I share that sympathy and compassion which extends beyond the barriers of class and race and form until it partakes of the universal whole. I am not ashamed to profess this emotion, nor will I call it a pathology. Only through this experience many times repeated and enhanced does man become truly human. Only then will his gun arm be forever lowered. I pray that it may sometime be so.

***

MAN has been seeking man for five thousand years: He has found him in blood, he has found him in dungeons, he has found him in the hermit's cell or in the torture chamber, he has waited for him by the cross, he has found him in meditation under the sacred Bô tree, he has found him in the laboratory or in the first mushroom cloud that lit the night at Roswell, N.M. He has never been found at all. The reason is simple. Men have been seeking Man, a capitalized, hypostatized thing: a god or a Frankenstein monster. Man does not exist, he is something sewn together in the laboratory of the human imagination, he may be beautiful, he may be as venomous as a serpent. Some men may thus perceive him, but they would be wrong: they are wrong so long as they say, having vitalized their creation, this is Man. There is no Man, there are only men: good, evil, uncountable millions, marred by genetic makeup, marred or perfected by environment. . . . Men should be the subject of human study, never Man; the moment we say Man we are lost in abstraction.

The Lost Notebooks of Loren Eiseley (1987)

CHRONOLOGY

NOTE ON THE TEXTS

NOTES

INDEX

Chronology

1907–12   Born Loren Corey Eiseley in Lincoln, Nebraska, on September 3, 1907, the only child of Clyde and Daisy Corey Eiseley. Mother, thirty-two, grew up in Iowa; deaf and reclusive, she takes in sewing to help support the family. Father, thirty-eight, works as a hardware-store clerk in Fremont, Nebraska; the son of a prosperous German-immigrant hardware merchant, as a younger man he managed the small Opera House in Norfolk, Nebraska (which his father owned), acted in plays, and committed much Shakespeare to memory, but his family's fortunes collapsed after a bad investment in a sugar beet factory. Fifteen-year-old half-brother Leo, from the first of father's two previous marriages, begins high school in January 1909 but soon drops out to work as a messenger boy. Family moves from Fremont to Lincoln in 1910; Leo leaves home the following year, becoming a telegraph operator.

1913–17   Begins school in Lincoln in 1913. Four years later, family moves to Aurora, Nebraska, where father continues to work as a hardware-store clerk.

1918–21   Paternal grandfather Charles Frederick dies on February 2, 1918, and maternal grandfather Milo on December 12. Taking a job as a traveling salesman, father moves family back to Lincoln; he survives a case of influenza. Loren visits the Museum at the University of Nebraska with his maternal uncle Buck, an attorney; for his twelfth birthday, receives a birdhouse and a copy of Jules Verne's *From the Earth to the Moon*. Leo, now married to Mamie Harris and the father of a two-year-old daughter, returns to Lincoln in 1920; Loren visits them often. Writes in an eighth-grade essay: "I have selected Nature Writing for my vocation because at this time in my life it appeals to me more than any other subject."

1922–24   Leo and family relocate to Colorado Springs. Loren enters Lincoln High School in September 1922; moves in with Aunt Grace and Uncle Buck. Unhappy in school, does not return the following January and takes a menial job. Uncle arranges for his enrollment at Teachers College High

School where he becomes a junior in September 1923; joins the football team. Buck gives him a copy of Henry Fairfield Osborn's *Origin and Evolution of Life*.

1925   During his senior year in high school, captains the football team and is elected class president; acts in the class play, Walter Ben Hare's *Kicked out of College*. With three friends, buys a 1919 Model T Ford and drives to Los Angeles in search of summer employment, camping along the way. Back in Lincoln, enters the University of Nebraska in September, living alternately with his parents and his aunt and uncle, his uncle helping to pay his tuition. Begins a correspondence with Mabel Langdon, a former student teacher at his high school who had taken her first job in rural Arnold, Nebraska; they exchange poems.

1926   Studies evolution and genetics, elementary paleontology, social psychology, and applied psychology in summer school, but takes only two courses in the fall; failing one, is forced to take a semester's leave. Travels widely throughout the western states on freight trains.

1927   Reenters the University of Nebraska in the summer. Becomes associate editor of the recently established literary magazine *Prairie Schooner*; publishes poems and prose sketches.

1928   Does not reenroll at the University of Nebraska in January; takes job on the graveyard shift at a chicken hatchery. Father, diagnosed with stomach and liver cancer, dies on March 31; unable to afford mortgage payments, mother moves in with uncle and aunt. Returns to school in September.

1929   Diagnosed with influenza and a pulmonary infection, is urged by doctors to take a rest cure; spends time in Manitou Springs, Colorado, to recuperate. Returns to Lincoln in the fall; his doctor recommends further rest in a dry climate. "The Deserted Homeland," a poem, appears in *Poetry* in December.

1930   Travels to California, where he becomes caretaker of a property in the Mojave Desert, near Lancaster. Mabel Langdon visits. His health improved, makes his way back to Lincoln by freight train, living as a hobo; arrives "in a dreadful state." Re-enrolls in classes in September, taking

Introductory Anthropology and Field & Museum Techniques.

1931   During spring vacation, joins a group of about a dozen students on an archaeological dig led by Dr. William Duncan Strong; they excavate remains of sixteenth-century Pawnee village in central Nebraska. On May 3 becomes president of the Poetry Society, part of the Nebraska Writers' Guild. Travels through Colorado, New Mexico, Arizona, and Utah in June, camping with Helen Hopt, a young teacher, and another couple. Spends August and September before classes as part of the Morrill Paleontological Expedition (the "South Party"); discovers a promising fossil site in Banner County, Nebraska.

1932   Returns to Banner County with the South Party in May, June, and July, unearthing bones of extinct mastodon, rhinoceros, camel, and saber-toothed cat.

1933   Graduates from University of Nebraska in June with a B.A. in anthropology and English; joins the South Party for third summer. Visits Taos, New Mexico, introducing himself to literary heiress Mabel Dodge Luhan. In August discovers remains of a large Brontothere and the skull of a saber-toothed cat. Moves to Philadelphia in the fall, beginning graduate school in anthropology at the University of Pennsylvania. Visits Harold Vinal, editor of the poetry magazine *Voices* and a longtime correspondent, in New York City, and his half-brother Leo and family, who now live in Philadelphia.

1934   Spends time on weekends hiking in the New Jersey Pine Barrens with Frank Speck, a favorite professor. In June, travels to Carlsbad, New Mexico, with another graduate student, Joseph B. Townsend Jr., on a fellowship; over six weeks they excavate a Native American burial site in a cave in what is now Guadalupe Mountains National Park. (At the end of the summer's work, celebrating in Ciudad Juárez, Mexico, the two are arrested for "doing something ridiculous," and are released only after the intervention of the American consul.) Visits Santa Fe with Mabel Langdon and Dorothy Thomas, a young writer from Lincoln; they attend the funeral of writer Mary Austin, and tour Pecos Pueblo and Chaco Canyon. Returning to Penn, moves into International House with two friends; is awarded a scholarship. Later remembers the time as "one of the happiest in my life." Completes master's thesis, "A Review of the Paleontological Evidence Bearing upon the Age of the Scottsbluff Bison Quarry and Its Assorted Artifacts."

1935   Part of thesis is accepted for publication in *American Anthropologist*. Receives master's degree on February 9. In June, joins eight others at the Lindenmeier archaeological site in Larimer County, Colorado, seeking skeletal remains of Folsom man; finds a Folsom arrowhead in the neck vertebrae of a Pleistocene bison, and gathers mollusk specimens hoping they may help date the party's finds. On August 20, learns of the death of his uncle Buck, and returns to Lincoln for his funeral. Unable to afford tuition at Penn without uncle's support, moves in with family in Lincoln; Mabel Langdon helps pay his tuition at the University of Nebraska, where he takes graduate courses in sociology and German.

1936   Takes a W.P.A. job in February, assisting the editor of *Nebraska: A Guide to the Cornhusker State*; writes essays on Nebraska geology, paleontology, and Native American culture. Returns to Penn in the fall with the aid of a fellowship. Goes on canoeing and hiking trips with Frank Speck, now his dissertation advisor. Maternal grandmother Malvina dies on December 17.

1937   Receives Ph.D. degree in June after Speck approves his dissertation, "Three Indices of Quaternary Time and Their Bearing on the Problems of American History: A Critique." Spends three weeks excavating eighteenth-century Native American burial sites in Doniphan County, Kansas, with a party from the National Museum of Natural History. Accepts position as assistant professor in the sociology department at the University of Kansas, moving to Lawrence; subsequently teaches courses on General Anthropology, the American Indian, Primitive Society, Peoples of the Pacific, the Evolution of Culture, and the Elements of Sociology. In August is reunited with Mabel in New Mexico; they stay with Dorothy Thomas at Mary Austin's house in Santa Fe, visit Taos, and meet Frieda Von Richtofen Lawrence, D. H. Lawrence's widow, at her home nearby.

1938   On weekends, investigates potential archaeological sites near Lawrence, and does some excavating work in early August. Visits Albuquerque with Mabel; they marry there on August 29 in a small informal ceremony. Lacking the funds to set up a joint household, she returns to her family in Lincoln, working as a curator at the university art gallery, he to bachelor quarters in Lawrence.

1939   In April, with Bert Schultz of Nebraska State Museum, announces archaeological discoveries in southeastern Nebraska; with Schultz, signs contract to write book "They Hunted the Mammoth: The Story of Ice Age Man" for Macmillan (later cancelled because of the war). Helps to prepare an exhibit on the works of Robinson Jeffers, the modern writer he claims most to admire, at the University of Nebraska Library. Beginning in the summer, is joined by wife Mabel in Lawrence. Signs a contract with publisher Thomas Y. Crowell to write a textbook on general anthropology.

1940   In June, begins postdoctoral fellowship in physical anthropology at Columbia University and the American Museum of Natural History. Mabel, arriving in New York in September, works part-time as a typist for crime writer Rex Stout. Registers for military service in October. Friend Dorothy Thomas rents a vacant apartment downstairs; she takes the Eiseleys to parties and introduces them to her circle.

1941   Working with Harry L. Shapiro at Columbia, undertakes biometric studies of Basket Maker skeletons excavated in Arizona. Returns to Kansas in September.

1942   In January, as second author with Frank Speck, publishes "Montagnais-Naskapi Bands and Family Hunting Districts of the Central and Southeastern Labrador Peninsula," in the *Proceedings of the American Philosophical Society*. Application for a commission in the Army Air Corps is rejected in July due to his weak eyesight; fears he may soon be drafted as an infantryman. Publishes "What Price Glory? The Counterplaint of an Anthropologist" in *American Sociological Review* in December, along with "The Folsom Mystery," the first of a series of articles in *Scientific American*.

1943   In the spring becomes an instructor in human gross anatomy in the short-handed University of Kansas Medical School; receives a draft deferment on the basis of his teaching responsibilities.

1944   Mabel, diagnosed with breast cancer, undergoes a radical mastectomy and radiation treatment. In May, accepts an offer to serve as a professor in the department of sociology at Oberlin College in Ohio; moves to Oberlin with Mabel in October.

1945   Teaches over the summer at the University of Pennsylvania. Becomes friendly with writer and photographer Wright Morris, a downstairs neighbor, who nicknames him "Schmerzie" (a diminutive of *Weltschmerz*), making light of his tendency to melancholia. Shares work in progress with Morris. Publishes essays "Myth and Mammoth in Archaeology" in *American Antiquity* and "There *Were* Giants" in *Prairie Schooner*.

1946   Is awarded tenure at Oberlin. Teaches over the summer at Columbia University. In November receives a grant of $3,500 from The Viking Fund for "A Survey and Investigation of Researches on Early Man in South Africa"; plans extensive travel, intending to visit South African anthropologists, museum collections, and archaeological sites.

1947   In January is appointed Professor of Anthropology at the University of Pennsylvania and assumes chairmanship of his department; with Mabel, rents a small apartment in suburban Rose Valley, outside Philadelphia. Works with Froelich Rainey, newly appointed director of the University Museum, to restore long-strained ties between the Museum and the Department of Anthropology; is named curator of early man at the Museum. Postpones South African trip after he learns that a group from the University of California will soon be en route with similar objectives; hopes to revise his itinerary and sail in a year.

1948   Travels to Oaxaca, Mexico, in the fall after human skeletal remains are unearthed near those of an ancient elephant, a potentially significant discovery that ultimately proves less important than it appears. Abandons South African travel plans, citing "the unsettled state of both departmental and family affairs." Suffers from partial deafness after an ear infection in the fall, recovering about six months later. Comes to an informal agreement with Harper & Brothers to publish a collection of essays; works with editor Jack Fischer. Presents paper on "Providence and the Death of Species" at meeting of the American Anthropological Association in December, in Toronto.

1949   Teaches at Berkeley over the summer, afterward traveling
        to Wyoming, near Cody, where he investigates a distinc-
        tive arrowhead site. Moves with Mabel to a new apartment
        in Wynnewood, a prosperous suburb on the Philadelphia
        Main Line. In September is elected president of the newly
        formed American Institute of Human Paleontology, after
        a meeting of leading physical anthropologists in New
        York, which he attends. Publishes "The Fire Apes" in
        Harper's; tells Ray Bradbury, who writes him in praise of
        the essay, that he is "contemplating doing a book."

1950   Delivers a eulogy for Frank Speck after his death on Feb-
        ruary 7.

1951   Toward the end of the year, travels briefly to England to
        arrange for the university museum's acquisition of an im-
        portant collection of plaster casts of paleontological speci-
        mens.

1952   Accepts a commission from the American Philosophical
        Society to write a book on the reception of Darwin's ideas
        in America, to be published on the centennial of the first
        edition of *On the Origin of Species*, in 1959. Helps the Society
        build its Darwin collections, locating books and manuscripts
        for acquisition. (Wright Morris occasionally accompanies him
        on buying trips.) Wife Mabel takes part-time job as secretary
        to the director of the Pennsylvania Academy of the Fine Arts,
        in subsequent years becoming an associate director of that
        institution. Begins a year's leave with fellowship support to
        work on a book about "the philosophical implications of
        human evolution."

1953   In July, meets with Jack Fischer to discuss essay collection.

1954   Essay "Man the Firemaker" appears in *Scientific American*
        in September.

1955   After discussions at the American Philosophical Society,
        abandons plans for a book on Darwinism in America, and
        agrees in December to edit two volumes of the letters of
        Darwin and his contemporaries. (Subsequently renegoti-
        ates this commitment, and instead of the proposed vol-
        umes works intensely on a book about the history of
        evolution tentatively titled "The Time Voyagers," solicited
        by Doubleday editor Jason Epstein.)

1956   Early in the year, meets with Hiram Haydn, editor-in-chief at Random House, who persuades him to work with the firm on a book gathering his essays; proposes to call it "The Great Deeps." Begins a year's sabbatical in September, freeing him to advance his American Philosophical Society publication projects. Presents Haydn with a manuscript of his essay collection, now titled "The Crack in the Absolute," on November 13; they ultimately agree on the title *The Immense Journey*. Readers for the press recommend substantial revisions to Eiseley's initial submission.

1957 In January, meets with Haydn in Philadelphia to discuss alterations to *The Immense Journey*; rewrites and reorganizes it, dropping several essays. Submits a manuscript of "The Time Voyagers" to Doubleday in June. *The Immense Journey* is published on August 26; travels to New York for radio interviews about the book. Joins editorial board of *The American Scholar*.

1958 Proposes to edit an anthology of naturalists' writings for Random House; at a meeting in New York in February, Haydn convinces him instead to sign a contract for a second book of his own essays. "The Time Voyagers," now titled *Darwin's Century: Evolution and the Men Who Discovered It*, is published by Doubleday in July.

1959 *Darwin's Century* wins Phi Beta Kappa science prize. Is appointed provost of the University of Pennsylvania in October; quickly realizes he has overcommitted himself. Mother dies on November 29. Delivers six public lectures at the College of Medicine of the University of Cincinnati.

1960 Publishes *The Firmament of Time*, gathering his University of Cincinnati lectures.

1961 In March, delivers the Montgomery Lectures on Contemporary Civilization at the University of Nebraska, addressing the life and significance of Francis Bacon on the four hundredth anniversary of his birth. On April 3, is awarded the John Burroughs Medal for *The Firmament of Time*. Resigns as provost in April, and in the fall becomes a fellow of the Center for Advanced Study in the Behavioral Sciences, in Palo Alto.

1962 Aunt Grace dies on January 17. In February, gives the fifth annual lecture of the John Dewey Society, in Chicago; it is published eight months later by Harper & Row, titled *The*

Mind as Nature. Visits New York in June, meeting with editors and publishers. Returning to Penn in July, is named University Professor in the Life Sciences.

1963 *Francis Bacon and the Modern Dilemma*, an expanded version of his 1961 Nebraska lectures, is published on January 15. Delivers address "The Divine Animal" on May 22, at a New York meeting of the National Institute of Arts and Letters. Accepts a position as director of the new Richard Prentice Ettinger Program for Creative Writing at Rockefeller University; on June 27, at ceremonies inaugurating the program, gives address "The Illusion of Two Cultures."

1964 Is awarded a Guggenheim Fellowship in April, enabling a year's leave of absence from Penn. Hires Caroline E. Werkley as his administrative assistant, to oversee the affairs of the Ettinger Program; she remains his assistant for the rest of his life. In November is relieved of his duties as Ettinger director, charged with neglecting them. Named to President's Task Force on the Preservation of Natural Beauty, travels to Washington, D.C., for a press conference, and meets with Lyndon Johnson.

1965 Addresses the Nebraska Academy of Sciences in April, on "The Inner Galaxy: A Prelude to Space."

1966 Gives address "Man, Time, and Prophecy" at the University of Kansas Centennial Celebration in April. Serves as host and narrator of *Animal Secrets*, an NBC television series that airs for about eighteen months.

1967 Vacations on Sanibel Island in Florida in February. Spends the fall semester at the University of Wisconsin, where he has been appointed Johnson Research Professor. Tours the countryside around Madison with Walter Hamady, an assistant professor of art and the proprietor of The Perishable Press, who proposes to publish some of Eiseley's works in an illustrated limited edition.

1968 In Dallas at a December meeting of the American Association for the Advancement of Science, gives an interview to the *Dallas Morning News* in which he questions the value of the space program, given its immense cost; many criticize his remarks.

1969   Spends three winter weeks in Aruba with Mabel; later in
        the year, she retires from her position at the Pennsylvania
        Academy of the Fine Arts. Delivers a series of lectures at
        the University of Washington in Seattle in the fall. *The
        Unexpected Universe* is published in October to widely
        positive reviews. Receives copies of *The Brown Wasps*, a
        collection of three essays published in a limited edition by
        Walter Hamady at The Perishable Press; "the books are
        beautiful," he writes Hamady. Begins a novel, never com-
        pleted, titled "The Snow Wolf."

1970   On February 21, *The New Yorker* publishes a warmly favor-
        able review by W. H. Auden of *The Unexpected Universe*;
        Eiseley and Auden subsequently correspond, and meet for
        lunch. *The Unexpected Universe* is nominated for a Na-
        tional Book Award. Invitation to deliver the commence-
        ment address at Kent State is cancelled by the university
        after the shooting of four students on May 4. Receives
        honorary doctor of science degree from St. Lawrence
        University in New York, one of many such honors. *The
        Invisible Pyramid* is published in October.

1971   Elected to the National Institute of Arts and Letters. *The
        Night Country* is published November 10.

1972   Suffers from viral pneumonia during the fall. In November
        publishes *Notes of an Alchemist*, a book of poems, with
        Scribner's; receives a note of praise from Auden, who asks
        permission to dedicate a poem to Eiseley.

1973   *The Man Who Saw Through Time*, a revised and expanded
        edition of *Francis Bacon and the Modern Dilemma*, is
        published by Scribner's in April. Over the summer spends
        ten days with Froelich Rainey of the Museum of Archaeol-
        ogy and Anthropology at Penn, doing archaeological
        fieldwork in Dawson County, Montana. Visits Walden
        Pond and Thoreau's grave in Masssachusetts; considers
        writing a book about Thoreau and discusses the project
        with Scribner's. *The Innocent Assassins*, a second collection
        of poetry, appears in October.

1974   On September 18, at the Kennedy Center in Washington,
        D.C., is presented with the Distinguished Nebraskan
        Award. Toward the end of the year, begins working inten-
        sively on a long-postponed volume of autobiography; it is
        initially titled "The Other Player."

1975   Publishes his autobiography, *All the Strange Hours: The Excavation of a Life*, in October.

1976   Sits for a portrait bust by sculptor Kappy Wells. Learns after routine prostate surgery in September that doctors have discovered a malignant bile duct tumor.

1977   Undergoes a pancreatectomy on January 27, returning home at the end of March. Receives galleys for *Another Kind of Autumn*, a book of poems to be published by Scribner's. Dictates a letter to Scribner's proposing an outline of the contents of his last book, published posthumously by Times Books as *The Star Thrower*. Returns to the hospital in June, his tumor having returned. Dies on July 9.

Note on the Texts

This volume—the first of a two-volume set, *Collected Essays on Evolution, Nature, and the Cosmos*, by Loren Eiseley—presents the complete texts of *The Immense Journey* (1957), *The Firmament of Time* (1960), and *The Unexpected Universe* (1969), along with a selection of prose writings that Eiseley did not include in any of his books. The second volume in the set presents *The Invisible Pyramid* (1970), *The Night Country* (1971), and essays gathered posthumously in *The Star Thrower* (1978). The texts of the books included here have been taken from the first printings; texts of previously uncollected items have been taken from the earliest published versions, as described below.

The Immense Journey. In January 1947 Eiseley published the first in a succession of well-received essays in *Harper's*, "The Long-Ago Man of the Future"; he followed it with "Obituary of a Bone Hunter" in October of the same year and "The Places Below" in June 1948. By July 1948, after a meeting with Harper & Brothers editor Jack Fischer, he had agreed to gather these essays along with others yet to be written into a book, for which he proposed the title "Manhunt." Declining subsequent inquiries and invitations from Ken McCormick and Jason Epstein at Doubleday, E. P. Swenson at W. W. Norton, and Kenneth Heuer at the Viking Press, he worked on the collection with Harper & Brothers for nearly a decade, sending an outline of its projected contents in 1949, begging Fischer's patience in 1950 ("the kind of book I hope to do cannot be rushed"), and at the end of 1952, now calling the book "The Great Deeps," promising a draft by the following summer.

Eiseley was still in the middle of this draft when he met with Fischer in New York in the summer of 1953. The editor attempted to push his author to organize and finish the book—sending a new outline and detailed notes after their long conversation—but Eiseley would not commit to a particular schedule and communicated little with Fischer thereafter. In the years that followed, *Harper's* rejected three of Eiseley's essays ("Is Man Alone in Space?," "The Flow of the River," and "The Judgment of the Birds"), and Eiseley found a more sympathetic editor, he felt, in Hiram Haydn at *The American Scholar*, who eagerly accepted two of the three (praising "The Judgment of the Birds" as "the most beautiful piece of writing I have read in several years").

Also editor-in-chief at Random House, Haydn let it be known that his firm would welcome the chance to publish Eiseley's first book if Harper & Brothers had no objection; the two met in New York early in 1956. Finally, in July of that year, Eiseley informed Fischer that he had decided to publish "The Great Deeps" with Random House. His completed first draft, collecting seventeen essays, arrived on Haydn's desk in mid-September.

Two months later, now with a book contract in hand and publication scheduled for the spring of 1957, Eiseley delivered a revised typescript, incorporating his responses to Haydn's initial comments and in places updating the essays to reflect changes in the scientific literature since they were written. Neither Haydn nor Eiseley was happy with the title "The Great Deeps." Eiseley tentatively suggested "Bones and Searches," which they agreed together to change to "The Crack in the Absolute" before Eiseley ultimately decided on *The Immense Journey*, a phrase he had recently noticed in the *Journal intime* of the Swiss philosopher Henri Frédéric Amiel.

Others at Random House turned out to be less enthusiastic about *The Immense Journey* than Haydn, who shared Eiseley's typescript with his colleagues. Robert Linscott recommended that several essays be cut and others altered. Robert D. Loomis, offering a second opinion, echoed Linscott's concerns in a number of particulars and made additional suggestions. Postponing publication until the fall, at the end of January 1957 Haydn traveled to Eiseley's home in Wynnewood, Pennsylvania, to discuss his fellow editors' recommendations. Given a few days to review and reflect on these, Eiseley was angered and upset: in a letter to Haydn, he threatened not only to tear up his contract but to give up nonscientific writing altogether ("when two editors in a good house want to go out and retch in the street after reading this stuff . . . this ought to be enough"). But he was soon persuaded not to abandon the book, and by February 28 he had submitted his final revisions. ("Big Eyes and Small Eyes," "The Places Below," and "The Brown Wasps"—essays Eiseley removed from the collection in the wake of his publisher's criticisms—were subsequently printed in *The Night Country*. "The Fire Apes" appeared posthumously in *The Star Thrower*.)

Further information on the composition and publication history of individual essays in *The Immense Journey* is provided in the list below. To varying degrees Eiseley revised all of the magazine versions of his essays before they appeared in the book.

"The Slit." First published in *The Immense Journey*.
"The Flow of the River." Early titles (from drafts among Eiseley's papers at the University of Pennsylvania) included "The Shape of

Water," "I Remember My Green Extensions," and "Men, Snow-flakes, and Green Leaves." Published in *The American Scholar*, Autumn 1953.

"The Great Deeps." Early titles included "The Abyss," "The Reaching Out," "The Abyssal Ooze," and "The Place Without Light." Published in *Harper's*, December 1951.

"The Snout." Early titles included "Things To Be Watched," "The Mud Skippers," "The Door in the Bog," "A Place of Low Life," "A Strange Door," "The Quagmire," "The Crawlers," "The Appearances," "The Moon Darkens," and "A Time of No Moon." Published in *Harper's*, September 1950.

"How Flowers Changed the World." An early draft bears the canceled title "The Green Twinkle." First published in *The Immense Journey*.

"The Real Secret of Piltdown." Some drafts are titled "The Question Darwin Never Answered." Published in *Harper's*, November 1955, as "Was Darwin Wrong About the Human Brain?"

"The Maze." Published in *Scientific American*, June 1956, as "Oreopithecus: Homunculus or Monkey?" Earlier rejected titles included "Dawn Ape," "Dawn Man," and "Oreopithecus: The Little Man."

"The Dream Animal." First published in *The Immense Journey*.

"Man of the Future." Published in *Harper's*, January 1947, as "The Long-Ago Man of the Future"; originally titled "The Man of the Future Lived in the Past."

"Little Men and Flying Saucers." Published in *Harper's*, March 1953; an earlier draft is titled "Of the Waiting for Ships."

"The Judgment of the Birds." Published in *The American Scholar*, Spring 1956.

"The Bird and the Machine." Published in *Harper's*, January 1956.

"The Secret of Life." Published in *Harper's*, October 1953.

Eiseley read author's proofs of *The Immense Journey* at least twice, in April and May 1957; his hand-corrected copies are now among his papers at the University of Pennsylvania. He is not known to have altered the text of the book when it was subsequently reprinted, or when it was published in London, by Victor Gollancz, in 1958. The text of *The Immense Journey* in the present volume is that of the August 26, 1957, Random House first printing.

The Firmament of Time. On April 6, 1959, Eiseley wrote to Bennett Cerf at Random House seeking to be released from his multivolume contract with the firm. Hiram Haydn—"an editor," he explained, "with whom I can work most satisfactorily, and with whom I wish to continue to work"—had joined several others in a new publishing venture, Atheneum, and Eiseley hoped to follow him there. Later the same year, between October 19 and November 5, Eiseley gave six lectures at the College of Medicine of the University of Cincinnati, where he had accepted an appointment as Visiting Professor of the Philosophy of Science. The following June 30, Atheneum published these lectures as *The Firmament of Time*.

None of Eiseley's Cincinnati lectures had previously appeared in print, but he had presented the fourth of the six, "How Human Is Man?," as part of the Benjamin Franklin Lecture Series at the University of Pennsylvania in 1958; originally titled "The Ethic of the Group," it was published separately by the University of Pennsylvania Press on July 11, 1960, in *Social Control in a Free Society*, a volume edited by Robert E. Spiller. A version of the final lecture, "How Natural is 'Natural'?," appeared in *Horizon* in July 1960 as "Nature, Man, and Miracle."

Eiseley read at least one set of proofs of *The Firmament of Time* before the book was published, but he is not known to have revised or corrected later printings or to have been involved in the preparation of an English edition, published by Victor Gollancz in London in 1961. The text of *The Firmament of Time* in the present volume has been taken from the Atheneum first printing of June 30, 1960.

The Unexpected Universe. In 1964, Eiseley followed his editor Hiram Haydn from Atheneum to Harcourt, Brace, signing contracts to produce a volume of autobiography and a new collection of essays. By September 1967, however, he confessed that he was encountering "a certain blockage" in his progress on both projects. ("Like the veriest amateur," he complained to William Jovanovich at Harcourt, "I have gone dry and am waiting for inspiration.") The autobiography seemed an impossible task, and he offered to return his advance. The essays, scheduled for publication in 1968, would need to be delayed by a year.

When Eiseley finally submitted a manuscript of *The Unexpected Universe* in February 1969, Haydn and Jovanovich both quickly reassured him that it had been worth the wait; they sought no significant alteration to the contents and made plans for a large initial printing. Eiseley took advantage of the considerable editorial control Haydn yielded to him, carefully reviewing proofs and even helping to draft the text on the dust jacket. He is not known to have been involved in the preparation of an English edition of *The Unexpected Universe*, published by Victor Gollancz in London in 1970.

Further information on the composition and publication history of individual essays in *The Unexpected Universe* is provided in the list below.

"The Ghost Continent." Incorporates material from "The Odyssean Voyage in Science and Literature," a lecture presented at the Dallas meeting of the American Association for the Advancement of Science, on December 26, 1968; a section appeared in *Science* on July 11, 1969, as "Activism and the Rejection of History." First published in its entirety in *The Unexpected Universe*.

"The Unexpected Universe." Rejected early titles included "The Mutable Cloud" and "Man, Nature, and Science." Presented as the fifth annual Grady Gammadge lecture at Arizona State University, February 8, 1966, under the title "Science and the Unexpected Universe," and published under the same title in *The American Scholar*, Summer 1966.

"The Hidden Teacher." Incorporates material from "Thoughts Provoked on the Nature of Man by a Visit to the Fair," an unpublished manuscript written for a 1964 Time, Inc. World's Fair souvenir book. Other draft titles included "Man: The Listener in the Web" and "The Fair and the Dreamer." Presented as a lecture on several occasions from 1965 to 1968. Published for the first time in *The Unexpected Universe*.

"The Star Thrower." First published in *The Unexpected Universe*.

"The Angry Winter." Draft titles included "The Doorway of Snow," "The Antagonist," and "The Lost Door." Presented as a lecture at Stanford University in October 1966 and at the University of Wisconsin in November 1967. Published as "Man Is an Orphan of the Angry Winter," *Life*, February 16, 1968, and in shortened form as "The Night the Shadows Whispered," *Reader's Digest*, May 1968.

"The Golden Alphabet." Early versions were titled "The Forfeit Paradise" and "The Forfeit Paradise: Darwin and Thoreau." Presented as a lecture on several occasions in 1968. First published in *The Unexpected Universe*.

"The Invisible Island." Draft titles included "The Isle of Voices," "Man: The Invisible Island," "The Rent in the Curtain," and "The Living Screen." First published in *The Unexpected Universe*.

"The Inner Galaxy." Draft titles included "The Bright Stranger: A History of Love," "New End for Inner Galaxy," "A Prelude to Space," "The Inner Galaxy: A Prelude to Space," and "Man Looks at Man." Presented as a lecture on numerous occasions from 1964 to 1969. First published in *The Unexpected Universe*.

"The Innocent Fox." Published in *Natural History*, October 1969.

"The Last Neanderthal." A draft is titled "The Season of the Leaves." First published in *The Unexpected Universe*.

The text of *The Unexpected Universe* in the present volume has been taken from the Harcourt Brace Jovanovich first printing of October 8, 1969.

Uncollected Prose. The final section of this volume presents a selection of prose writings that Eiseley did not choose to include in any of his books: two sketches written while he was an undergraduate at the University of Nebraska, and one written shortly thereafter; a lecture delivered at a scientific conference in 1957; and twenty-nine entries from his private notebooks, written over a span of approximately two decades. The texts of these prose writings have been taken from the earliest published sources, listed below.

"Autumn—A Memory." *Prairie Schooner,* October 1927.
"Riding the Peddlers." *Prairie Schooner,* Winter 1933.
"The Mop to K. C." *Prairie Schooner,* Winter 1935.
"Neanderthal Man and the Dawn of Human Paleontology." *Quarterly Review of Biology,* December 1957.
"FROM *The Lost Notebooks.*" *The Lost Notebooks of Loren Eiseley,* ed. Kenneth Heuer (Boston: Little, Brown, 1987), 52, 81–82, 82, 83, 84, 85, 86, 87–88, 89, 89–90, 91, 92–93, 93–94, 96, 96–97, 98, 98–99, 100, 102, 103–4, 116, 127–28, 128–30, 136, 138, 141–42, 150, 155, 156.

This volume presents the texts of the essays and other writings chosen for inclusion here, but it does not attempt to reproduce features of their typographic design, such as the display capitalization of chapter openings. The texts are reprinted without change, except for the correction of typographical errors. Spelling, punctuation, and capitalization are often expressive features, and they are not altered, even when inconsistent or irregular. The following is a list of typographical errors corrected, cited by page and line number: 30.29, and and; 89.20, him; 196.24, ice lie; 235.8, uncreate; 272.25, exceeds; 403.4, Quarternary; 412.9, aways; 420.22, "No"; 420.27, "Cause; 426.12, Chamber's; 427.2, compete; 427.10, Schaffahausen's; 431.30, the; 432.15, Archibishop; 435.10, horozontal.

Notes

In the notes below, the reference numbers denote page and line of this volume (the line count includes chapter headings but not blank lines). No note is generally made for material included in standard desk-reference works. Quotations from Shakespeare are keyed to *The Riverside Shakespeare*, ed. G. Blakemore Evans (Boston: Houghton Mifflin, 1974), and biblical references to the King James Version. For further information about Eiseley's life and works, and references to other studies, see Gale E. Christianson, *Fox at the Wood's Edge: A Biography of Loren Eiseley* (New York: Henry Holt, 1990); Peter Heidtmann, *Loren Eiseley: A Modern Ishmael* (Hamden, CT: Archon Books, 1991); Tom Lynch and Susan N. Maher, eds., *Artifacts and Illuminations: Critical Essays on Loren Eiseley* (Lincoln: University of Nebraska, 2012); and Mary Ellen Pitts, *Toward a Dialogue of Understandings: Loren Eiseley and the Critique of Science* (Bethlehem, PA: Lehigh University Press, 1995).

THE IMMENSE JOURNEY

4.1–4 "*Man can* . . . THOREAU] From an entry in Thoreau's journals dated March 23, 1853.

4.5–7 "*Unless all* . . . TEMPLE] See "Revelation and Its Mode," lecture 12 of Temple's *Nature, Man, and God; Being the Gifford Lectures Delivered in the University of Glasgow in the Academic Years 1932–33 and 1933–34* (1934).

7.29 "Whirl . . . Aristophanes] See *The Clouds* (423 B.C.E.), line 828.

8.23–26 "In habitat . . . restricted."] See Frederik Barth, "On the Relationships of Early Primates," *American Journal of Physical Anthropology*, June 1950.

14.1–2 MacKnight Black . . . pole."] See Black's poem "Apart," first collected in *Machinery* (1929).

16.15–16 Thoreau . . . animalized water"] See *Walden* (1854), chapter 16: "They are not green like the pines, nor gray like the stones, nor blue like the sky; but they have, to my eye, if possible, yet rarer colors, like flowers and precious stones, as if they were pearls, the animalized *nuclei* or crystals of the Walden water."

21.7–9 "Urge the use . . . seas."] See Darwin's letter of February 11, 1857, to Charles Lyell.

21.20–24 "It was like . . . monster."] See chapter 4 of *The Depths of the Sea; an Account of the General Results of the Dredging Cruises of H.M.S.S. 'Porcupine' and 'Lightning' During the Summers of 1868, 1869, and 1870* (1873), by C. Wyville Thomson.

22.37–38 Thorfinn the Skull Cleaver] Thorfinn Torf-Einarsoon (?–c. 963), an Earl of the Orkney Islands.

23.12 Conan Doyle's . . . reverse.] In Conan Doyle's 1912 novel, an expedition to South America discovers an Amazon plateau inhabited by dinosaurs and otherwise extinct prehistoric mammals.

24.38–25.3 Sir Charles Tomlinson . . . protoplasm."] See *The Depths of the Sea*, chapter 9.

25.13–17 Mr. Buchanan . . . sea water.] See J. Y. Buchanan, "Preliminary Report to Professor Wyville Thomson, F.R.S., Director of the Civilian Scientific Staff, on Work (Chemical and Geological) Done on Board H.M.S. 'Challenger,'" *Proceedings of the Royal Society of London*, January 1875.

27.2–9 Thomas Huxley . . . *age.*] See "The Problems of the Deep Sea," first published in *The Popular Science Monthly*, August 1873.

27.11–13 Agassiz . . . periods."] From a letter of January 16, 1872, to Benjamin Peirce, Superintendent of the Coast Survey.

27.29–30 "an infinite capacity . . . direction."] See chapter 9 of Thomson's *The Depths of the Sea* (1873).

27.30–33 "Our ardor . . . the world."] See chapter 22 of H. N. Moseley's *Notes by a Naturalist on the "Challenger"* (1879).

37.37–38 Cope . . . unspecialized,"] See chapter 3 of *The Primary Factors of Organic Evolution* (1896), by Edwin Drinker Cope.

42.15–16 Charles Darwin . . . mystery,"] See Darwin's letter of July 22, 1978, to Joseph Dalton Hooker.

42.20–21 Francis Thompson . . . a star.] See "The Mistress of Vision," first collected in Thompson's *New Poems* (1897).

51.21–23 Wallace . . . if anything!"] See Wallace's letter of August 26, 1913, to Mr. E. Smedley, of Dorset.

52.26–29 Darwin . . . the same country."] Eiseley quotes Darwin indirectly, by way of a retranslation from the French of Pierre Trémaux's *Origine et transformations de l'homme et des autres êtres* (1865) in Robert Mackenzie Beverly's *The Darwinian Theory of the Transmutation of Species* (1867). The original text in *The Origin of Species* (1859) reads as follows: "Natural selection tends only to make each organic being as perfect as, or slightly more perfect than, the other inhabitants of the same country with which it comes into competition."

53.32–36   "How, then . . . learned societies."] See Wallace's review, in the *London Quarterly Review* for April 1869, of new editions of Charles Lyell's *Principles of Geology* and *Elements of Geology*.

54.14–17   "If you . . . sorry for it."] See Darwin's letter of April 14, 1869, to Alfred Russel Wallace.

55.31–36   "If," . . . *development*."] See Wallace's Presidential Address to the Biological Section of the British Association, Glasgow, September 6, 1876.

59.25–28   the words of Wallace . . . to do."] See "The Limits of Natural Selection as Applied to Man," chapter 10 of Wallace's *Contributions to the Theory of Natural Selection* (second edition, 1875).

61.22–24   Solly Zuckerman . . . other Primate"] See "Evolution of the Human Brain," *The Advancement of Science*, October 1939.

63.4–8   as St. George Jackson Mivart . . . know it."] See part 2 of Mivart's essay "Man and Apes," first published in *The Popular Science Review* in 1873 and collected in *Lessons from Nature, as Manifested in Mind and Matter* (1876).

65.5–10   Wood Jones . . . great apes."] See the final sentence of Frederic Wood Jones's *Hallmarks of Mankind* (1948).

65.21–25   Henry Fairfield Osborn . . . pro-human limbs."] See "Recent Discoveries Relating to the Origin and Antiquity of Man," *Science*, May 1927.

65.38–66.2   William L. Straus . . . anthropoid-like."] See William L. Straus Jr., "The Riddle of Man's Ancestry," *Quarterly Review of Biology*, September 1949.

66.13   Hurzeler's recent announcement] See Johannes Hürzeler, "*Oreopithecus bambolii* Gervais: A Preliminary Report" (*Verhandlungen der Naturforschenden Gesellschaft Basel*, vol. 69, 1958), and earlier papers.

70.35–40   Dr. Tilly Edinger . . . extinct."] See Edinger's "Paleoneurology Versus Comparative Brain Anatomy," *Confinia Neurologica*, January 1949.

75.14–17   Professor Frederick Zeuner . . . actually be.] See "Pre-history Lost in the Wash?," *New Scientist*, February 14, 1957.

76.16   "red in tooth and claw."] See canto 56 of *In Memoriam A.H.H.* (1850) by Alfred, Lord Tennyson.

85.8   Weena . . . Time Machine.] The protagonist of H. G. Wells's novel *The Time Machine* (1895), arriving in the year 802,701, rescues an Eloi girl, Weena, from drowning in a river, and befriends her.

86.6–12   Dr. Ronald Singer . . . discussing.] Singer's paper "The Boskop 'Race' Problem" was first published in *Man* in November 1958, after Eiseley's *The Immense Journey* appeared in print, but he had presented a preliminary version at the Fifth International Congress of Anthropological and

Ethnological Sciences in Philadelphia in September 1956, while Eiseley was writing the book.

94.32–36 Louis Agassiz . . . external forms."] See the "Essay on Classification" with which Agassiz begins his *Contributions to the Natural History of the United States of America* (1857).

95.10–14 the Reverend Mr. Kirby . . . him?"] See chapter 1 of William Kirby's *On the Power, Wisdom, and Goodness of God, as Manifested in the Creation of Animals, and in Their History, Habits, and Instincts* (1836).

96.3–5 "Geology . . . later."] See chapter 11 of *Typical Forms and Special Ends in Creation* (1857) by James McCosh and George Dickie.

99.4–6 one author . . . *Worlds?*] See *The Mystery of Being* (1863) by schoolmaster Nicholas Odgers (1839–1889).

114.8–13 "What . . . lesser machines."] See the Introduction to Thomas Hobbes's *Leviathan* (1651).

121.1–3 "What Next . . . Themselves."] See Waldemar Kaempffert, "What Next in the Attributes of Machines? It Might Be Power to Reproduce Themselves," *New York Times*, May 3, 1953.

121.5–8 "It does not seem . . . other hand . . ."] See Jonathan Cohen, "Can There Be Artificial Minds?," *Analysis*, December 1955.

124.23–26 Darwin . . . pond."] See Darwin's letter of February 1, 1871, to Joseph Dalton Hooker.

125.35–126.3 Woodger . . . happen."] See chapter 9 of Joseph Henry Woodger's *Biological Principles: A Critical Study* (1929).

129.5–11 *New York Times* . . . artificially."] No such headline or article is known to have appeared in *The New York Times*, but Lepeshinskaya's claims were widely quoted in U.S. newspapers in December 1952; see, for instance, "Creation of Life a Goal," *Kansas City Times*, December 3, 1952.

129.39–130.3 "To grasp . . . capacity."] See Ludwig von Bertanlanffy's "An Outline of General System Theory," published in *The British Journal for the Philosophy of Science* in August 1950.

132.16–17 Hardy . . . behind."] See the final lines of Thomas Hardy's poem "The Last Chrysanthemum," first collected in *Poems of the Past and Present* (1901).

THE FIRMAMENT OF TIME

134.1–11 *The splendours* . . . SHELLEY] See Shelley's "Adonais" (1821), stanza 4.

135.3–6 *That then* . . . DONNE] See Donne's commentary on Genesis 1:1 in his *Essayes in Divinity* (1651).

136.18–20 A scientist . . . world of illusions.] Probably Arthur James Balfour, who writes in his "Reflections Suggested by the New Theory of Matter" (*Popular Science Monthly*, October 1904): "It is presumably due to these circumstances that the beliefs of all mankind about the material surroundings in which it dwells are not only imperfect but fundamentally wrong. It may seem singular that down to, say, five years ago, our race has, without exception, lived and died in a world of illusions; and that its illusions, or those with which we are here concerned, have not been about things remote or abstract, things transcendental or divine, but about what we can see and handle, about those 'plain matters of fact' among which common sense daily moves with its most confident step and most self-satisfied smile."

138.20–24 Sir Thomas Browne . . . designs."] See section 14 of Browne's *Religio Medici* (1643).

142.1–3 Kepler . . . the gears."] See Kepler's letter to J. G. Herwart von Hohenburg, February 10, 1605.

142.34–36 Kant . . . the Void."] See Kant's *Allgemeine Naturgeschichte und Theorie des Himmels* (*Universal Natural History and Theory of Heaven*, 1755).

144.40–145.2 he states . . . an end."] See Hutton's *The Theory of the Earth* (1788).

147.23–32 Hutton . . . in another."] See *The Theory of the Earth* (1788).

150.2–10 La Fontaine . . . conversation."] See the testimony of M. de Liancourt, quoted in *Mémoires pour servir à l'histoire de Port-Royal*, volume 2 (1736), by Nicolas Fontaine (1625–1709).

151.3–4 *The world* . . . AGASSIZ] See "America the Old World," *The Atlantic Monthly*, March 1863.

152.28–30 Ray . . . produced."] See Ray's letter of October 8, 1695, to Edward Lhwyd.

153.3–7 Thomas Jefferson . . . broken."] See *Notes on the State of Virginia* (1785), chapter 6.

153.15–17 "To suppose . . . seventeenth-century naturalist] See William Cole's letter of March 27, 1685, to John Ray, collected in *Philosophical Letters between the Late Learned Mr. Ray and Several of His Ingenious Correspondents, Natives and Foreigners* (1718).

154.9–10 began to be whispered . . . Seas."] See John Ray's letter of October 22, 1684, to Dr. Tancred Robinson, collected in *Philosophical Letters between the Late Learned Mr. Ray and Several of His Ingenious Correspondents, Natives and Foreigners* (1718).

155.9–12 Peter Collinson . . . depended on."] See Collinson's letter to John Bartram of September 20, 1736 (collected in *Memorials of John Bartram and Humphry Marshall*, 1849).

155.13–17 Bartram . . . Natural History."] See John Bartram's letter of December 15, 1746, to Johann Friedrich Gronovius.

155.20–21 "The frogs . . . the King."] See John Fothergill to William Bartram, in an undated letter, c. 1772.

155.22–31 "I received . . . extremely ignorant."] See John Fothergill's letter to John Bartram of July 8, 1774.

155.32–34 "That all petrifactions . . . another correspondent] See Johann Friedrich Gronovius's undated letter of 1754 to John Bartram.

155.36–156.2 "Every uncommon . . . to America. . . ."] See Peter Collinson's letters to John Bartram in *Memorials of John Bartram and Humphry Marshall* (1849): an undated letter of 1736, a letter of February 3, 1736 or 1737, and a letter of May 17, 1768.

156.3–9 Collinson . . . *annihilated*."] See Collinson's letter to John Bartram, May 17, 1768.

157.31–34 "At certain periods . . . ridiculous."] See Broad's "The New Philosophy: Bruno to Descartes," *Cambridge Historical Journal*, January 1944.

158.3–10 "Half a century . . . science."] See *Natural Science and Religion: Two Lectures Delivered to the Theological School of Yale College* (1880), lecture 1 ("Scientific Beliefs").

158.22–28 Cuvier . . . the world."] See *Essay on the Theory of the Earth* (1813), Robert Kerr's translation of Cuvier's *Recherches sur les ossements fossils de quadrupèdes* (1812).

159.8–16 Adam Sedgwick . . . another planet."] See *A Discourse in the Studies of the University* (1833).

159.19–24 "It can be shown . . . obvious . . ."] See "A Period in the History of Our Planet," *Edinburgh New Philosophical Journal*, July–October 1843.

160.33–36 William Smith . . . pursuit of a hare,"] See the Introduction to Smith's *Stratigraphical System of Organized Fossils, with Reference to the Specimens of the Original Geological Collection in the British Museum* (1817).

161.11–16 Charles Gillespie . . . other way about."] See *Genesis and Geology: A Study in the Relations of Scientific Thought, Natural Theology, and Social Opinion in Great Britain, 1790–1850* (1951), chapter 6.

162.34–39 As early as 1829 . . . maintained."] See Lyell's letter of February 7, 1829, to his father, also Charles Lyell.

167.3–6 *If we can* . . . 1836] See "A Visit to the Zoological Gardens," *New Monthly Magazine*, August 1836.

173.25–26 the apt phrase . . . departure"] See Wallace's essay "On the Tendency of Varieties to Depart Indefinitely from the Original Type," published in the *Journal of the Proceedings of the Linnean Society* in 1859; Darwin received an advance copy from Wallace on June 18, 1858.

176.8–10   Professor Lovejoy . . . life."] See Arthur O. Lovejoy's *The Great Chain of Being: A Study in the History of an Idea* (1933), chapter 7.

176.14–17   Buckland . . . balance of creation."] See William Buckland's paper "On the Discovery of Coprolites, or Fossil Fæces, in the Lias at Lyme Regis, and in Other Formations," originally presented on February 6, 1829, and published in the *Transactions of the Geological Society of London*, volume 3 (1835).

177.22–28   In 1835 . . . qualities of a species."] See Blyth's article "An Attempt to Classify the 'Varieties' of Animals, with Observations on the Marked Seasonal and Other Changes Which Naturally Take Place in Various British Species, and Which Do Not Constitute Varieties," *The Magazine of Natural History*, January 1835.

177.29–40   by 1837 . . . *common heritage?*"] See Blyth's "On the Psychological Distinctions between Man and All Other Animals," *The Magazine of Natural History*, January 1837.

183.3–5   *Here below . . .* NEWMAN] See Newman's *An Essay on the Development of Christian Doctrine* (1845), chapter 1, section 1.

184.32–185.4   "If species . . . discovered."] See note 158.22–28.

185.26–30   "Man . . . contriving power."] The original text, from Sedgwick's *A Discourse on the Studies of the University* (1833), reads as follows: "Geology, like every other science when well interpreted, lends its aid to natural religion. It tells us, out of its own records, that man has been but a few years a dweller on the earth; for the traces of himself and of his works are confined to the last moments of its history. Independently of every written testimony, we therefore believe that man, with all his powers and appetencies, his marvellous structure and his fitness for the world around him, was called into being within a few thousand years of the day in which we live—not by transmutation of species, (a theory no better than a phrensied dream), but by a provident contriving power."

186.5–13   Sir James Geikie . . . mile after mile . . ."] See *The Founders of Geology* (1901), lecture 6.

187.7–14   W. E. Baker . . . that of man."] See W. E. Baker and H. M. Durand, "Sub-Himálayan Fossil Remains of the Dádúpur Collection," *Journal of the Asiatic Society of Bengal*, November 1836.

187.16–19   Falconer's . . . exhibits now."] See P. T. Cautley and H. Falconer, "Notice on the Remains of a Fossil Monkey from the Tertiary Strata of the Sewalik Hills in the North of Hindoostan," presented on June 14, 1837, and subsequently published in the *Transactions of the Geological Society of London*, volume 5 (1840).

187.22–30　"was clear evidence . . . terrestrial life."] See Hugh Falconer, "On the Asserted Occurrence of Human Bones in the Ancient Fluviatile Deposits of the Nile and Ganges; with Comparative Remarks on the Alluvial Formation of the Two Valleys," *Quarterly Journal of the Geological Society of London*, March 22, 1865.

191.1–5　"We do not . . . his ancestors."] See *The Descent of Man, and Selection in Relation to Sex* (1871), chapter 2.

192.1–3　its describer . . . *barbarous races*,"] See D. Schaaffhausen, "On the Crania of the Most Ancient Races of Man," *The Natural History Review*, April 1861; originally published in German in 1858. (*Homo neanderthalensis* was formally named and described by geologist William King in 1864.)

194.2–3　Thomas Hardy . . . oblivion."] See "Heredity," in Hardy's *Moments of Vision and Miscellaneous Verses* (1917).

194.33–35　nineteenth-century writers . . . ferocious."] See *The Development Theory: A Brief Statement for General Readers* (1884), by Joseph Y. Bergen and Fanny D. Bergen.

197.2–25　I made these remarks . . . the ape."] See "Neanderthal Man and the Dawn of Human Paleontology," included in the present volume on pages 422–34.

199.3–5　*Be not under* . . . BROWNE] See section 14 of Browne's *Christian Morals*, published posthumously in 1716.

199.10–14　"He who fights . . . meet."] See Kierkegaard's *Edifying Discourses in Diverse Spirits* (1847).

202.6–7　In Browning's words . . . they it."] See "Cleon," first published in *Men and Women* (1855).

202.36–38　"Whatever interrupts . . . animal."] See *Scepticism and Animal Faith* (1923), chapter 20.

205.18–22　"Trial . . . it now is."] See the opening sentence of *The Great Instauration* (1620).

206.12–20　"It is not fear . . . the species."] See "The Mythic Past," chapter 2 of *The Territory Ahead* (1958).

211.22　Berdyaev] Nikolai Alexandrovich Berdyaev (1874–1948), Russian religious and political philosopher.

219.3–5　*The very design* . . . EMERSON] See "Poetry and Imagination," first collected in *Letters and Social Aims* (1875).

222.1–5　"In astronomy . . . transformations."] See Emerson's lecture "Natural History of Intellect," as published posthumously in 1893.

222.18–19　"He can be . . . first creation."] See note 135.3–6.

222.38–40 one astute biologist . . . ingenuity."] See Robert Gesell's lecture "The Wisdom of the Body and the Wisdom of the Mind," *The Quarterly Review of the Michigan Alumnus*, Autumn 1953.

222.40–223.4 Pascal . . . we do not destroy."] See Pascal's *Pensées* (1670), 377.

230.28–30 "The special value . . . knower."] See James Street Fulton, *Science and Man's Hope* (1954).

233.23–27 Kierkegaard . . . understood."] From an undated entry in Kierkegaard's journals of 1849.

234.1–2 words of Pascal . . . found me."] See "Le Mystère de Jésus," *Pensées*, 553.

234.18–21 "With reference . . . transcendentalists."] From Thoreau's journals, June 7, 1851.

THE UNEXPECTED UNIVERSE

242.1–3 *The universe* . . . HALDANE] See "Possible Worlds" in Haldane's *Possible Worlds and Other Essays* (1927).

242.4–6 *If you do not* . . . HERACLITUS] See fragment 7 of Heraclitus's "On Nature" (c. 535–475 B.C.E.).

243.3–5 *The winds* . . . BURTON] See "Democritus Junior, to the Reader," an introductory section of Burton's *Anatomy of Melancholy* (1621).

246.2–3 the words of Kazantzakis . . . heart."] See Kazantzakis's *The Odyssey: A Modern Sequel* (1938), book 13, line 1218.

246.31–33 It was once . . . winds."] See Eratosthenes's *Geography* (c. 245–40 B.C.E.), book 1.

249.8–10 Cook . . . high latitude."] From Cook's journals, March 31, 1770.

249.18–19 "exhibited such . . . represented by it."] From Cook's journals, December 26, 1772.

249.28–30 one of his scientist . . . being shipwrecked."] See the entry for December 27, 1773, in George Forster's *A Voyage Round the World, in His Britannic Majesty's Sloop, Resolution* (1777).

249.40–250.2 "I can be bold . . . have done."] See volume 2 of Cook's *A Voyage Towards the South Pole and Round the World, Performed in H.M.S. Resolution and Adventure* (1777).

254.26–27 "God is dead . . . permitted."] See Fyodor Dostoyevsky's novel *The Brothers Karamazov* (1880), part 4, book 11, chapter 4.

255.5–14 Plato . . . with confidence."] See Plato's *Phaedo* (c. 360 B.C.E.).

259.3–7 *Imagine God . . .* DONNE] See Donne's sermon on Isaiah 65:20, "Preached to the King, at White-Hall" (c. 1624–27).

261.23–24 Francis Bacon . . . made to do."] See Bacon's *Novum Organum* (1620), book 2, section 10.

262.7–11 John Donne . . . are not.] See "A Nocturnal upon St. Lucy's Day, Being the Shortest Day," first published posthumously in 1633.

262.18–22 a little book of essays . . . Natural?] See *The Firmament of Time* (1960); its sixth chapter, "How Natural Is 'Natural'?," appears on pages 219–35 in the present volume.

264.16–23 "In every nature . . . neglected."] See the conclusion of *Natural Theology; or, Evidences of the Existence and Attributes of the Deity, Collected from the Appearances of Nature* (1802).

264.29 "natural government . . . John Hunter.] See "Observations in Natural History," collected posthumously in *Essays and Observations on Natural History, Anatomy, Physiology, Psychology, and Geology* (1861).

265.9–10 "no vestige . . . end."] See note 144.40–145.2.

265.38 Alfred Russel Wallace . . . departure."] See note 173.25–26.

268.25–26 Einstein . . . the universe.] See Einstein's letter of December 4, 1926, to Max Born.

269.5–8 "The most important . . . anticipation."] See *The Principles of Mechanics, Presented in a New Form* (1899).

272.33–34 "Force maketh . . . once written.] See Bacon's essay "Of Nature in Men" (1617).

273.8–12 "It is very unhappy . . . directly."] See Emerson's "Experience," collected in *Essays: Second Series* (1844).

273.19–20 "We have learned . . . same thinker] See Emerson's "Illusions," collected in *The Conduct of Life* (1860): "What terrible questions we are learning to ask!"

284.12–13 a student . . . anthropologist] Frank G. Speck (1881–1950), subsequently Eiseley's dissertation advisor at the University of Pennsylvania.

293.11–12 "Certain coasts . . . shipwreck."] See Robert Louis Stevenson's essay "A Gossip on Romance" (1882).

299.18–19 Things . . . being calculated.] See the final chapter of Chesterton's novel *The Return of Don Quixote* (1927).

299.30–31 "The dangerous gift . . . termed it] See Goethe's "Probleme" (1823).

301.37–39 "Love not the world . . . in the world."] See I John 2:15.

302.15–16 "There is no boon . . . written harshly] See William Graham Sumner's essay "The Boon of Nature," first published in *The Independent*, October 27, 1887.

305.20–21 Bacon's forgotten . . . life."] See section 49 of Bacon's *Novum Organum* (1620).

307.3–6 *As to what happened . . .* XENOPHON] See Xenophon's *Hellenica* (c. 393–357 B.C.E.), book 7.

309.35 "Nature . . . Thoreau once said.] From Thoreau's journals, May 27, 1841.

310.3–9 "The human brain . . . out of it."] From Thoreau's journals, January 30, 1854.

313.13–15 Alfred Russel Wallace . . . impoverished world."] See *The Geographical Distribution of Animals* (1876), volume 1, chapter 7.

313.31–35 "The association of unusual . . . evolve together."] See *The Meaning of Evolution: A Study of the History of Life and of Its Significance for Man* (1949).

317.4–9 J. K. Charlesworth . . . tectonic history."] See volume 2 of Charlesworth's *The Quaternary Era, with Special Reference to Its Glaciation* (1957).

320.5–7 Leonard Silliman . . . providence."] See the opening sentence of "The Genesis of Man," *The International Journal of Psycho-Analysis*, January 1953.

320.18–24 the legendary cycles . . . their dreams."] See "The Blackfoot Genesis" in George Bird Grinnell's *Blackfoot Lodge Tales: The Story of a Prairie People* (1907).

322.3–6 "hazardously supplied . . . higher centers."] See F. A. Gibbs, "The Most Important Thing," *American Journal of Public Health*, December 1951.

325.3–5 "A creature without . . . SANTAYANA] See "The Realm of Truth" (1937), the third book of Santayana's *Realms of Being* (1927–40).

325.27–28 "There is no circulation . . . protested.] From Thoreau's journals, March 19, 1841.

326.26–29 Darwin . . . shall be mine."] See Darwin's letter of July 24, 1834, to John Stevens Henslow.

326.29–31 Thoreau . . . sunbeam."] From Thoreau's journals, March 17, 1841.

326.33–37 Ellery Channing . . . I judge."] Channing wrote these lines in his notebook after Thoreau's death on May 6, 1862.

327.12–14 "But . . . living person can tell."] See L. Frank Baum's *The Wonderful Wizard of Oz* (1900), chapter 10.

327.29–31 "It was a lonely . . . while before."] See *The Wonderful Wizard of Oz*, chapter 4.

330.23–25 Darwin writing . . . search for "facts."] See Darwin's letter to Lyell of March 17, 1863.

330.30–31 "Forfend me . . . Scotch prudence."] See Darwin's letter of May 12, 1847, to Joseph Dalton Hooker.

331.31–36 "Lord . . . queer cases of variation."] See Darwin's letter of November 12, 1862, to Joseph Dalton Hooker.

331.37–38 "I sat one evening . . . pigeon fanciers."] See Darwin's letter of November 27, 1859, to Thomas Henry Huxley.

332.1–13 "How awfully flat . . . nothing new."] See Darwin's letters of August 11, 1860, to Charles Lyell ("The Bishop makes a very telling case"); November 18, 1856, to Joseph Dalton Hooker ("A seed has just germinated"); March 26, 1863, to Hooker ("I am like a gambler"); November 12, 1862, to Hooker ("I am horribly afraid" and "I trust to a sort of instinct"); and May 7, 1855, to W. D. Fox ("All nature is perverse").

335.3–5 "What a book . . . works of Nature."] From a letter of July 13, 1856, to Joseph Dalton Hooker.

335.31–336.1 Stanley Hyman . . . savage tribes."] Hyman discusses the anonymous review, published in *The Spectator* on March 12, 1871, in *The Tangled Bank: Darwin, Marx, Frazer, and Freud as Imaginative Writers* (1962).

336.36–337.1 One contemporary . . . animal."] See the recollections of John Weiss in "Thoreau," first published anonymously in the *Christian Examiner*, July 1865.

337.24–27 In one passage . . . outrage at most."] From Thoreau's journals, September 3, 1841.

337.40–338.1 "If we see nature . . . decays."] From Thoreau's journals, March 13, 1842.

338.11–14 "It appeared to have . . . beneath it."] See *Walden; or, Life in the Woods* (1854), chapter 17.

338.15–16 "fox belongs . . . the village."] From an undated entry in Thoreau's journals, c. 1842–44.

338.23–25 "Some . . . happened to the universe."] See the chapter title "Thursday" in Thoreau's *A Week on the Concord and Merrimack Rivers* (1849).

338.38–339.1 "I am sensible . . . kind of fiction."] See *Walden*, chapter 5.

339.11–13 "for if the voices . . . I hear."] From Thoreau's journals, February 11, 1853.

339.20–21 "It is ebb . . . reports."] From Thoreau's journals, March 5, 1858.

339.31–36 "I did not formerly . . . unrecognized service."] See note 191.1–5.

340.13–14 "consists not . . . his instincts"?] From Thoreau's journals, July 16, 1850.

340.19–24 Viking Eddas . . . other spare.] See the tenth-century *Völuspá* or "Wise-Woman's Prophecy."

340.36–38 "If the condition . . . substitute?"] See *Walden*, chapter 18.

341.15–16 The soul . . . never seen.] A view attributed to the Alaskan shaman Najagneg in H. Ostermann's *The Alaskan Eskimos, as Described in the Posthumous Notes of Dr. Knud Rasmussen* (1952).

341.29–30 what Coleridge . . . the universe.] See "The Destiny of Nations. A Vision," first published in its entirety in 1817: "For all that meets the bodily sense I deem / Symbolical, one mighty alphabet / For infant minds."

343.3–5 *I say* . . . CALIBAN] See *The Tempest*, III.ii.52–53.

343.28–29 Herman Melville . . . any map."] See *Moby-Dick* (1851), chapter 12, on Queequeg's birthplace, Kokovoko: "It is not down in any map; true places never are."

346.40–347.2 the young man . . . *species*."] From an entry in Darwin's *Ornithological Notes*, c. September–October 1835.

348.14–15 Everything . . . everything else.] See *My First Summer in the Sierra* (1911), chapter 6: "When we try to pick out anything by itself, we find it hitched to everything else in the universe."

348.38–349.4 Sir Francis Bacon . . . confined.] See book 2 of Bacon's *De Augmentis Scientiarum* (1623): "however it be that vast and strange swellings . . . take place occasionally in nature,—whether of the sea or the clouds or the earth or any other body,—nevertheless all such exuberances and irregularities are by the nature of things caught and confined in an inextricable net, and bound down as with a chain of adamant."

350.36–40 Darwin . . . origin of species."] See Darwin's letter to Wagner, October 13, 1876.

352.31–36 a Brocken specter . . . evasion.] De Quincey discusses the phenomenon of the Brocken spectre—the magnified shadow of an observer, cast upon distant clouds or fog—in "The Apparition of the Brocken," first published in 1845 and later retitled "Dream Echoes Fifty Years Later": "Make the sign of the cross, and observe whether he repeats it, (as on Whitsunday he surely ought to do.) Look! he *does* repeat it; but these driving April showers perplex the images, and *that*, perhaps, it is which gives him the air of one who acts reluctantly or evasively."

354.6–12 "The crabs . . . deliberation."] From a letter of July 31, 1910, by

Henry "Birdie" Bowers, quoted in Cherry-Garrard's *The Worst Journey in the World: Antarctic, 1910–1913* (1922).

355.5–8 Thoreau . . . acorns."] From Thoreau's journals, December 21, 1840.

355.22–25 Sherwin Carlquist . . . obsolescent."] See Carlquist's *Island Life: A Natural History of the Islands of the World* (1965).

356.37–38 the isle . . . mortal business."] See *The Tempest*, I.ii.407.

358.35–37 "This thing . . . acknowledge mine."] See *The Tempest*, V.i.275–76.

358.38–40 "There is no . . . service to man."] See the opening chapter of Wilson's *Signs and Wonders upon Pharaoh: A History of American Egyptology* (1964).

359.6–10 spirit Ariel . . . follow it"?] See *The Tempest*, III.ii.148–49.

361.3–6 *There is strong* . . . McGLASHAN] See *The Savage and Beautiful Country* (1967), chapter 7.

365.19–20 Dante . . . other stars."] See the final line of Dante's *Paradiso* (1320).

365.31–32 "The conviction . . . plague of man."] See the "Apology for Raymond Sebond" in Montaigne's *Essais* (1580).

366.30 "Double in ourselves . . . Montaigne.] See "Of Glory" in Montaigne's *Essais* (1580).

368.8–10 John Donne . . . beare."] See Donne's sermon on Psalms 51:7 ("Purge me with hyssop").

368.28–33 Emerson . . . worthless?"] See "The Over-Soul," from *Essays: First Series* (1841).

369.5–8 "Cosmic nature . . . the tiger."] See Huxley's "Evolution and Ethics," first presented as the Romanes Lecture at Oxford on May 18, 1893.

374.8–9 "the bright stranger . . . Emerson] See the entry for August 29, 1849, in Emerson's journals: "Love is the bright foreigner, the foreign self."

375.3–8 *Only to a magician* . . . BEAGLE] See Beagle's novel *The Last Unicorn* (1968).

382.37–39 "I am the thing . . . Charles Williams] See Williams's play *Thomas Cranmer of Canterbury* (1936).

386.38–40 Thoreau . . . Royal Society.] From Thoreau's journals, October 4, 1859: "Your greatest success will be simply to perceive that such things are, and you will have no communication to make to the Royal Society."

390.16–17 "We must regard . . . John Joly] See "The Abundance of Life," a paper presented before the Royal Dublin Society on November 19, 1890, and published in *The Scientific Proceedings of the Royal Dublin Society*, volume 7 (1891–92).

397.40–398.2 Sir Francis Bacon . . . into view."] See the "Aphorisms on the Composition of the Primary History" in Bacon's "Preparative towards a Natural and Experimental History," first published with the *Novum Organum* in 1620.

UNCOLLECTED PROSE

410.29 U.P.] Union Pacific.

414.5 "The Big Rock Candy Mountains"] A folk song first recorded by Harry McClintock (1882–1957) in 1928.

435.6 Petrie] Flinders Petrie (1853–1952), English Egyptologist.

440.3–4 Richard Jeffries' . . . winter.] See Jeffries's "Outside London," collected in *The Open Air* (1905).

452.20–21 "There has been . . . once wrote] See "Monday," in *A Week on the Concord and Merrimack Rivers* (1849): "There was but the sun and the eye from the first."

Index

Abyssal fauna, 21–30

Adaptation, 38, 47, 52, 97, 100, 177, 196

Adaptive radiation, 171, 267, 351

Africa, 154–55, 366, 392; Bushmen in, 85–87; early hominids in, 56–57, 74, 81–82, 84–88, 190, 195–96; early primates in, 65, 234, 319; effect of ice age in, 316–18; Hottentots in, 63, 425, 430, 432

Agassiz, Louis, 27, 94, 151, 158–59, 187–88

Ainu, 431

Alaska, 341

Alexandria, Egypt, 397

Algeria, 211

Alps, 192

Amazon River, 157

Amphibians, 19, 31, 43, 93, 96, 155, 164–65, 343

Amundsen, Roald, 250

Anderson, William, 251

Andes Mountains, 22

Angiosperms, 41–50

Antarctic, 29, 192, 243, 247, 249–50, 252, 436

Anthropology, 51, 62, 76, 82, 260

Antimatter, 266

Ants, 438–39

Apes, 49–50, 53, 56–67, 70, 72, 74, 78–79, 94, 186–87, 190–91, 195–96, 234, 243, 245, 319, 368, 422–28, 430, 433

Arboreal primates, 8–9, 49–50, 64–65, 195, 319, 422, 431

Archaeopteryx, 44, 48, 50

Archean (Archeozoic) eon, 164, 266

Archencephala, 63–64

Archeology, 261, 268, 282, 429–30, 432

Archeozoic (Archean) eon, 164, 266

Arctic, 312–13, 317, 322

Arcturus, 362

Argos (dog), 256–57

Argyll, Duke of (George Campbell), 190, 430

Aristophanes, 7

Arrhenius, Svante, 30, 126

Asia, 55–56, 64, 155, 193, 215, 317–21, 366

Athens, Greece, 397

Atlantic Ocean, 21–22, 24, 26, 73, 193, 354

Atom, primordial, 266

Auschwitz concentration camp, 271

Australia, 33, 75, 154, 190, 247, 312, 397, 425, 428

Australopithecus, 74

Azoic era, 22

Aztecs, 407

Bacon, Francis, 11, 261, 272, 299, 305, 348–49, 398; *Novum Organum*, 205; on earthly paradise, 204–6, 208, 211, 217

Badlands, 107

Baker, W. E., 187

Balance of creation, 173–76, 179, 263–65

Balfour, Arthur James, 136

Banks, Joseph, 251–52

Bartram, John, 155

Bathybius haeckelii, 24–25

Beagle, H.M.S., 158, 171, 177–78, 250, 256, 326, 329–31, 333, 346

Beagle, Peter, 375

Belgium, 194

Belsen concentration camp, 273

Berdyaev, Nikolai, 211

Bering Strait, 251, 316

Berlin, Germany, 192, 366

Bernard, Claude, 239

Bertalanffy, Ludwig von, 130

Bible, 138–41, 146–47, 156–57, 159–60, 263, 275, 280, 283, 295, 301, 333, 387

"Big Rock Candy Mountain," 414

Bipedalism, 42, 159, 195–96, 318–19

Birds, 7, 94, 297, 341, 370–71, 443, 446, 454; evolution of, 43–44, 47–48, 50; judgment of, 103–10; and machines, 113–20; and snakes, 231–33

Bison, 48, 50, 71, 307, 316

Black, MacKnight, 14

491

Blackfoot, 320
Blacksnake, 128, 231–33, 451
Blake, Carter C., 427–28
Blake, William, 451
Blyth, Edward, 177–78
Borek, Ernest: *Man, the Chemical Machine*, 440
Borneo, 190, 195, 426
Boskopoid people, 85–88
Brain, 42; hominid, 70, 319–20, 370; human, 51–61, 63–65, 67, 69–71, 74–80, 82–88, 102, 132, 192, 194, 228, 266–67, 277–78, 314, 321–22, 353, 370; Neanderthal, 192, 194, 197, 427–28, 431–33, 436–37
Brazil, 337
Brewster, David: *More Worlds Than One*, 98–99
Britain, 26, 51, 74, 147, 151–52, 154, 159, 170–71, 188, 249, 252, 269, 316–17, 322, 325, 329–31, 333, 339, 428–30
British Association for the Advancement of Science, 63, 333
Broad, C. D., 157
Bronze age, 247
Browne, Thomas, 138, 199
Browning, Robert, 202
Brückner, John, 174–76, 178–79; *Philosophical Survey of the Animal Creation*, 174
Buchanan, John Young, 25
Buchenwald concentration camp, 208, 273
Buckland, William, 176
Buddhism, 289
Burial of the dead, 197, 261, 314, 397, 433
Burton, Robert, 243
Bushmen, 85–87
Byzantium, 345–46

Calcium, 6, 202, 446
California, 45, 413–14
Cambrian period, 163, 311–12
Cambridge University, 329–30, 333
Canada, 316
Cape Horn, 329, 336
Cape of Good Hope, 63
Carbon, 30, 124
Carbon-14 dating technique, 72–73, 75
Caribbean Sea, 73

Carlquist, Sherwin, 355
Catastrophism, 146–47, 157–63, 169, 174, 176, 185–86, 189, 266
Catfish, 17–18, 20
Caucasians, 85–86, 432
Cautley, Proby, 186–87
Cave paintings, 57, 77, 87, 244, 341, 397
Cecropia moths, 363
Celebes (Sulawesi), 422
Celestial machine, 142–44, 149–50, 157, 229, 263–64, 336
Cenozoic era, 187
Challenger expedition, 25–28
Chambers, Robert, 168
Chance, M. R. A., 55
Channing, William Ellery, 326
Charlesworth, J. K., 317
Cherry-Garrard, Apsley, 354
Chesterton, G. K., 299
Chichén Itzá, 244
Chile, 326
Chimpanzees, 64, 190–91, 194
China, 318, 320–21, 342
Chipmunks, 8
Choukoutien skull, 321
Christianity, 93, 97, 140, 145–46, 151–52, 167, 175, 184, 204, 210–11, 263, 326, 329, 422–23, 431
Cicadas, 439
Clarke, Howard, 255
Climate, world, 317–18
Coleridge, Samuel Taylor, 167–68, 174, 178, 341
Collinson, Peter, 155–56
Columbus, Christopher, 389
Computers, 281
Conan Doyle, Arthur: *The Lost World*, 23
Concentration camps, 208, 271, 273
Concord, Mass., 337
Congo, 86
Contingency evolution, 270–71, 295
Conus spurius atlanticus, 342
Cook, James, 243, 246–52, 257, 329, 351, 397
Cope, Edward Drinker, 37
Copernicus, Nicolaus, 97
Costabel beach, 289–93, 297, 300–301, 305
Crabs, 180–81, 354

Creation: balance of, 173–76, 179, 263–65; biblical, 140, 146, 159; evolutionary, 153, 158–60, 162, 165, 169, 173; special, 424
Cretaceous period, 16, 44, 46
Crickets, 448, 454
Crossopterygian, 35–36, 38–39, 354
Crows, 105–6, 183
Cummings, Ray, 99
Curaçao, 387–88, 399
Cuvier, Georges, 94, 158–60, 184–85, 424
Cyprus, 211

Dalrymple, Alexander, 248–50
Dante Alighieri, 256, 365
Dart, Raymond, 196
Darwin, Charles, 21, 23, 93, 137, 158–59, 167–68, 173, 175, 178–79, 185, 215, 229, 246, 254, 256–57, 302, 305, 347, 369–70, 373, 423, 441–42; *The Descent of Man*, 190, 335, 350, 423, 428; evolutionary theory of, 52, 61–62, 76, 99–101, 124, 169, 171–72, 188, 265–66, 299–300; on humans, 51–55, 57–58, 63, 69, 71, 189–92, 194–98, 426–32; on islands, 353–56; *On the Origin of Species*, 53–54, 97, 163, 168, 178, 213, 296, 299, 330, 332, 337, 339, 350, 423, 426, 429; on plants, 42, 177; in South America/Galápagos Islands, 250–51, 253, 326, 329, 331, 333, 341, 346, 350–51, 354; as "voyager," 325–42
Death, 151–65, 176
De Beer, Gavin, 79, 178
Degeneration, 429, 431–32
Deism, 156
De Quincey, Thomas, 352
Descartes, René, 149, 440
Devonian period, 33, 37, 164
Dinosaurs, 42, 46–47, 49, 117, 159, 225, 282, 367
Divergent evolution, 97, 100, 267
DNA, 279
Dogs, 240, 256–57, 307–9, 322, 380–81, 387–88, 444
Donne, John, 135, 221–22, 234, 259, 262, 368
Drennan, Matthew, 85–86
Dunsany, Lord (Edward Plunkett), 373, 446

Durand, H. M., 187
Dyersville, Iowa, 301

Eden, 80, 140, 146, 151, 154, 159, 190–91, 195, 198, 204–8, 211–12, 217, 220, 279, 344
Edinburgh University, 330–31
Edinger, Tilly, 60, 70
Egypt, ancient, 107, 363, 372, 429, 435
Einstein, Albert, 136, 299
Eiseley, Clyde (father), 2
Eiseley, Daisey Corey (mother), 297, 300–301
Eiseley, Mabel Langdon (wife), 33–34, 123, 438
Electron microscope, 127
Elements, 10, 30, 124, 200
Elephants, 154
Elihu, 275, 280
Elizabeth age, 165, 200, 269, 349, 429
Emerson, Ralph Waldo, 219, 222, 229, 273, 368, 378, 384, 452
Emiliani, Cesare, 72–74
Empiricism, 145
Endeavor, H.M.S., 250
Enlightenment, 148, 245, 330
Epeira, 112
Erosion, 141, 147–48, 154, 230
Eskimos, 58, 325, 339–41
Europe, 55–56, 63, 84, 87, 154, 184–88, 191–94, 316–17, 422, 427, 429, 431–32
Evolution, 20, 27–28, 38, 296–97, 330–31, 348, 367, 369–70, 374, 387, 442–44; of birds, 43–44, 47–48, 50, contingency, 270–71, 295; development of theory, 99–101, 143, 148, 151–53, 162, 167–79, 229, 427, 430; divergent, 97, 100, 267; of flowering plants, 43–50; human, 51–88, 183–98, 266–67, 340, 424–33; of mammals, 6–9, 43–44, 48, 55
Experiments, on animals, 149–50
Explorers Club, 33
Extinction, 9, 28, 48, 56, 71, 96, 151–63, 170, 173–77, 188, 282, 313, 316, 320, 344, 425, 431

Falconer, Hugh, 186–88
Falcons, 118–20
Faust, 256

Ferns, 8, 41, 93, 109, 127
Fertilization, 41, 43–48
Fetalization, 58–59, 76, 78, 82–88
Fire, 320–22
Fish, 33–39, 188
Fitzroy, Robert, 329–31, 333
Flood, biblical, 139, 146, 157, 263
Florence, Italy, 250
Flowering plants, 41–50
Fontaine, Nicolas, 150
Fontechevade skull, 74
Forbes, Edward, 22
Forster, Johann, 252
Fossils, 6, 16, 28, 91, 96, 233, 263, 276–77, 300, 307, 366, 395; and evolutionary theory, 140–41, 148, 152–54, 156, 158, 160–64, 174, 179, 184–85, 188; hominid, 5–6, 191, 423, 425, 428, 430–32; human, 51–55, 61–62, 64–67, 74–75, 86, 189–91, 423, 425, 428, 430–32; living, 21, 23, 75–76, 190; primate, 186–87
Foxes, 385–86
France, 57, 74, 143, 147, 149, 152, 158, 171, 174, 193, 197, 245, 330, 433
Francis of Assisi, 366, 372
French Revolution, 147, 152, 171, 174, 330
Freud, Sigmund, 136, 214, 254, 299
Frogs, 19, 31, 43, 93, 155
Fuhlrott, Johann, 191, 426

Galápagos Islands, 250–51, 253, 331, 333, 341, 346, 350–52, 354–55
Galaxies, 297
Galileo Galilei, 199
Geikie, James, 186
Gelli, Giovanni Battista, 250
Genetics, 52, 137, 189, 266, 282–83, 351, 451
Geological prophecy, 100–101, 158–59, 165, 169, 179
Geological Survey, U.S., 312
Geological time, 21, 23, 25, 41–42, 47, 60, 71–72, 87, 143, 149, 151–52, 159, 162–63, 172–73, 179, 189, 202, 266
Geology, historical, 143, 152
Geostrophic cycle, 147–48
Germany, 27, 156, 158, 191–94
Gervais, Paul, 66
Gesell, Robert, 222

Gibbon ape, 319
Gibbs, Frederick, 322
Gibraltar, 430
Gillispie, Charles, 161
Glaciers, 186, 188, 192, 310–14
Goethe, Johann Wolfgang von, 296
Gomulka, Wladyslaw, 445
Gorillas, 70, 83, 190–91, 194–96, 368, 427
Grant, Robert, 330
Grasses, 45, 48–50, 313, 319
Gravitation, 141–42, 229
Gray, Asa, 158
Great Depression, 271
Great Lakes, 155
Greece, ancient, 10, 93, 139, 185, 212, 246–47, 250, 253, 255, 285, 370, 397, 422
Gregory, William King, 65
Gulf of Mexico, 13–15
Gulls, 372–73
Gunz glaciation, 74

Haeckel, Ernst, 24–25, 129
Haldane, John Burdon, 79, 242
Halley, Edmund, 142, 150, 171–72
Halley's comet, 293
Hannibal, 154
Hardy, Thomas, 132, 194
Hawaii, 250, 252
Hawkesworth, John, 252
Henslow, John, 251, 329
Heraclitus, 242
Hertz, Heinrich, 269
Hibernation, 43, 355
Hieroglyphs, 282–83, 342
Himalaya Mountains, 16, 186–87, 319, 349, 450
Hinduism, 281
Historical geology, 143, 152
Hobbes, Thomas, 114
Homer: *Odyssey*, 243–58
Hominids, 64, 179, 243, 312, 354, 356, 366, 368; brains of, 70, 319–20, 370; at end of ice age, 318–22; fossils of, 5–6, 191, 423, 425, 428, 430–32; missing links, 53–54, 63, 69, 126, 426; Neanderthals, 191–94, 196–97, 423–33
Homo alalus, 425
Homo diluvii testis, 96

Homo faber, 223, 235
Homo sapiens, 50, 94, 100, 203, 234, 250, 312–13, 394–95, 400, 438; bipedalism of, 42, 159, 195–96, 318–19; brain of, 51–61, 63–65, 67, 69–71, 74–80, 82–88, 102, 132, 192, 194, 228, 266–67, 277–78, 314, 321–22, 353, 370; burial of the dead, 197, 261, 314, 397, 433; Darwin on, 51–55, 57–58, 63, 69, 71, 189–92, 194–98, 426–32; evolution of, 51–88, 183–98, 266–67, 340, 424–33; fossils of, 51–55, 61–62, 64–67, 74–75, 86, 189–91, 423, 425, 428, 430–32; migrations of, 56, 71, 316; speech of, 53–54, 76, 321, 354, 370–71; tools of, 56–58, 74, 87, 135, 185, 188, 192–93, 196, 223, 235, 243, 261, 270, 298, 321, 339, 429–30
Hopis, 272
Horses, 48, 66–67
Hottentots, 63, 425, 430, 432
Hudson, W. H., 445
Humboldt Current, 343
Hummingbirds, 47–48
Hunt, Leigh, 167
Hunter, John, 264–65
Hurzeler, Johannes, 62, 66–67
Hutton, James, 143–49, 152, 154, 157, 161–63, 173, 175, 179, 186, 188, 229, 263–64
Huxley, Thomas H., 24–25, 27, 63–64, 190, 192, 333, 369, 428; *Evidence as to Man's Place in Nature*, 64
Hydrogen, 30, 128

Ice age, 14, 54–55, 84–85, 89, 148, 154, 179, 183–84, 186, 188–89, 279, 307, 309–12, 314, 322, 395, 425, 436–37; and cave paintings, 57, 77, 87, 244, 341, 397; dating of, 72–75; development of human brain during, 58, 74, 78; effect in Africa, 316–18; impact on mammals, 50, 56, 71, 313, 322; migrations during, 56, 71; Neanderthals in, 192–94, 196, 215
India, 186–87, 319
Indiana, 411, 414
Industrialization, 326
Insects, 48, 438–39, 448, 454
Instinct, 42, 77–78, 202, 234, 370

Ireland, 427, 432
Irish elk, 70, 156
Iron, 202, 446
Iroquois, 380
Ishi, 366
Islands, 351–56, 358, 442
Italy, 62, 69, 194, 250, 253, 396
Ithaca, 245, 251, 253, 255–56

Jackrabbits, 323–24
Janson, H. W., 422
Japan, 431
Java man, 54–55, 64, 366
Jefferies, Richard, 440
Jefferson, Thomas, 22, 153–54
Jesus, 295, 365
Job, 275, 280, 387
Judaism, 139
Jupiter, 98–99, 141

Kalahari Desert, 85, 87
Kangaroo rat, 8
Kansas, 316, 327
Kansas City, Mo., 415, 417, 420
Kant, Immanuel, 142
Kazantzakis, Nikos, 246, 256
Keilor skull, 75
Keith, Arthur, 85–86
Kelvin, Baron (William Thomson), 126
Kepler, Johannes, 142
Kestrels, 118–20
Khrushchev, Nikita, 445
Kierkegaard, Søren, 199, 201, 208, 210, 215
King, James, 252
King, Philip, 312
Kirby, William, 95, 99
Korean War, 207
Krishna, 281

La Chapelle-aux-Saints cave, 197, 433
Lake, Philip, 184
Lamarck, Jean-Baptiste, 157, 170–71, 330
La Naulette skull, 428, 430
Language, 53–54, 76, 321, 354, 370–71
Laplace, Pierre-Simon, 144, 229; *Treatise on Celestial Mechanics*, 142
Lapland, 317
Lascaux cave paintings, 244
Leakey, Louis B., 196

Leavenworth, Kans., 415, 417, 419–20
Lemaître, Georges, 254
Lemurs, 67, 371
Lepeshinskaya, Olga, 129, 131
Lhuillier, Ruz, 283
Lifespan, 83
Light, speed of, 141
Linnaeus, Carolus (Carl von Linné), 172
Living fossils, 21, 23, 75–76, 190
Lizards, 388–90
London, England, 196–97, 251–52, 331, 427
Louisiana Purchase Exposition, 431
Lovejoy, Arthur O.: *The Great Chain of Being*, 424
Lovejoy, Owen, 176
Lubbock, John, 430
Lungfish, 35–36, 38–39, 354
Lyell, Charles, 152, 158, 161–63, 165, 172–73, 176–78, 186, 330, 427–28; *Principles of Geology*, 161, 184

Macusi people, 384
Madagascar, 318, 352
Malthus, Thomas, 174–76, 178, 336, 442; *Essay on the Principle of Population*, 174, 178
Mammals, 6–9, 43–44, 48, 55, 313, 316, 348, 356
Mammoths, 15, 18, 22, 48, 50, 56, 71, 154–56, 184–85, 192, 277, 313, 315, 322, 326, 392
Mars, 41, 100, 447
Marxism, 216
Massingham, H. J., 259, 262
Mastodons, 22, 156, 316
Mayans, 244–45, 283
McCosh, James, 96
McGee, W. J., 431
McGlashan, Alan, 361
Mead, A. P., 55
Mechanics, 142–44, 149–50, 229
Mediterranean Sea, 192–93, 244, 247, 253, 396–97
Melville, Herman, 343; *Moby-Dick*, 201, 248, 252
Mendel, Gregor, 136–37
Mesopotamia, 429
Metabolism, 42–43, 48
Meteorites, 126

Mias papan, 426–27
Mice, 114–15, 124, 440
Microscope, 97–98, 125, 127, 140, 200, 261
Middle Ages, 98, 138–40, 204, 265
Migrations, human, 56, 71, 316
Milky Way, 59, 142, 450
Miller, Hugh: *The Testimony of the Rocks*, 95
Miocene epoch, 58, 65–66, 319
Missing links, 53–54, 63, 69, 126, 426
Mississippi River, 429
Missouri River, 14, 420
Mivart, St. George Jackson, 63
Moe, Henry Allen, 241
Mohegans, 284
Mollusks, 33, 161
Mongols, 432
Monkeys, 63–68, 71, 167, 422, 424–25, 446
Montaigne, Michel de, 365–66
Moon, 243
Morning Chronicle (London), 252
Morris, Wright, 206
Moseley, Henry Nottidge, 27–28
Moses, 263
Mosses, 41
Mount Monadnock, 310
Mudskippers, 33, 39
Muir, John, 348
Muncie, Ind., 419
Museums, 447–48
Mushrooms, 351–52
Muskrats, 220–21, 234
Mutation, 265–67, 349, 371

Napoléon Bonaparte, 174
Nationalism, 211
Native Americans, 155, 185; Aztecs, 407–8; Blackfoot, 320; Eskimos, 58, 325, 339–41; Hopis, 272; Iroquois, 380; Mayans, 244–45, 283; Mohegans, 284; Pequots, 284; of Tierra del Fuego, 329, 336
Natural History magazine, 178
Natural law, 138–39, 149
Natural selection, 20, 51–54, 76, 79, 171–72, 175–79, 333, 350, 369–71
Natural theology, 424
Neander River, 423, 432

Neanderthals, 64, 85, 191–94, 197, 215, 282, 368, 391, 393–94, 397, 400, 422–33
Nebraska, 390
Nebulas, 142
Negroes, 85–86, 427, 432
Neolithic period, 267, 371, 428
Neptune, 99
New Guinea, 190
New Jersey, 316
Newman, John Henry, 183
Newton, Isaac, 141–44, 146, 148–49, 170–71, 229, 254, 263–65, 332, 336
New York City, 62, 103, 117, 316, 442
New York Herald Tribune, 62
New York Times, 62, 113, 129
Niger River, 37
Nile River, 429
Nitrogen, 30
North America, 75, 154–55, 184, 187–88, 279, 316, 322
Nuclear energy, 379
Nuclear weapons, 113, 129, 205, 217, 455

O'Brien, Fitz-James, 99
Oceans, depths of, 21–30
Octopuses, 33, 291
Odysseus, 243–58, 331
Ohio, 155
Olduvai Gorge, 196
Oligocene epoch, 65–66
Ooze: pond, 34–37; primordial, 23–28, 124, 129, 164–65
Orangutans, 64, 187, 190, 319, 425, 430
Orb spider, 275–80
Oreopithecus, 62, 66, 69
Orinoco River, 157
Osborn, Henry Fairfield, 65
Osman Hill, W. C., 66
Ospreys, 454
Owen, Richard, 63–64, 159
Owls, 312
Oxygen, 30, 35–37, 39, 43, 266, 277, 354
Oxygen-18 dating technique, 72–73
Oz, 327–28, 336–37, 340

Pacific Ocean, 243, 247–48, 251
Palenque, 283

Paleocene epoch, 7–9, 340
Paleolithic period, 185, 267, 371
Paleontology, 422–33
Paleozoic era, 21, 28, 38, 93, 101, 314
Paley, William, 264
Palomar Observatory, 31, 278, 361
Parapithecus, 66
Paris, France, 158, 192
Pascal, Blaise, 222–23, 229, 234–35
Pascoli, Giovanni, 253, 256–57
Peddlers, 409–14
Pedomorphism, 58–59, 76, 78, 82–88
Peking man, 366, 368
Pequots, 284
Periophthalmus, 39
Permian period, 311–13
Petrie, Flinders, 435
Phagocytes, 279
Pheasants, 231–33
Philosophes, 245
Phosphorus, 202, 446
Photosynthesis, 266, 272, 277
Picard, Max, 205, 210–11
Pigeons, 103–5, 443
Piltdown man, 51–52, 54–57, 60–62, 71
Pines, 43–47, 109, 116
Pithecanthropus, 60, 64, 320, 368, 426
Planets, 97–100, 263, 350
Plato, 96–97, 147, 160, 170, 176, 204, 255, 363, 437
Platte River, 14, 16–18
Playfair, John, 161
Pleistocene epoch, 74–75, 309, 311–13, 316–20, 432
Pliocene epoch, 317, 319
Plotinus, 257
Pluto, 350
Pluvials, 316
Point-extinction, 162
Poland, 445
Pollen, 43–48
Polynesia, 248
Pompeii, 244
Prairie dogs, 8
Prairies, 5, 14, 18
Prairie Schooner, 408, 414, 421
Primates, 8–9, 49–50, 53, 56–67, 70, 72, 74, 78–79, 186–87, 195–96, 243–45, 318–20, 422–33
Proconsul africanus, 65

Progeria, 69
Progress, 136, 207, 211–12, 215, 223, 265, 299, 336, 429
Progressionism, 157–58, 161, 163, 169–70, 173, 176
Propithecus, 371
Proteins, 127–28
Protozoans, 126
Pruner, Franz, 427
Pterodactyls, 44, 282
Ptolemy, 248

Quarterly Review of Biology, 434

Railroads, 409–21
Ramsay, A. C., 312
Rasmussen, Knud, 341
Rationalism, 148, 156–57, 229
Rats, 8–9
Rattlesnakes, 115
Ravens, 109–10
Ray, John, 152; *The Wisdom of God Manifested in the Works of the Creation*, 141
Redwoods, 44–45, 47
Rembrandt van Rijn, Harmensz, 244
Renaissance, 165, 200, 232, 269, 349, 429, 445
Reno, Nev., 414
Reptiles, 7, 9, 42–44, 46, 48, 94–96, 99, 115–17, 128, 155, 159, 217, 225, 231–33, 282, 297, 348, 352, 356, 388–90, 441, 451
Resolution, H.M.S., 250–51
Rhinoceroses, 392, 395
Ritson, Joseph, 425
Robots, 113
Rockets, 30, 205
Rock pigeon, 103–5, 443
Rocky Mountains, 14, 391
Rodents, 8–9
Roemer, Olaus, 141
Rome, ancient, 154, 185, 208, 270–71, 427, 430
Roswell, N.M., 455
Rousseau, Jean-Jacques, 366
Royal Geographical Society, 24
Russia, 129, 201, 208, 211, 445, 448

Saber-toothed cat, 48, 50, 56, 70, 277, 319, 367

Sacramento Valley, 413
St. Louis, Mo., 431
Santayana, George, 202, 325, 376
Scale of nature concept, 424
Schaaffhausen, Hermann, 191–92, 426–27
Scheuchzer, Johann Jakob, 96
Schmid, Rudolf, 430
Science, 269–70, 298; Bacon on, 205, 211, 261; history of, 136–50; limits of, 167–68; modern, 209–12, 216–17, 222–23, 233, 244, 256, 267, 344, 378; progress of, 136; and skepticism, 142; value of, 230
Scientific method, 398
Scott, Robert F., 250, 436
Seashells, 342
Sea urchins, 21, 26
Seccho, 289
Sedgwick, Adam, 159, 185
Semmelweis, Ignaz, 137
Sequoias, 44, 50
Shakespeare, William, 148, 269; *The Tempest*, 343, 356, 358–59, 370
Shaw, Bernard, 448
Shelley, Percy Bysshe, 134
Shensi skull, 320
Siberia, 156, 208
Sierras, 413–14
Sillman, Leonard, 320
Silurian period, 36
Simpson, George Gaylord, 313
Singer, Ronald, 86
Siwalik Hills, 186–87, 319
Slit (excavation site), 5–11
Smith, William, 160
Snakes, 115, 128, 231–33, 441
Snowflakes, 20, 309
Snowy owl, 312
Society, 58, 77, 88, 196, 203, 210, 214, 217, 267, 368, 371
Solar system, 141, 150, 171
Song sparrow, 132
South Africa, 57, 63, 65, 74, 81–82, 84–88
South America, 22, 154, 157, 316
South Trinidad Island, 354
Soviet Union, 129, 201, 208, 211, 445, 448
Space exploration, 243–44, 362
Spain, 194, 250

Sparrows, 117, 132
Special creation, 424
Specialization, biological, 37–38, 48, 65, 67, 70, 79, 190, 278
Species, 172, 176–79, 184, 282, 299–300
Speck, Frank G., 284
Sperm whale, 28, 248
Spiders, 110–12, 275–80, 442, 448, 454
Squids, 25–26
Starfish, 290–92, 304
Stars, 140, 142, 150, 297
Stone age, 267, 371, 428
Strait of Magellan, 329
Strandlooper people, 81–82, 84
Straus, William L., Jr., 65–67
Struggle for existence, 162, 173, 175–77, 190–92, 265, 296, 300, 302, 305, 338, 351, 442
Succession, 159, 162, 173
Sun, 128, 142
Swanscombe skull, 74, 321–22
Switzerland, 186

Tahiti, 248
Tanzania, 196
Tarsiers, 64–66
Technology, 208–11, 216, 244, 261, 267, 269–70, 378
Telegraph, 333
Telescope, 97–98, 140, 199, 261
Television, 201
Temple, William, 4
Tennyson, Alfred, 256
Terns, 370–71
Tertiary period, 65–67, 91, 186, 312, 317–19, 392
Thompson, D'Arcy, 270
Thompson, Francis, 42
Thomson, Charles, 21–26, 28; *The Depths of the Sea*, 23
Thoreau, Henry David, 16, 234, 314, 325, 327–29, 355, 386, 452; and glaciation, 310; journals, 325, 337–38, 445; and nature, 4, 309, 326, 336–42; *Walden*, 201, 326, 340
Tierra del Fuego, 329, 336
Times (London), 208
Tollund man, 253
Tolstoy, Leo, 201

Tools, 56–58, 74, 87, 135, 185, 188, 192–93, 196, 223, 235, 243, 261, 270, 298, 321, 339, 429–30
Tornados, 294–95
Transcendentalism, 234, 452
Trilobites, 161, 164, 202, 226, 446
Tristram, H. B., 429
Trojan War, 245, 247
Turtles, 155, 217, 352
Tuscarora Deep, 22
2001: A Space Odyssey, 244
Tyrannosaurus rex, 42

Uniformitarianism, 161, 265
Union Pacific Railroad, 409–10
Universe, expanding, 91, 355
University of Chicago, 72
University of Kansas, 443
Ussher, James, 146
Utah, 414

Valparaiso, Chile, 326
Variation, 173, 177, 179, 333, 371
Velikovsky, Immanuel: *Worlds in Collision*, 135
Venezuela, 388
Venus, 30
Victorian age, 53, 76–77, 190, 273, 299, 318, 330, 336, 339, 373, 432
Vikings, 340–41
Virchow, Rudolph, 427–28
Virginia, 22, 154
Viruses, 127
Vogt, Carl, 428

Wagner, Moritz, 350
Walden Pond, 16–17
Wallace, Alfred Russel, 51–60, 71, 76, 79, 173, 179, 265, 313, 318, 427, 431
Warblers, 108
Wasps, 446
Water, 13–20
Water lilies, 441
Watt, James, 149
Weapons, 113, 129, 192–93, 205, 208, 217
Weasels, 441
Wells, H. G.: *The Time Machine*, 81, 85
Wenner-Gren Foundation for Anthropological Research, 3, 62
Whales, 28, 248, 343

Whately, Richard, 429, 432
Whewell, William: *The Plurality of Worlds*, 98
White, Gilbert, 366
Whitehead, Alfred North, 138–39
White Sands proving grounds, 30
Wilberforce, Samuel, 333
Wilderness, 11, 103–4, 112, 201, 269, 452
William of Ockham, 261
Williams, Charles, 382
Williams, William: *The Universe No Desert, the Earth No Monopoly*, 99
Wilson, John, 358–59

Wolf (dog), 240, 307–9
Wollaston, T. Vernon: *On the Variation of Species*, 168
Wolves, 48
Woodchucks, 43
Woodger, Joseph, 125–26
Wood Jones, F., 65–66
World War II, 207
Wright, Thomas, 142

Xenophon, 307

Zeuner, Frederick, 75
Zuckerman, Solly, 61

*This book is set in 10 point ITC Galliard, a face
designed for digital composition by Matthew Carter and based
on the sixteenth-century face Granjon. The paper is acid-free
lightweight opaque that will not turn yellow or brittle with age.
The binding is sewn, which allows the book to open easily and lie flat.
The binding board is covered in Brillianta, a woven rayon cloth
made by Van Heek–Scholco Textielfabrieken, Holland.
Composition by Dedicated Book Services.
Printing and binding by Edwards Brothers Malloy, Ann Arbor.
Designed by Bruce Campbell.*

THE LIBRARY OF AMERICA SERIES

The Library of America fosters appreciation of America's literary heritage by publishing, and keeping permanently in print, authoritative editions of America's best and most significant writing. An independent nonprofit organization, it was founded in 1979 with seed funding from the National Endowment for the Humanities and the Ford Foundation.

1. Herman Melville: Typee, Omoo, Mardi
2. Nathaniel Hawthorne: Tales & Sketches
3. Walt Whitman: Poetry & Prose
4. Harriet Beecher Stowe: Three Novels
5. Mark Twain: Mississippi Writings
6. Jack London: Novels & Stories
7. Jack London: Novels & Social Writings
8. William Dean Howells: Novels 1875–1886
9. Herman Melville: Redburn, White-Jacket, Moby-Dick
10. Nathaniel Hawthorne: Collected Novels
11 & 12. Francis Parkman: France and England in North America
13. Henry James: Novels 1871–1880
14. Henry Adams: Novels, Mont Saint Michel, The Education
15. Ralph Waldo Emerson: Essays & Lectures
16. Washington Irving: History, Tales & Sketches
17. Thomas Jefferson: Writings
18. Stephen Crane: Prose & Poetry
19. Edgar Allan Poe: Poetry & Tales
20. Edgar Allan Poe: Essays & Reviews
21. Mark Twain: The Innocents Abroad, Roughing It
22 & 23. Henry James: Literary Criticism
24. Herman Melville: Pierre, Israel Potter, The Confidence-Man, Tales & Billy Budd
25. William Faulkner: Novels 1930–1935
26 & 27. James Fenimore Cooper: The Leatherstocking Tales
28. Henry David Thoreau: A Week, Walden, The Maine Woods, Cape Cod
29. Henry James: Novels 1881–1886
30. Edith Wharton: Novels
31 & 32. Henry Adams: History of the U.S. during the Administrations of Jefferson & Madison
33. Frank Norris: Novels & Essays
34. W.E.B. Du Bois: Writings
35. Willa Cather: Early Novels & Stories
36. Theodore Dreiser: Sister Carrie, Jennie Gerhardt, Twelve Men
37. Benjamin Franklin: Writings (2 vols.)
38. William James: Writings 1902–1910
39. Flannery O'Connor: Collected Works
40, 41, & 42. Eugene O'Neill: Complete Plays
43. Henry James: Novels 1886–1890
44. William Dean Howells: Novels 1886–1888
45 & 46. Abraham Lincoln: Speeches & Writings
47. Edith Wharton: Novellas & Other Writings
48. William Faulkner: Novels 1936–1940
49. Willa Cather: Later Novels
50. Ulysses S. Grant: Memoirs & Selected Letters
51. William Tecumseh Sherman: Memoirs
52. Washington Irving: Bracebridge Hall, Tales of a Traveller, The Alhambra
53. Francis Parkman: The Oregon Trail, The Conspiracy of Pontiac
54. James Fenimore Cooper: Sea Tales
55 & 56. Richard Wright: Works
57. Willa Cather: Stories, Poems, & Other Writings
58. William James: Writings 1878–1899
59. Sinclair Lewis: Main Street & Babbitt
60 & 61. Mark Twain: Collected Tales, Sketches, Speeches, & Essays
62 & 63. The Debate on the Constitution
64 & 65. Henry James: Collected Travel Writings
66 & 67. American Poetry: The Nineteenth Century
68. Frederick Douglass: Autobiographies
69. Sarah Orne Jewett: Novels & Stories
70. Ralph Waldo Emerson: Collected Poems & Translations
71. Mark Twain: Historical Romances
72. John Steinbeck: Novels & Stories 1932–1937
73. William Faulkner: Novels 1942–1954
74 & 75. Zora Neale Hurston: Novels, Stories, & Other Writings
76. Thomas Paine: Collected Writings
77 & 78. Reporting World War II: American Journalism
79 & 80. Raymond Chandler: Novels, Stories, & Other Writings

81. Robert Frost: Collected Poems, Prose, & Plays
82 & 83. Henry James: Complete Stories 1892–1910
84. William Bartram: Travels & Other Writings
85. John Dos Passos: U.S.A.
86. John Steinbeck: The Grapes of Wrath & Other Writings 1936–1941
87, 88, & 89. Vladimir Nabokov: Novels & Other Writings
90. James Thurber: Writings & Drawings
91. George Washington: Writings
92. John Muir: Nature Writings
93. Nathanael West: Novels & Other Writings
94 & 95. Crime Novels: American Noir of the 1930s, 40s, & 50s
96. Wallace Stevens: Collected Poetry & Prose
97. James Baldwin: Early Novels & Stories
98. James Baldwin: Collected Essays
99 & 100. Gertrude Stein: Writings
101 & 102. Eudora Welty: Novels, Stories, & Other Writings
103. Charles Brockden Brown: Three Gothic Novels
104 & 105. Reporting Vietnam: American Journalism
106 & 107. Henry James: Complete Stories 1874–1891
108. American Sermons
109. James Madison: Writings
110. Dashiell Hammett: Complete Novels
111. Henry James: Complete Stories 1864–1874
112. William Faulkner: Novels 1957–1962
113. John James Audubon: Writings & Drawings
114. Slave Narratives
115 & 116. American Poetry: The Twentieth Century
117. F. Scott Fitzgerald: Novels & Stories 1920–1922
118. Henry Wadsworth Longfellow: Poems & Other Writings
119 & 120. Tennessee Williams: Collected Plays
121 & 122. Edith Wharton: Collected Stories
123. The American Revolution: Writings from the War of Independence
124. Henry David Thoreau: Collected Essays & Poems
125. Dashiell Hammett: Crime Stories & Other Writings

126 & 127. Dawn Powell: Novels
128. Carson McCullers: Complete Novels
129. Alexander Hamilton: Writings
130. Mark Twain: The Gilded Age & Later Novels
131. Charles W. Chesnutt: Stories, Novels, & Essays
132. John Steinbeck: Novels 1942–1952
133. Sinclair Lewis: Arrowsmith, Elmer Gantry, Dodsworth
134 & 135. Paul Bowles: Novels, Stories, & Other Writings
136. Kate Chopin: Complete Novels & Stories
137 & 138. Reporting Civil Rights: American Journalism
139. Henry James: Novels 1896–1899
140. Theodore Dreiser: An American Tragedy
141. Saul Bellow: Novels 1944–1953
142. John Dos Passos: Novels 1920–1925
143. John Dos Passos: Travel Books & Other Writings
144. Ezra Pound: Poems & Translations
145. James Weldon Johnson: Writings
146. Washington Irving: Three Western Narratives
147. Alexis de Tocqueville: Democracy in America
148. James T. Farrell: Studs Lonigan Trilogy
149, 150, & 151. Isaac Bashevis Singer: Collected Stories
152. Kaufman & Co.: Broadway Comedies
153. Theodore Roosevelt: Rough Riders, An Autobiography
154. Theodore Roosevelt: Letters & Speeches
155. H. P. Lovecraft: Tales
156. Louisa May Alcott: Little Women, Little Men, Jo's Boys
157. Philip Roth: Novels & Stories 1959–1962
158. Philip Roth: Novels 1967–1972
159. James Agee: Let Us Now Praise Famous Men, A Death in the Family, Shorter Fiction
160. James Agee: Film Writing & Selected Journalism
161. Richard Henry Dana Jr.: Two Years Before the Mast & Other Voyages
162. Henry James: Novels 1901–1902
163. Arthur Miller: Plays 1944–1961
164. William Faulkner: Novels 1926–1929
165. Philip Roth: Novels 1973–1977

166 & 167. American Speeches: Political Oratory

168. Hart Crane: Complete Poems & Selected Letters

169. Saul Bellow: Novels 1956–1964

170. John Steinbeck: Travels with Charley & Later Novels

171. Capt. John Smith: Writings with Other Narratives

172. Thornton Wilder: Collected Plays & Writings on Theater

173. Philip K. Dick: Four Novels of the 1960s

174. Jack Kerouac: Road Novels 1957–1960

175. Philip Roth: Zuckerman Bound

176 & 177. Edmund Wilson: Literary Essays & Reviews

178. American Poetry: The 17th & 18th Centuries

179. William Maxwell: Early Novels & Stories

180. Elizabeth Bishop: Poems, Prose, & Letters

181. A. J. Liebling: World War II Writings

182. American Earth: Environmental Writing Since Thoreau

183. Philip K. Dick: Five Novels of the 1960s & 70s

184. William Maxwell: Later Novels & Stories

185. Philip Roth: Novels & Other Narratives 1986–1991

186. Katherine Anne Porter: Collected Stories & Other Writings

187. John Ashbery: Collected Poems 1956–1987

188 & 189. John Cheever: Complete Novels & Collected Stories

190. Lafcadio Hearn: American Writings

191. A. J. Liebling: The Sweet Science & Other Writings

192. The Lincoln Anthology

193. Philip K. Dick: VALIS & Later Novels

194. Thornton Wilder: The Bridge of San Luis Rey & Other Novels 1926–1948

195. Raymond Carver: Collected Stories

196 & 197. American Fantastic Tales

198. John Marshall: Writings

199. The Mark Twain Anthology

200. Mark Twain: A Tramp Abroad, Following the Equator, Other Travels

201 & 202. Ralph Waldo Emerson: Selected Journals

203. The American Stage: Writing on Theater

204. Shirley Jackson: Novels & Stories

205. Philip Roth: Novels 1993–1995

206 & 207. H. L. Mencken: Prejudices

208. John Kenneth Galbraith: The Affluent Society & Other Writings 1952–1967

209. Saul Bellow: Novels 1970–1982

210 & 211. Lynd Ward: Six Novels in Woodcuts

212. The Civil War: The First Year

213 & 214. John Adams: Revolutionary Writings

215. Henry James: Novels 1903–1911

216. Kurt Vonnegut: Novels & Stories 1963–1973

217 & 218. Harlem Renaissance Novels

219. Ambrose Bierce: The Devil's Dictionary, Tales, & Memoirs

220. Philip Roth: The American Trilogy 1997–2000

221. The Civil War: The Second Year

222. Barbara W. Tuchman: The Guns of August, The Proud Tower

223. Arthur Miller: Plays 1964–1982

224. Thornton Wilder: The Eighth Day, Theophilus North, Autobiographical Writings

225. David Goodis: Five Noir Novels of the 1940s & 50s

226. Kurt Vonnegut: Novels & Stories 1950–1962

227 & 228. American Science Fiction: Nine Novels of the 1950s

229 & 230. Laura Ingalls Wilder: The Little House Books

231. Jack Kerouac: Collected Poems

232. The War of 1812

233. American Antislavery Writings

234. The Civil War: The Third Year

235. Sherwood Anderson: Collected Stories

236. Philip Roth: Novels 2001–2007

237. Philip Roth: Nemeses

238. Aldo Leopold: A Sand County Almanac & Other Writings

239. May Swenson: Collected Poems

240 & 241. W. S. Merwin: Collected Poems

242 & 243. John Updike: Collected Stories

244. Ring Lardner: Stories & Other Writings

245. Jonathan Edwards: Writings from the Great Awakening

246. Susan Sontag: Essays of the 1960s & 70s

247. William Wells Brown: Clotel & Other Writings

248 & 249. Bernard Malamud: Novels & Stories of the 1940s, 50s, & 60s

250. The Civil War: The Final Year

251. Shakespeare in America

252. Kurt Vonnegut: Novels 1976–1985

253 & 254. American Musicals 1927–1969

255. Elmore Leonard: Four Novels of the 1970s

256. Louisa May Alcott: Work, Eight Cousins, Rose in Bloom, Stories & Other Writings

257. H. L. Mencken: The Days Trilogy

258. Virgil Thomson: Music Chronicles 1940–1954

259. Art in America 1945–1970

260. Saul Bellow: Novels 1984–2000

261. Arthur Miller: Plays 1987–2004

262. Jack Kerouac: Visions of Cody, Visions of Gerard, Big Sur

263. Reinhold Niebuhr: Major Works on Religion & Politics

264. Ross Macdonald: Four Novels of the 1950s

265 & 266. The American Revolution: Writings from the Pamphlet Debate

267. Elmore Leonard: Four Novels of the 1980s

268 & 269. Women Crime Writers: Suspense Novels of the 1940s & 50s

270. Frederick Law Olmsted: Writings on Landscape, Culture, & Society

271. Edith Wharton: Four Novels of the 1920s

272. James Baldwin: Later Novels

273. Kurt Vonnegut: Novels 1987–1997

274. Henry James: Autobiographies

275. Abigail Adams: Letters

276. John Adams: Writings from the New Nation 1784–1826

277. Virgil Thomson: The State of Music & Other Writings

278. War No More: American Antiwar & Peace Writing

279. Ross Macdonald: Three Novels of the Early 1960s

280. Elmore Leonard: Four Later Novels

281. Ursula K. Le Guin: The Complete Orsinia

282. John O'Hara: Stories

283. The Unknown Kerouac: Rare, Unpublished & Newly Translated Writings

284. Albert Murray: Collected Essays & Memoirs

285 & 286. Loren Eiseley: Collected Writings on Evolution, Nature, & the Cosmos